6th International Conference on Mechatronics and Mechanical Engineering (ICMME 2019)

MATEC Web of Conferences Volume 306 (2020)

Osaka, Japan
27 – 30 November 2019

Editor:

Mohd Hamdi Abdul Shukor

ISBN: 978-1-7138-0292-1

Printed from e-media with permission by:

Curran Associates, Inc.
57 Morehouse Lane
Red Hook, NY 12571

Some format issues inherent in the e-media version may also appear in this print version.

For additional information, please contact EDP Sciences – Web of Conferences
at the address below.

EDP Sciences – Web of Conferences
17, Avenue du Hoggar
Parc d'Activité de Courtabœuf
BP 112
F-91944 Les Ulis Cedex A
France

Phone: +33 (0) 1 69 18 75 75
Fax: +33 (0) 1 69 28 84 91

contact-edps@webofconferences.org

Additional copies of this publication are available from:

Curran Associates, Inc.
57 Morehouse Lane
Red Hook, NY 12571 USA
Phone: 845-758-0400
Fax: 845-758-2633
Email: curran@proceedings.com
Web: www.proceedings.com

TABLE OF CONTENTS

Preface

We are honored to bring you this collection of articles from 2019 The 6th International Conference on Mechatronics and Mechanical Engineering (ICMME 2019), which has been held in Osaka, Japan on November 27-30, 2019. The primary focus of this conference is to provide you opportunity to meet with academicians as well as practitioners in the fields of mechatronics and mechanical engineering from all over the world, and get the latest insights from every area of mechatronics and mechanical engineering theory and practice.

The general theme of the conference is Mechatronics and Mechanical Engineering, papers were submitted from all over the world in the domains of Power Engineering and Applied Mechanics, Mechanical Design and Manufacturing System, Control Theory and Control Engineering, Power Electronics and Transmission Technology and Aerospace Engineering. Out of around 60 submitted papers, 32 papers were accepted for publication after peer review by international reviewers who were drawn from the technical committee members as well as other external reviewers invited from all over the world. Hereby we would like to thank all the reviewers for their time and effort in reviewing the documents.

Meanwhile, 4 speeches were delivered by eminent researchers from Japan, Russia, and Malaysia. They are Prof. Susumu Hara from Nagoya University, Japan with the talk "Spacecraft Landing Using Base-Extension Separation Mechanisms"; Prof. Dr. Mohd Hamdi Abdul Shukor from University of Malaya, Malaysia with the speech title "Engineering Evolution of Functional Devices from Fundamental Research to Infrared Sensors and Blended wing Body Aircraft"; Prof. Alexander Pesin from Nosov Magnitogorsk State Technical University, Russia, talked about "Numerical Modeling and Development of New Technical Solutions in Metallurgy" as well as Assoc. Prof. Mikio Ito, Osaka University, Japan on the topic of "Rapid and Energy-Saving Densification of Metallic and Semiconductive Powders by Using Directly Applied Current Sintering". All the talks were very impressive for the high level of professionalism, and in many cases original ideas and activities have been accomplished or proposed. Overall, the conference was a grand success. The participants were immensely benefitted by presenting and interacting among themselves as well as with the experts.

The proceedings of this conference, including all 32 full papers are compiled and published by MATEC Web of Conferences. Finally, it is appropriate that we record our thanks to our fellow members of the technical organizing committee for their work, we are also indebted to those who served as conference chairs, program chairs, steering committee. Without their support, the conference could not have been the success that it was. We also acknowledge the authors themselves, without whose expert input there would have been no conference.

Regarding the continuing success of this conference series, we are looking forward to the 7th International Conference on Mechatronics and Mechanical Engineering next year, we hope that it would be more interesting, enjoying and significant.

Program Chair
Prof. Dr. Mohd Hamdi Abdul Shukor

University of Malaya, Malaysia
Dec 12, 2019.

edp sciences

Web of Conferences

Statement of Peer review

In submitting conference proceedings to *Web of Conferences*, the editors of the proceedings certify to the Publisher that

1. They adhere to its **Policy on Publishing Integrity** in order to safeguard good scientific practice in publishing.
2. All articles have been subjected to peer review administered by the proceedings editors.
3. Reviews have been conducted by expert referees, who have been requested to provide unbiased and constructive comments aimed, whenever possible, at improving the work.
4. Proceedings editors have taken all reasonable steps to ensure the quality of the materials they publish and their decision to accept or reject a paper for publication has been based only on the merits of the work and the relevance to the journal.

Title, date and place of the conference

The 6th International Conference on Mechatronics and Mechanical Engineering

Osaka, Japan

November 27-30, 2019

Proceedings editor(s): Prof. Dr. Mohd Hamdi Abdul Shukor

Date and editor's signature

13/12/2019

MATEC Web of Conferences **306**, 01001 (2020) https://doi.org/10.1051/matecconf/202030601001
ICMME 2019

Investigation on Dynamic Interaction of Multiple Clearance Joints for Flap Actuation System

Qi Wan, *Geng* Liu*, *Shangjun* Ma and *Ruiting* Tong

Shaanxi Engineering Laboratory for Transmissions and Controls, Northwestern Polytechnical University, Xi'an 710072, China

Abstract. A dynamic simulation model of the flap actuation system is developed in this paper to analyze the dynamic interaction behaviours of multiple clearance joints. The nonlinear contact force model and modified Coulomb friction model are adopted in the four clearance joints to capture their motion modes, including free flight mode, impact mode and continuous mode. The results show that there exists a strong dynamic interaction between different clearance joints and one impact motion mode in a joint will immediately affect other joints. And when the system reaches a stable state, the four clearance joints almost appears the similar tendency due to the rigid connection. Therefore, in order to accurately predict the dynamic responses of multibody system, it is essential for all joints to be modelled as imperfect ones.

1 Introduction

As flap actuation system is widely employed in the field of aircrafts, the demands of predicting the system's dynamic characteristics accurately are becoming increasingly essential. The flap actuation system mainly consists of an actuator, linkage mechanism, flight control surface and control system, which is a high precise position servo system [1]. The actuator receives the control command of flight attitude of aircrafts, and then produces force to drive linkage mechanism connected with control surface. At last, the aim of flight direction and attitude of aircrafts is accomplished. The output angle of control surface is the main control target of the flap actuation system, and its follow-up of the command signal directly determines the flight quality of aircrafts. However, as the typical multibody mechanism with clearances, the linkage mechanism of flap actuation system has great influence on dynamic behaviors of aircrafts. Therefore, it is necessary to analyze the influence of joint clearance in the linkage mechanism on the dynamic responses of flap actuation system.

Due to the manufacturing, assembly processes, and wear errors, joint clearance is inevitable and always leads to extraordinary negative effect on the dynamic characteristic of multibody system. A number of researchers pointed that there is a strong dynamic interaction between different clearance joints and all joints should be considered as non-ideal to accurately predict the dynamic performance of aircrafts and spacecrafts. Li [2] established the dynamic model of the spacecraft with large deployable solar arrays considering multiple clearance joints and studied the effect of joint clearance on the satellite attitude under deployment disturbance. The results showed that system response with two clearance joints is not a simple superposition of that in a mechanism with one clearance joint. Bai [3] studied the dynamic responses of satellite antenna dual-axis driving mechanism with two planar revolute clearance joints and developed an optimization method to reduce the undesirable vibrations due to impact forces in clearance joints. Li [4] investigated the effects of the number of clearance joints on the dynamic characteristics of planar deployable structure based on scissor-like element, and the results indicated that the motion in one revolute clearance joint will affect the motion in the other clearance joint. However, the combination motion modes between multi-clearance joints and some inner rules existing those modes are not revealed. Muvengei [5] and Ma [6] studied the interaction and combination motion modes between clearance joints, but the slider-crank mechanism only including four simple parts is always studied and there has been less investigated on the combination motion modes and inner rules of aircrafts with imperfect revolute joints.

Therefore, in order to achieve the relationship between system responses and combination motion modes as well as dynamic interaction of clearance joints, the dynamic characteristics of flap actuation system with multiple clearance joints are investigated in this paper, which is significant for the dynamic simulation and analysis of transmission system in aerospace field.

2 Modelling of flap actuation system with joint clearance

2.1 Model of revolute joint with clearance

The perfect revolute joint assumes that the centers of journal and bearing are coincident all the time with only

* Corresponding author: npuliug@nwpu.edu.cn

relative rotation allowed, while the imperfect one means that there is a clearance between the centers of journal and bearing.

The clearance can be expressed by

$$c = r_B - r_J \quad (1)$$

where c denotes the radial clearance and r_B and r_J are the radii of bearing and journal, respectively.

The mathematical model of the revolute clearance joint is shown in Fig. 1. XOY donates the generalized coordinate system. O_B and O_J are the centers of bearing and journal, respectively.

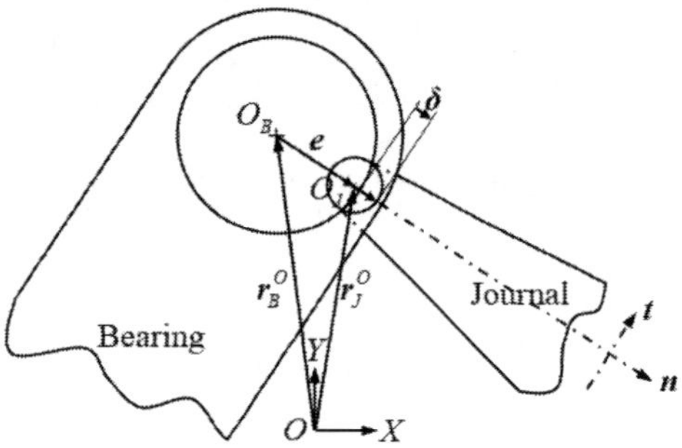

Fig. 1. Mathematical model of revolute clearance joint.

As shown in Fig. 1, the eccentricity vector which connects the centers of bearing and journal (O_B and O_J) can be written as [3-4]:

$$\boldsymbol{e} = \boldsymbol{r}_J^O - \boldsymbol{r}_B^O \quad (2)$$

where $\boldsymbol{r}_B^O$ and $\boldsymbol{r}_J^O$ are the positional vectors of centers of bearing and journal in generalized coordinate system (XOY), respectively.

The magnitude of the vector $\boldsymbol{e}$ can be described as

$$e = \sqrt{\boldsymbol{e}^{\mathrm{T}}\boldsymbol{e}} \quad (3)$$

And the normal unit vector $\boldsymbol{n}$ at collision point can be expressed as

$$\boldsymbol{n} = \boldsymbol{e} / e \quad (4)$$

The tangential unit vector at the collision point can be obtained by reversing n for 90°.

When the journal and the bearing come into contact with each other, the relative penetration depth can be calculated with respect to e and c

$$\delta = e - c \quad (5)$$

The motion of a journal in a bearing can be divided into three categories: the free flight mode, the contact mode and the impact mode. The relative motion relationship between the journal and the bearing can be accurately determined by the relative penetration depth δ. There are several situations as follows: (a) If δ<0, there is no collision between two bodies and the journal moves freely within the bearing's boundaries, which means the system is in the free flight mode and the contact force between journal and bearing is equalled to zero; (b) If δ>0, the journal and bearing contact with each other and the system is in the continuous contact mode; (c) If δ changes from the negative to the positive, it means the two bodies are gradually approaching. Particularly, when the relative penetration depth is equalled to zero, there is an impact force between the journal and bearing and the system is in the impact mode at that moment; (d) If δ changes from the positive to the negative, it indicates the two are gradually separating.

2.2 Contact force model

The "Impact-function" in multi-body dynamic software, ADAMS [7-8], which is a nonlinear spring-damper model considering the effect of energy dissipation process and is widely employed by many researchers. Ma compared the simulation results of a slider-crank mechanism with clearance joints, which the normal contact forces are respectively calculated by "Impact-function" and the presented model, and the simulation results of those two models have a good agreement with each other [6]. Khemili studied the dynamic behavior of a planar slider-crank mechanism with clearance by "Impact-function" and verified the effectiveness of contact force model by some experiments [9]. Chen analyzed the dynamic responses of a closed high speed and heavy load press and the "Impact-function" is employed to describe the contact response between the journal and bearing. As the numerical results agree well with the experimental data, it indicates that the "Impact-function" can accurately describe the dynamic performance of the multibody system [10]. Therefore, the "Impact-function" is used to predict the dynamic characteristics of contact-impact process for the flap actuation system, and its equations can be given by [11]

$$F_n = \begin{cases} K\delta^{1.5} & \delta \le 0 \\ K\delta^{1.5} + c_{\max}\left(\dfrac{\delta}{\delta_{\max}}\right)^2\left(3 - 2\dfrac{\delta}{\delta_{\max}}\right)\dot{\delta} & 0 < \delta < \delta_{\max} \\ K\delta^{1.5} + c_{\max}\dot{\delta} & \delta > \delta_{\max} \end{cases} \quad (6)$$

where δ is the relative penetration depth; $\delta_{\max}$ is the maximal penetration depth; $c_{\max}$ is the maximal damping coefficient, which is about one percent of the contact stiffness; K is the generalized contact stiffness and relate to the material property and geometries of bearing and journal, which can be expressed as

$$K = \frac{4}{3\left(\dfrac{1-\nu_B}{E_B} + \dfrac{1-\nu_J}{E_J}\right)}\sqrt{\frac{R_B R_J}{R_B - R_J}} \quad (7)$$

where R_B and R_J are the radii of bearing and journal, respectively; and the material parameters of bearing and journal are associated with Poisson's ratio, ν_B, ν_J, and Young's modulus, E_B, E_J, respectively.

The frictional characteristics of the revolute clearance joint is depicted by a modified Coulomb friction force model, and its formula is shown as [11]

$$F_t = \begin{cases} -\mu_d \cdot \mathrm{sign}(v_t)F & |v_t| > v_d \\ -\mathrm{step}(|v_t|, v_d, \mu_d, v_s, \mu_s) \cdot \mathrm{sign}(v_t) & v_s \le |v_t| \le v_d \\ \mathrm{step}(v_t, -v_s, \mu_s, v_s, -\mu_s) & |v_t| < v_s \end{cases} \quad (8)$$

where v_t is the relative sliding velocity at contact point; v_s is the critical stiction transition velocity; v_d is the friction transition velocity; μ_s is the static friction coefficient; μ_d is the dynamic friction coefficient.

3 Simulations and results

3.1 System properties

The dynamic model of flap actuation system with four clearance joins is shown in Fig. 2. The material properties of the system are shown in Table 1 and the parameters used in the dynamic simulations are presented in Table 2. The "Impact-function" is adopted in the contact force model of clearance joints, and Gear Stiff integrator (GSTIFF) is chosen in ADAMS/Solver. During the simulation, the integration tolerance is 0.001 and the step size is 0.001s.

3.2 Dynamic interaction

This section presents the dynamic interaction phenomenon by analyzing the influences of clearance on the dynamic responses of the flap actuation system. The command of actuator's output displacement is defined as 10 mm with a constant frequency equal to 4 Hz, and size of the radial clearance equals 0.1 mm with the bearing diameter as 15 mm. The total simulation time is 0.5 s.

Fig. 3 shows that the existence of joint clearance has extraordinary obvious effects on the angular velocity and angular acceleration, which is obviously shaky around the ideal curve. The constant value in angular velocity correspond to the zero angular acceleration and the impulse type of angular acceleration means the abrupt increase of angular velocity, which indicates that joint clearance will cause oscillation of the system movement and lead to the worse system stability.

Table 1. Material properties.

Parameters / Parts	Young's modulus (GPa)	Poisson's ratio (/)	Mass density (kg/m^3)
Actuator Rod1 Bearing house 1,2,3,4 Inertia plate	209	0.269	7890
Rod2 Lever arm1 Lever arm2	45	0.33	2700
Shaft	211	0.28	7700

Table 2. Parameters used in the dynamic simulations.

Parameter	Description	Value	Unit
δ_{max}	Maximal penetration depth	0.01	mm
μ_s	Static friction coefficient	0.01	/
μ_d	Dynamic friction coefficient	0.01	/
v_s	Stiction transition velocity	0.1	mm/s
v_d	Friction transition velocity	10	mm/s

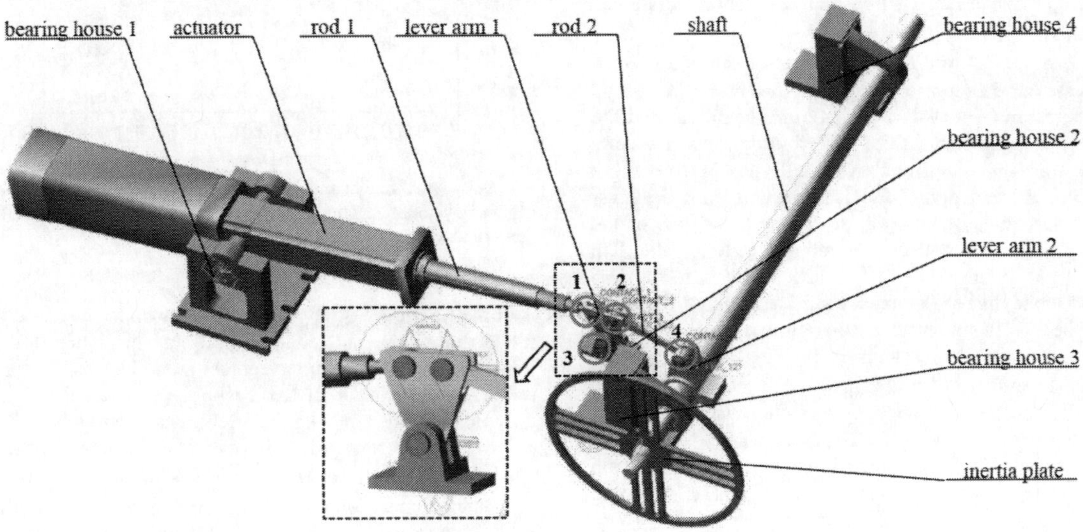

Fig. 2. Dynamic model of flap actuation system with four clearance joints.

 https://doi.org/10.1051/matecconf/202030601001

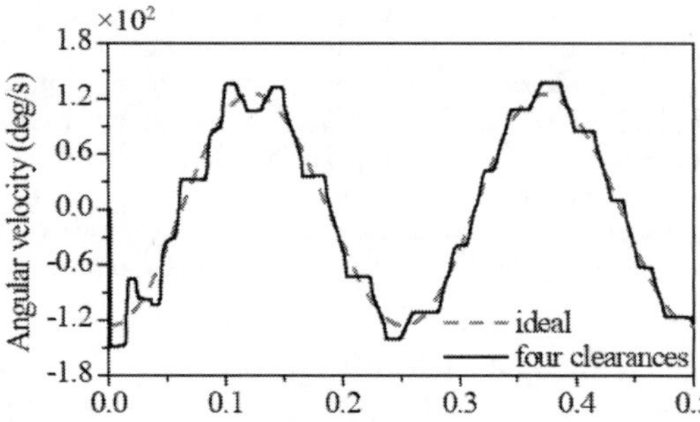

(a) Angular velocity

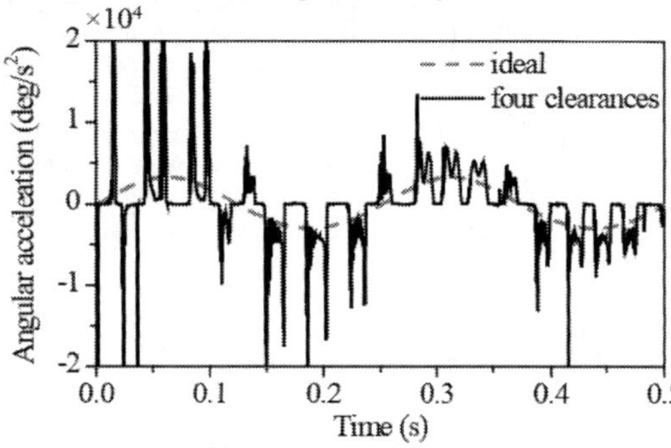

(b) Angular acceleration

Fig. 3. Dynamic responses of the system with four clearances.

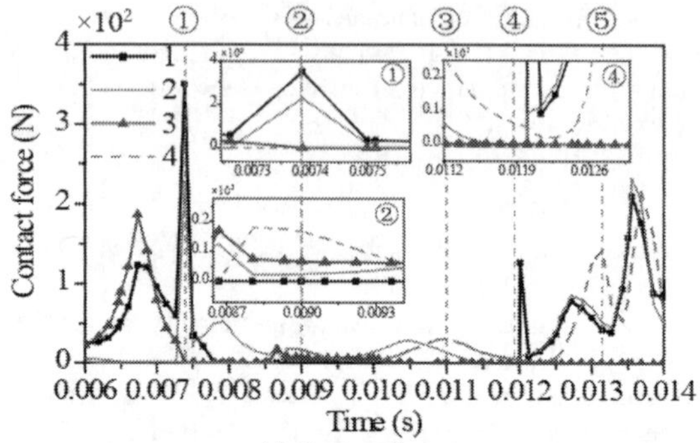

(a) Contact force

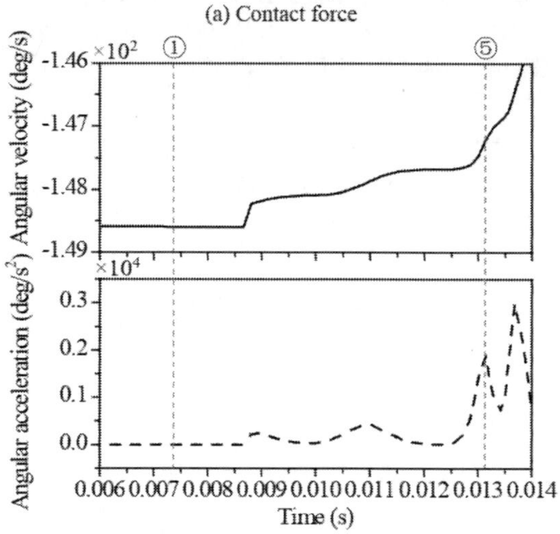

(b) Angular velocity and angular acceleration of cases ①⑤

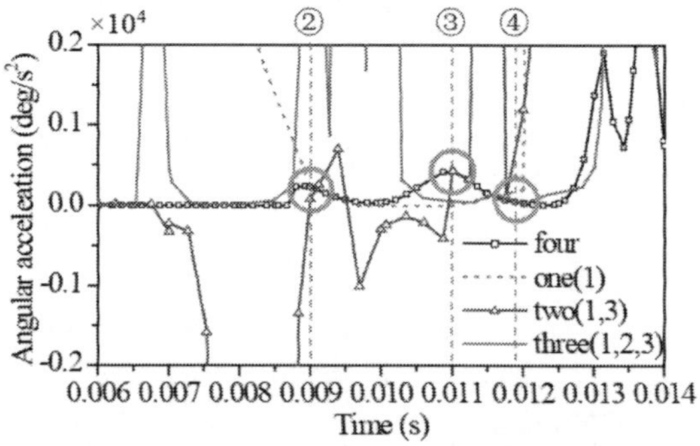

(c) Angular acceleration of cases ②③④

Fig. 4. Dynamic responses of the system at the beginning of numerical simulation

Fig. 4 shows the dynamic responses of the system during the period time between 0.006 and 0.014 seconds. Contact forces at four clearance joints with five marked time indicate five combination mode, which are presented as follows.

Case ① Impact-impact-free-free combination mode

Case ② Free-contact-contact-contact combination mode

Case ③ Free-contact-free-contact combination mode

Case ④ Free-free-free-contact combination mode

Case ⑤ Contact-contact-free-impact combination mode

(1) Cases ①⑤

Case ① shows that there occurs an unexpected increase at the contact forces of the first and second joints, meaning that they are in impact motion, and the contact forces at the third and fourth joints are zero, indicating that the latter two joints are in free flight motion. As the fourth joint is connected with the lever arm2 and shaft of the system, the motion mode of the fourth joint can reflect the motion state of the flap actuation system. When the fourth joints is in the free flight mode, there is no reaction force acted on the shaft, and hence, the angular acceleration of the system is equalled to zero and its angular velocity keeps a constant value, as shown in the Fig. 4(b). Case ⑤ shows that the angular acceleration of the system changes abruptly and the angular velocity keeps increasing, since the impact force occurred in the fourth joint with the assumption that the lever arm2 and shaft are modelled as two rigid parts.

Therefore, the change tendency of system responses can judged by the motion mode of the joint connected with the output component.

(2) Cases ②③④

Case ② shows that the first joint is in the free flight mode due to the null contact force and the latter three joints are in the continuous contact mode because of the stable contact forces. It can be observed from the Fig. 4(c) that the system angular acceleration curve almost coincides with that of only the first clearance joint situation. Similar to case ②, it can be observed in case ③. When contact motion occurs in the second and fourth joints, the system angular acceleration curve almost overlap that of cases which only considers two clearance joints including the first and third joints. Similarly, this

https://doi.org/10.1051/matecconf/202030601001

is evident in case ④, the system angular acceleration curve with four clearance joints case almost coincides with the one with three clearance joints, namely the first, second and third one.

Fig. 5 presents the contact forces of the four clearance joints after the system reaches a steady state, which appears that the clearance joint located closer to the input component undergoes the higher contact force and the corresponding conclusion can be obtained from the reference [6]. In addition, the second and fourth clearance joints, connected by a rigid rod, present almost the same contact force. The four clearance joints experience similar changing trends of contact force since impact mode occurred in one joint will immediately affect other joints in the rigid system.

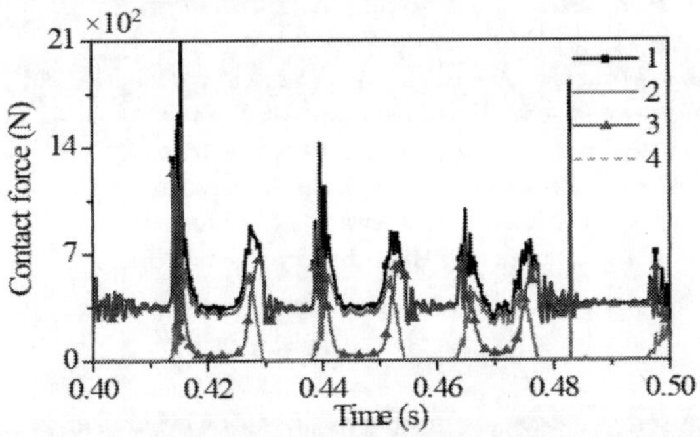

Fig. 5. Contact force of the four clearance joints at the end of numerical simulation

4 Conclusions

In this paper, the dynamic characteristics of flap actuation system with four clearance joints are investigated using computational methodology, where the contact force model is described by the "Impact-function", which consists of a nonlinear normal contact force model and a modified Coulomb friction model. And then the influence of clearances on the system dynamic interaction phenomenon is further analysed and some combination modes are illustrated.

From the numerical simulation results, it can be concluded that the existence of joint clearances will lead to the worse dynamic behaviours of the flap actuation system, namely, angular velocity, angular acceleration and contact force. 1) There exists a strong dynamic interaction between four clearance joints, where the impact mode occurred in one joint will immediately affect other joints. 2) It is clear that the clearance joint nearer to the input component appears higher contact force and four joints present the similar trend of contact force when the system reaches stable, particularly, the contact forces of two joints connected by a rigid rod are almost the same. 3) The system stability can judged by the motion mode of the joint connected with the output component. 4) When one joint is in the continuous contact motion mode, the angular acceleration of this system will coincide with that in the occasion of the joint is considered as ideal.

Acknowledgments

This research was supported by the National Natural Science Foundation of China (Grant No. 51675429), Key Project of National Natural Science Foundation of China (Grant No. 51535009), the Fundamental Research Funds for the Central Universities (Grant No.31020190504003) and the 111 Project (Grant No. B13044). The authors would like to express their appreciation to the agencies.

References

1. D. Arriola, F. Thielecke, Model-based design and experimental verification of a monitoring concept for an active-active electromechanical aileron actuation system. *Mech Syst Signal Pr*, **94**, 322-345 (2017)
2. Y. Y. Li, Z. L. Wang, C. Wang, W H Huang, Planar rigid-flexible coupling spacecraft modeling and control considering solar array deployment and joint clearance. *Acta Astronaut*, **142**, 138-151 (2018)
3. Z. F. Bai, J. J. Zhao, J. Chen, Y. Zhao, Design optimization of dual-axis driving mechanism for satellite antenna with two planar revolute clearance joints, *Acta Astronaut*, **144**, 80-89 (2018)
4. B. Li, S. M. Wang, V. Makis, X. Z. Xue, Dynamic characteristics of planar linear array deployable structure based on scissor-like element with differently located revolute clearance joints, *Proc IMechE Part C: J Mechanical Engineering Science*, **232** (10), 1759-1777 (2018)
5. O. Muvengei, J. Kihiu, B. Ikua, Dynamic analysis of planar rigid-body mechanical systems with two-clearance revolute joints. *Nonlinear Dynam*, **73**, 259-273 (2013)
6. J. Ma, L. F. Qian, Modeling and simulation of planar multibody systems considering multiple revolute clearance joints. *Nonlinear Dynam*, **90**, 1907-1940 (2017)
7. C. Birleanu, S. Talu, Machine elements. Designing and computer assisted graphical representations, Cluj-Napoca, Romania, Victor Melenti Publishing house, 2001. ISBN 973-99539-6-4.
8. T. Nitulescu, S. Talu, Applications of descriptive geometry and computer aided design in engineering graphics, Cluj-Napoca, Romania, Risoprint Publishing house, 2001. ISBN 973-656-102-X.
9. I. Khemili, L. Romdhane, Dynamic analysis of a flexible slider-crank mechanism with clearance. *Eur J Mech A-Solid*, **27**, 882-898 (2008)
10. Y. Chen, Y. Sun, C. Chen, Dynamic analysis of a planar slider-crank mechanism with clearance for a high speed and heavy load press system. *Mech Mach Theory*, **98**, 81-100 (2016)
11. MDI. Building models in ADAMS/View, 2012, ADAMS/Solver online help.

MATEC Web of Conferences **306**, 01002 (2020)
ICMME 2019
https://doi.org/10.1051/matecconf/202030601002

A Study of Hybrid Energy Storage System for Electric vehicle Air Conditioning System

Phucha Veerathanaporn [1,*], *Danai* Phaoharuhansa[1], and *Masaki* Yamakita[2]

1Department of Mechanical Engineering, King Mongkut's University of Technology Thonburi, Bangkok, Thailand
[2]Department of Mechanical and Control Systems Engineering, Tokyo Institute of Technology, Japan

Abstract. This paper introduces improvement efficiency of battery for air-conditioning (A/C) system. Super-capacitor (SCs) is mounted with Lithium-ion battery called hybrid energy storage system (HESS). The system modeling of heat system in passenger room and battery pack are derived to represent system in mathematic model. The characteristic of power loss relative to discharge rate are examined using several models of Li-Ion battery. The relationship between discharge rate and power loss in battery is tested by experiment kits. A/C system of passenger car is simplified to demonstration kit. Single energy storage system (SESS) and HESS batteries are used as energy storage. The both typed of energy storage are tested with on-off control. HESS outperforms rather than SESS. The efficiency of HESS battery is greater than SESS about 57.57% at peak load and 14.34% at rated load.

1 Introduction

Nowadays, the pollution problem is increasing which causes global warming, so the amount of electric vehicle (EV) are increasing to reduces the pollution problem [1]. Lithium-ion (Li-Ion) battery pack is usually used in commercial EV. The single type of battery cell is well-known as single energy storage system (SESS). Energy consumption of EV is consumed by two main systems such as powertrain system and air-conditioning system (A/C). A/C system is an essential system of electric vehicle for driver's comfort.

The papers [2-4] has introduced that peak load was occurred when A/C system and powertrain system are enabled at the same time. Energy consumption of A/C system decreases driving range about 36% [2]. Discharge efficiency will be reduced when discharge current is greater than 1 C-rate or high fluctuation [5]. The peak load of battery pack is main cause which reduce the efficiency of energy consumption in A/C system.

There are many types of controller for automotive A/C system such as on-off controller, PID controller and Fuzzy controller. On-off controller is usually used for conventional automotive

Fig. 1 presents the diagram of on-off control system. The temperature in passenger room (Ti) is feedback to the on-off controller by temperature sensor. On-off controller will compare with desired temperature (Td) . Then, it will perform A/C system, if temperature in passenger is higher than desired temperature.

Hybrid energy storage system (HESS) consist of different kind of battery types. Super-capacitor (SCs) is mounted with conventional battery such as lead-acid or Li-Ion. Supercapacitor obtains discharge rat greater than others battery types so that it can reduce the peak current of battery pack. They have many advantages such as large discharge current in short time, low internal resistance, wide range operation temperature, and long-life cycle time. Moreover, the supercapacitor has significant effect to reducing temperature of battery and extend battery lifespan [6-7]. Thus, supercapacitor is one of device to improve discharge current and efficiency of battery.

Therefore, this research will introduce heat balanced of automotive A/C system, characteristic of battery pack and improvement of efficiency and power loss of battery at peak. on-off control system is used to demonstrate energy consumption of SESS and HESS.

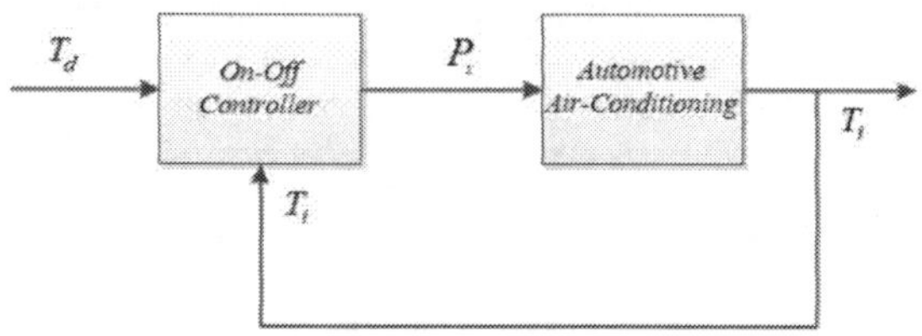

Fig. 1. Schematic diagram of on-off control system.

2 System modeling development

Air-condition load is considered to derive the mathematical model of passenger room. Load of A/C system consists of battery temperature, solar radiation, metabolic and ambient temperature. The models will be presented as follows

* Corresponding author: phucha.v@mail.kmutt.ac.th

2.1. Passenger model

Passenger model describes the relationship between total heat input ($\Sigma\dot{Q}_{ip}$), heat output ($\dot{Q}_{op}$), and the variation heat in passenger room ($\dot{Q}_p$). The equation is given as

$$\dot{Q}_p = \Sigma\dot{Q}_{ip} - \dot{Q}_{op} \quad (1)$$

Total heat input is calculated by summation of heat flow from battery pack ($\dot{Q}_{bp}$), solar ($\dot{Q}_{sp}$), and metabolic ($\dot{Q}_{mp}$). It is expressed that

$$\dot{Q}_{ip} = \dot{Q}_{bp} + \dot{Q}_{sp} + \dot{Q}_{mp} \quad (2)$$

Heat transfer between battery pack and passenger room is occurred when temperature of battery pack (T_b) and passenger room (T_i) are different level. It is written as

$$\dot{Q}_{bp} = U_p A_{bp}\left(T_b - T_i\right) \quad (3)$$

where, U_p is overall heat transfer coefficient between battery pack and passenger room. A_{bp} is area on top surface of battery pack. T_b is temperature of battery pack. T_i is temperature in passenger room.

The heat radiation from solar to passenger room is expressed as Eq.(4). An average of solar irradiation ($\dot{Q}_{si}$) in Thailand is approximately about 1000 W/m^2. The heat radiation enters to the passenger room pass though the surface such as roof and windshield. Which depends on heat transfer area and absorptivity of material.

$$\dot{Q}_{sp} = \dot{Q}_{si}\left(A_r\beta_r + A_g\beta_g\right) \quad (4)$$

where, A_r is area of roof. A_g is area of windshield. β_r is absorptivity of roof. β_g is absorptivity of glass.

When the driver or passenger is inside the vehicle, the heat is generated by metabolism from body. The heat from metabolic is given as

$$\dot{Q}_{mp} = \sum_{passenger} M_m A \quad (5)$$

where, M_m is constant of passenger metabolic heat. A is constant, which estimate from body. It depends on weight and height.

The heat output of the system is occurred from the air-conditioning system. The equation is written as

$$\dot{Q}_{op} = \eta_{COP} P_z \quad (6)$$

where, η_{COP} is coefficient of performance of electric compressor in A/C system. P_z denotes power of air conditioning system, which equals with discharge power from battery pack

As above sentences, the equation of passenger model can be derived to equation as show in Eq. (7)

$$\dot{T}_i = \left(\frac{\dot{Q}_{sp} + \dot{Q}_{mp} - \left(\eta_{COP} P_z\right)}{m_a c_{pa}}\right) \quad (7)$$

where, m_a is mass of air. c_{pa} is specific heat of air

3 Experiment on battery efficiency

The efficiency of battery is usually related to discharge current when the discharge current of battery increases, the battery capacities is decreased more than discharge energy because the power loss has been occurred as in battery.

Experiment is configured as shown in Fig. 2, it consists of 5 main components: motor, speed controller, battery pack, receiver, and remote controller. The motor is load of the system, which is adjusted sped of motor by controller.

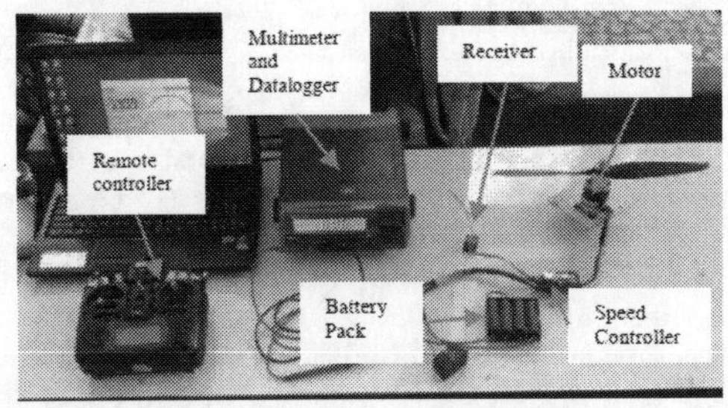

Fig. 2. Demonstration kit for efficiency of battery.

The battery models in this experiment are LG M26, LG HG2 and LG MH1 which have capacities such as 2,600 mAh, 3,000 mAh, and 3,200 mAh, respectively. The cells are packed as 4-series and 1-parallel (4S1P) which obtain energy is 38.48, 44.4, and 47.36 Wh, respectively. The battery pack is discharged at constant discharge current in each discharge rate until the ESC stop supplying power supply from battery with A/C system. The result of discharge current, efficiency, and the coefficient of power loss are shown in Table 1. The coefficient of power loss is calculated by equation which express as

$$C_p = \frac{1-\eta}{n_p} \quad (8)$$

where, C_p is coefficient of power loss, η is efficiency of battery. n_p denotes number of parallel can be found from specification of battery in Table 2.

Table 2. Efficiency of 18650 Li-Ion battery at several discharge current.

Battery model	Discharge Rate [A]	Discharge Current [A]	η	C_p
LG M26 2.6Ah	1.0	0.94	1.10	0.024
	2.6	2.41	0.81	
	5.0	4.98	0.51	
LG HG2 3Ah	1.0	0.93	0.72	0.029
	3.0	3.02	0.71	
	5.0	4.80	0.54	
LG MH1 3.2Ah	1.0	0.99	0.70	0.026

Battery model	Discharge Rate [A]	Discharge Current [A]	η	C_p
	3.2	2.85	0.65	
	5.0	5.07	0.47	

Table 3. Specification of batteries and supercapacitor.

	LG M26	LG HG2	LG MH1	SCs
Maximum capacity (Ah)	2.6	3	3.2	-
Fully charged voltage (V)	4.2	4.2	4.2	2.85
Energy maximum (Wh)	10.92	12.6	13.44	0.1
Rated Capacitance (F)	-	-	-	100
Maximum continuous discharge current (A)	10	20	10	-
Nominal battery and SCs discharge current (A)	2.2	2.5	2.7	-
Internal resistance (Ω)	0.060	0.017	0.007	0.012
Capacity (Ah) at nominal voltage	2.2	2.5	2.8	-
Nominal voltage (V)	3.65	3.6	3.7	2.7
Rated capacity (Ah)	2.6	3	3.2	-
Number of parallel (n_p)	22	19	18	-

According to the experiment result, the power had not supplied to the motor before the voltage of battery emptied because the current control was not permitting to discharge current. The cause of discharge permission is insufficient supply power from battery to meet power requirement of motor

The result presents efficiency and power loss of battery relative to discharge rate as shown in Fig. 3 and Fig. 4. The efficiency of battery decreases when discharge current enlarges. It affects an amount of power loss. The efficiency of batteries can be presented relative to discharge current as follow;

$$\eta_{dis.batt,LGM26} = \left(-0.14 \cdot I_{dis,LGM26}\right) + 1.21 \quad (9)$$

$$\eta_{dis.batt,LGHG2} = \left(-0.05 \cdot I_{dis,LGHG2}\right) + 0.78 \quad (10)$$

$$\eta_{dis.batt,LGMH1} = \left(-0.04 \cdot I_{dis,LGMH1}\right) + 0.79 \quad (11)$$

where, $\eta_{dis.batt}$ is efficiency of batteries. I_{dis} is discharge current of batteries. The relationships between power loss with power supply are calculated by the equations, which are expressed as

$$P_{l,LGM26} = \left(0.024 \cdot P_{s,LGM26}\right) + 5e^{-05} \quad (12)$$

$$P_{l,LGHG2} = \left(0.028 \cdot P_{s,LGHG2}\right) - 3e^{-05} \quad (13)$$

$$P_{l,LGMH1} = \left(0.026 \cdot P_{s,LGMH1}\right) - 3e^{-05} \quad (14)$$

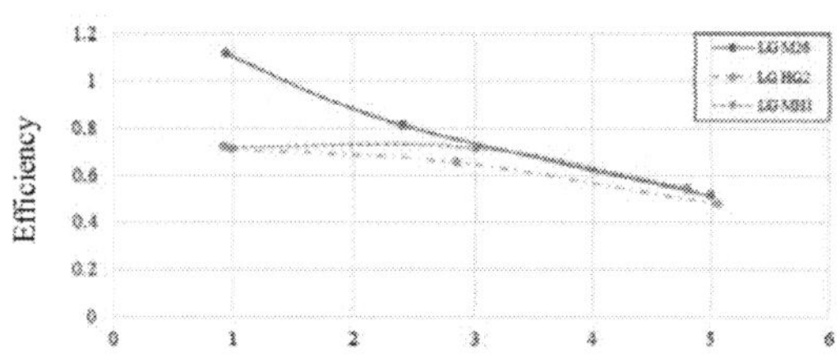

Fig. 3. Efficiency with discharge current of Li-Ion battery.

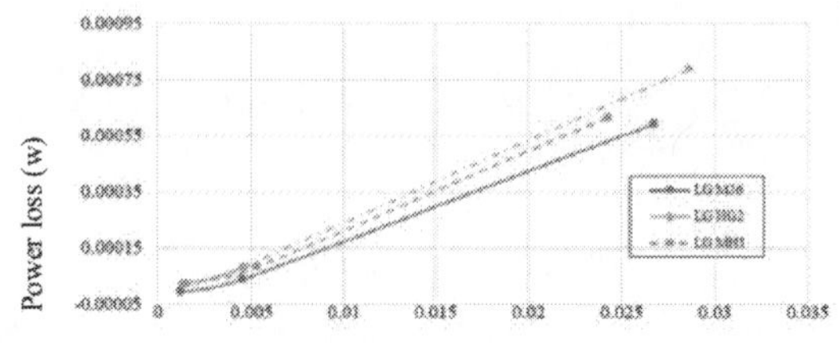

Fig. 4. Power loss with power discharge of Li-Ion battery.

4 Experiment of energy consumption

This topic presents experiment of single energy storage system (SESS) and hybrid energy storage system (HESS) to control temperature in control room. 4S1P LG M26 is used as SESS. HESS consist of 4S1P LG M26 and SCs. The super-capacitor is mounted with SESS as parallel. The energy of battery pack is 38.48 Wh and SCs is 0.55 Wh. This experiment is designed to find efficiency of HESS that affect to A/C system.

To simplify passenger model in Eq. (7) which neglect the heat of battery pack, the ratio of heat energy output with input passenger model is 2.49. Fig.5 illustrates block diagram of energy consumption experiment. It consists of energy storage, on-off control system and A/C system. The Eq. (7) is represented for experiment system that $\dot{Q}_{ia}$ of control room is 7 watts, $\dot{Q}_{sp}$ and $\dot{Q}_{mp}$ are replaced by heat input from heater which is defined as 12.45 watts, and the heat output of demonstration kit ($\dot{Q}_o$) is 48.2 watts. Then, Eq. (7) is transform as follow;

$$(\dot{Q}_h + \dot{Q}_{ia}) - \dot{Q}_o = m_{a/c} c_{p,a/c} \dot{T}_{i,a/c} \quad (15)$$

where, $\dot{Q}_h$ is the heat energy input which consist of $\dot{Q}_{sp}$ and $\dot{Q}_{mp}$. $\dot{Q}_{ia}$ is the heat energy input from ambient. $\dot{Q}_o$ is the heat energy output by demonstration A/C system. $m_{a/c}$ is mass of air in control room. $c_{p,a/c}$ is specific heat of air in control room. $\dot{T}_{i,a/c}$ is changing temperature in control room.

Battery and supercapacitor are connected in parallel. When A/C system turn on, battery and supercapacitor discharge current collaboratively to electronic speed control (ESC). Then, when A/C system turn off, supercapacitor will be recharged by battery.

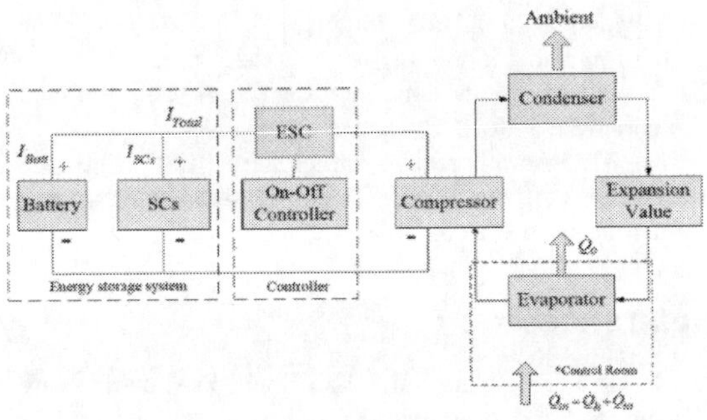

Fig. 5. Block diagram of demonstration kit.

5 Result

The results present variation of temperature in control room using SESS and HESS, energy consumption using SESS and HESS, the efficiency of battery using SESS and HESS at peak, power loss using SESS and HESS at peak. The boundary condition is defined that the desired temperature (T_d) is 25±1 °C , initial temperature (T_{int}) is 30°C, and ambient temperature (T_a) is 26 °C. The on-off control is bases on Hysteresis control system. If temperature is higher than 26 °C, the A/C system turn on. On the other hand, if temperature is lower than 24 °C, the A/C system turn off.

Temperature in control room using SESS and HESS is shown in Fig. 6. The temperature in control room using SESS and HESS achieve the desired temperature about 549 second and 367 second, respectively. Thus, using SESS achieve desired temperature slower than using HESS about 31.28 %.

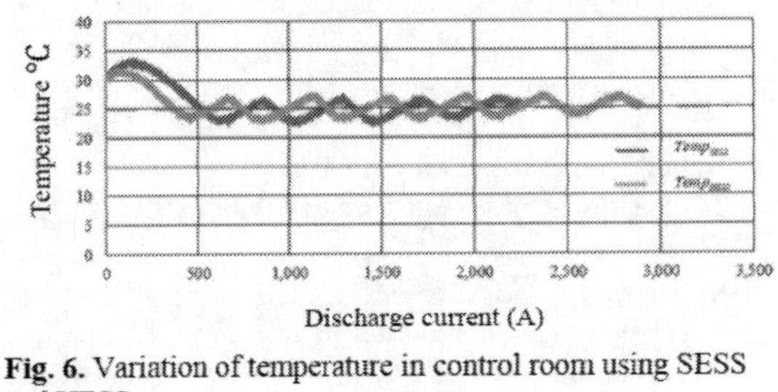

Fig. 6. Variation of temperature in control room using SESS and HESS

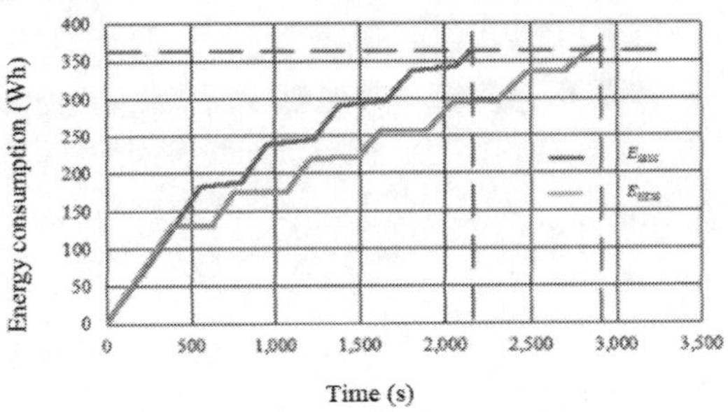

Fig. 7. Energy consumption of SESS and HESS.

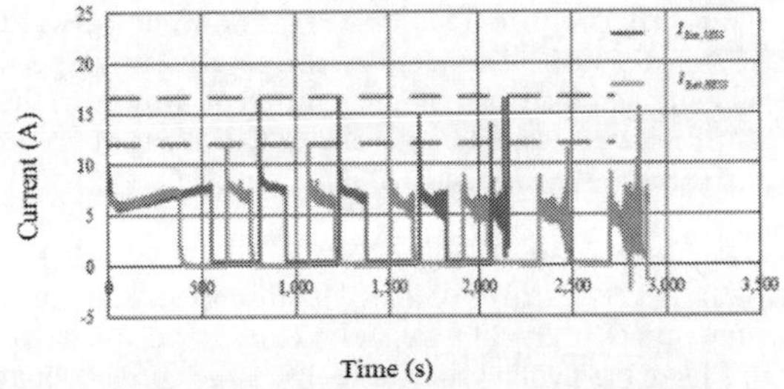

a) Comparison discharge current of Li-Ion in SESS and HESS

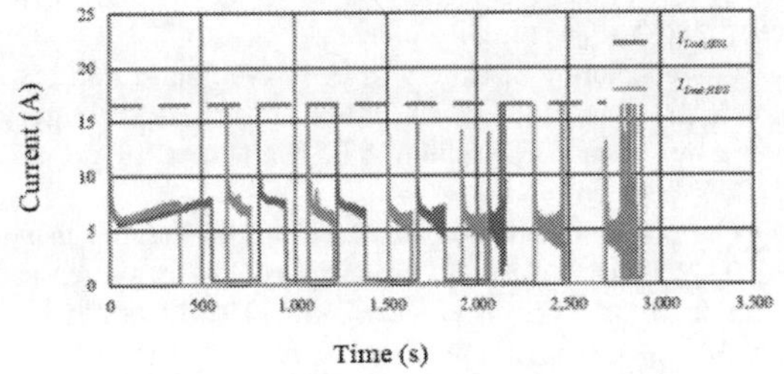

b) Total discharge current of SESS and HESS to A/C system.

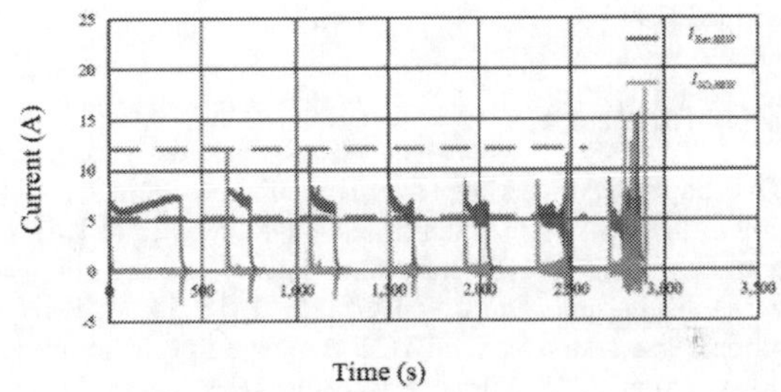

c) Discharge current of Li-Ion in HESS and discharge current of SCs in HESS to A/C system.

Fig. 8. Discharge current of single energy storage system (SESS) and hybrid energy storage system (HESS).

The energy consumption of SESS and HESS is not much difference because the air-conditoing system using SESS and HESS is stop suppling power at 22.7 % of SOC battery. SOC is calculated in Eq. (16). The operation period of control temperature using HESS and SESS are 2900 and 2253 second, respectively. Thus, the operation period of HESS is longer than SESS about 22.3% of SESS as shown in Fig. 7.

Fig. 8 presents the discharge current using single energy storage system (SESS) and hybrid energy storage system (HESS) during temperature control at 25±1 °C. The result of peak load and rated load are considered. The peak load will occur while electric compressor is starting, and rated load will occur while electric compressor consumes stable power consumption. The peak load will occur, while electric compressor is starting, the load will surge high discharge current. After electronic compressor is starting, the current will decrease nearly to constant which it is called rated load.

The discharge current of Li-Ion battery pack in SESS and HESS to A/C system are compared as show in Fig. 8a. At peak load, the discharge current is 16.38A and 11.78A, respectively. Thus, the discharge current battery of HESS can be decreased about 28.07%.

Fig. 8b presents total discharge current between SESS and HESS. The total discharge current at peak load and rated load are not much different. However, the total discharge current at starting A/C system of SESS and HESS are 9.57A and 10.735A, respectively, which increase about 10.85%.

Fig. 8c presents discharge current of HESS between Li-Ion and SCs. At peak load, the discharge current of Li-Ion and SCs are 11.98A and 4.68A, respectively. At rated load, the Li-Ion battery still discharge current about 6.49A to A/C while SCs can't discharge current. Moreover, the SCs is recharged by battery in short time when A/C system turn off.

The efficiency of battery at peak load and rated load is calculated by Eq. (9). The efficiency of HESS battery is greater than SESS about 57.57% at peak load and 14.34% at rated load.

The power discharge of battery can be derived using Eq. (12) thtat the power discharge of Li-Ion battery at peak load of SESS and HESS are 198.88 watts and 160.30 watts, respectivly. The power discharge of HESS is lower than SESS about 19.36%. Moreover, the power loss of HESS is improved about 16.59% comparison with SESS.

6 Conclusion

This paper develops improvement of the efficiency and power loss in Li-Ion battery. The characteristic of Li-Ion batteries such as efficiency and power loss are related with discharge current. LGM26, LGHG2, LGMH1 models are tested and LGM26 model outperforms than other batteries. Single energy storage system (SESS) contains only LGM26 Li-Ion battery and hybrid energy storage system (HESS) consists of LGM26 Li-Ion battery and super-capacitor (SCs).

The temperature in control room using HESS achieves the desired temperature faster than SESS about 31.28% because SCs supplies additional current to A/C system. The total discharge current of HESS is higher than SESS during beginning period, but the total current is slightly lower than SESS during another period.

The efficiency and power loss at peak of Li-Ion battery in HESS is better than SESS about 57.57% and 16.59% because SCs can reduced the discharge current as shown in Fig. 8a. The fluctuation of HESS has high amplitude than SESS.

Especially, SOC of batteries become low. The fluctuation also become higher than the previous because the battery in HESS isn't supplied to A/C system but also super-capacitor.

The HESS outperforms rather than SESS. The SCs can reduce with surge current at peak load which occur frequently in on-off control system.

References

1. Yong Li, Jie Yang, and Jian Song, Design structure model and renewable energy technology for rechargeable battery towards greener and more sustainable electric vehicle, Renewable and Sustainable Energy Reviews, vol. **74**, pp. 19-25, (2018).
2. R. Farrington and J. Rugh, *Impact of Vehicle Air-Conditioning on Fuel Economy, Tailpipe Emissions, and Electric Vehicle Range*, the Earth Technologies Forum, (2000).
3. Ziqi Zhang, Dandong Wang, Chenquan Zhang, and Jiangping Chen, Electric vehicle range extension strategies based on improved AC system in cold climate, International Journal of Refrigeration, vol. **88**, pp. 141-150, (2018).
4. Brahim Mebarki, Boumediene Allaoua, Draoui Belkacem, Elhadj Benachour, and Lakhdar Rahmani, Impact of Air-Conditioning System on the Power Consumption of an Electric Vehicle Powered by Lithium-Ion Battery, Modelling and Simulation in Engineering, vol. **2013**, (2013).
5. Chelladurai A/L Sinkaram, Vijanth S. Asirvadam, Nursyarizal Bin Mohd Nor, and Mumtaj Begam, *Battery characteristics due to various discharging current and temperatures: A simulation approach*, IEEE Student Conference on Research and Development (SCOReD), (2012).
6. Rebecca Carter, Andrew Cruden, and Peter J. Hall, Optimizing for Efficiency or Battery Life in a Battery/Supercapacitor Electric Vehicle, IEEE Transactions on Vehicular Technology, vol. **61**, (2012).
7. S. Pay, and Y. Baghzouz, *Effectiveness of battery-supercapacitor combination in electric vehicles*, IEEE Bologna Power Tech Conference Proceedings, (2003).

Dynamic Model of Planetary Roller Screw Mechanism with Considering Torsional Degree of Freedom

Linping Wu, *Shangjun* Ma*, *Qi* Wan and *Geng* Liu

Shaanxi Engineering Laboratory for Transmissions and Controls, Northwestern Polytechnical University, Xi'an 710072, China

Abstract. To predict accurately the dynamics performance of planetary roller screw mechanism, it is necessary to establish its streamline and engineering-compliant dynamic model, which is the basis of mechanical design and precision control of the system. In this paper, the relative displacement between roller and ring gear along the line of action is deduced and the relationship between nature frequencies and the number of rollers is discussed. Considering the torsional stiffness of all components and the thread mesh stiffness based on the Hertzian contact theory, the purely torsional model for planetary roller screw mechanism is presented to reveal the natural frequencies and vibration mode characteristics of the system. The results show that the natural properties of undamped system in planetary roller screw mechanism are mainly reflected by two typical vibration modes: rotational mode and roller mode.

1 Introduction

Planetary roller screw mechanism (PRSM) is a rolling screw transmission device that can convert screw's rotational motion into nut's linear motion or turn the nut's rotational motion into screw's linear motion, which the former one is called the standard PRSM and the latter one is defined as the inverse PRSM. As shown in Fig. 1, the PRSM mainly consists of a screw, a nut, rollers, carriers and ring gears. Owing to its higher speed, larger bearing capacity, higher transmission accuracy, stronger environmental adaptability and longer working life, the PRSM is widely employed in the aerospace and voyage fields, even the medical equipment and precision machine tools. Therefore, it is essential to build a streamlined and engineering-compliant dynamic model of PRSM in this paper, which is helpful to investigate its transmission performance and improve the mechanical design and precision control of the system.

In recent years, there are many researches on PRSM, which are primarily focused on load distribution [1-3], parameter matching [4], friction and wear behaviours [5], axial stiffness [6-7] and kinematic analysis [8-10]. Instead, only few publications on the dynamic characteristic of the PRSM are mentioned.

Jones and Velinsky [9] established the kinematical contact model at the screw/roller and nut/roller interfaces on the basis of the principle of conjugate surfaces. And Jones et al. [11] also proposed the dynamics equations of motion for the screw, rollers, and nut of PRSM by Lagrange's Method and derived their steady-state angular velocities and screw/roller slip velocities. Ma and his colleagues [12] studied the contact characteristics of threaded surfaces under the combined action of normal pressure and tangential friction forces, and derived not only the velocities of the contact points in the contact ellipse but also the normal pressure and friction force distributions. Fu et al. [13] developed a nonlinear six degrees of freedom dynamic model, in which the load distribution coefficient is introduced to describe the load distribution among threads of PRSM, and studied the transient and steady-state behaviours of the PRSM under a heavy and light external load, respectively. However, the natural frequencies and vibration mode characteristics of the PRSM are not studied in recent researches.

In this paper, a dynamic model of PRSM considering torsional stiffness and mesh stiffness is established, in which the thread mesh stiffness is calculated by using Hertzian contact theory. Then the nature frequencies and the vibration modes of PRSM are investigated, which is used to determine the dynamic characteristics of the system, and the relationship between nature frequencies and the number of rollers is discussed. At last, this paper proposes the direction for further research.

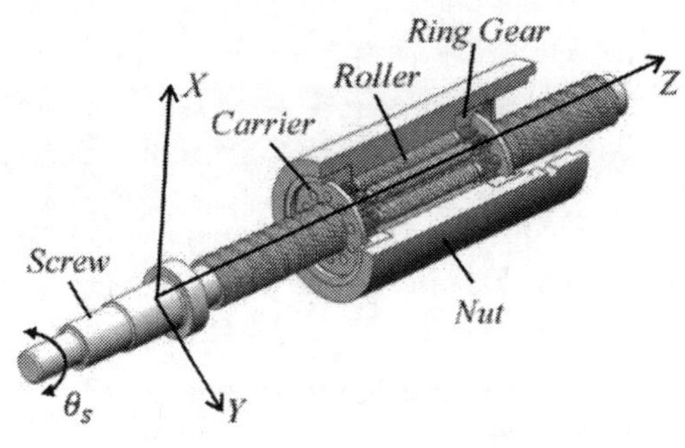

Fig. 1. Mechanical structure of PRSM.

* Corresponding author: mashangjun@nwpu.edu.cn

https://doi.org/10.1051/matecconf/202030601003

2 Modelling

2.1 Introduction of PRSM

As shown in Fig. 1, due to the similarities between roller and planetary wheel, the transmission principles of PRSM is similar to that of planetary gear. When the screw rotates, the roller revolves both round the screw and on its own axis, and at the same time, the nut moves in a straight line. The motion and power are transmitted through the threaded contact surfaces.

2.2 Purely torsional model

The purely torsional model of PRSM is established and some assumptions and definitions are described as follows: 1) The torsional line displacement is regarded as the product of the torsional angular displacement of each major component and its gyration radius; 2) The gyration radius of the screw is its nominal radius, which is a half of the pitch diameter of the thread, and the definition of gyration parameters for the nut is the same as that of the screw; 3) The gyration radius of the inner ring gear is a half of its standard pitch circle diameter; 4) The gyration radius of the carrier is considered as the distance between the roller's axis and the geometric center of the carrier, and that of roller is the same. As shown in Fig. 2, the purely torsional model of PRSM mainly considers the effects of the torsional support stiffness, rotational inertia, mass properties of the main components, the thread contact stiffness of the screw/roller and the nut/roller and the gear mesh stiffness between the spur gear located at the outer end of the rollers and the ring gear.

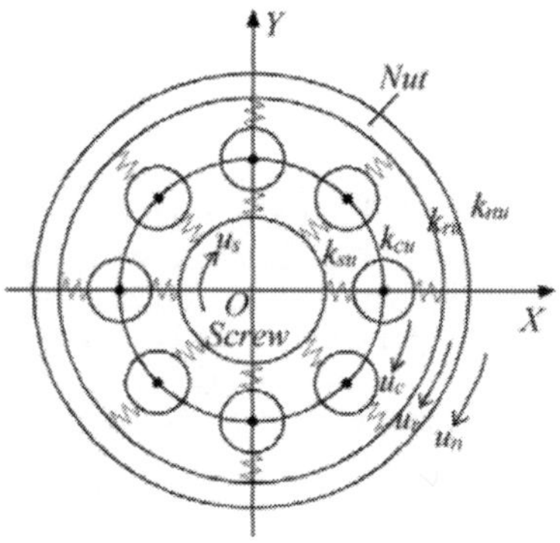

Fig. 2. Pure torsion model.

where K_{su}, K_{Nu}, K_{cu} and K_{ru} is the torsional support stiffness of the screw, nut and carrier, respectively; u_s, u_N, u_c and u_r are torsional line displacements of the screw, nut, carrier and ring gear, respectively.

According to the Newton's second law, the undamped free vibration differential equations of system motion are obtained

$$\begin{cases} \dfrac{J_s}{r_s^2}\ddot{u}_s + \sum_{n=1}^{8} K_{sn}d_{sn} + K_{su}u_s = 0 \\ \dfrac{J_N}{r_N^2}\ddot{u}_N + \sum_{n=1}^{8} K_{Nn}d_{Nn} + K_{Nu}u_N = 0 \\ \dfrac{(J_c + \sum_{n=1}^{8} m_{pn}r_c^2)}{r_c^2}\ddot{u}_c + K_{cu}u_c = 0 \\ \dfrac{J_r}{r_r^2}\ddot{u}_r + \sum_{n=1}^{8} K_{rn}d_{rn} + K_{ru}u_r = 0 \\ \dfrac{J_{R1}}{r_{R1}^2} + K_{s1}d_{s1} - K_{N1}d_{N1} - K_{r1}d_{r1} = 0 \\ \vdots \\ \dfrac{J_{Rn}}{r_{Rn}^2} + K_{sn}d_{sn} - K_{Nn}d_{Nn} - K_{rn}d_{rn} = 0 \end{cases} \tag{1}$$

where n is the number of the roller; J_s, J_N, J_c and J_r are the rotational inertia of the screw, nut, carrier and ring gear, respectively; r_s, r_N, r_c, r_r are the gyration radii of the screw, nut, carrier and ring gear, respectively; J_{Rn}, m_{Rn}, r_{Rn} are the rotational inertia, mass and gyration radius of the roller; d_{sn} is the relative displacement between the roller n and screw along the z-axis, which can be expressed as

$$d_{sn} = \frac{\theta_s L_s}{2\pi} = \frac{u_s L_s}{r_s \cdot 2\pi} \tag{2}$$

where θ_s is the rotational angular of the screw; L_s is the screw's lead; d_{Nn} is the relative displacement between the roller n and nut along the z-axis, which can be formed as

$$d_{Nn} = \frac{\theta_N L_N}{2\pi} = \frac{u_N L_N}{r_N \cdot 2\pi} \tag{3}$$

where θ_N is the rotational angular of the nut; L_N is the nut's lead; d_{rn} is the relative displacement between the roller n and ring gear along the mesh line, which can be written as

$$d_{rn} = u_r - u_n \tag{4}$$

The Eq. (1) can also be formed as

$$[M]\{\ddot{u}\} + [K_b + K_m]\{u\} = 0 \tag{5}$$

where u is the torsional line displacements; $[M]$ is the mass matrix; $[K_b+K_m]$ is the stiffness matrix, and thier formulas can be expressed as

$$\{u\} = \{u_s, u_N, u_c, u_r, u_1, \cdots, u_n\}^{\mathrm{T}} \tag{6}$$

$$[M] = \mathrm{diag}\left[\frac{J_s}{r_s^2}, \frac{J_N}{r_N^2}, \frac{J_c + \sum_{n=1}^{8} m_{pn}r_c^2}{r_c^2}, \frac{J_r}{r_r^2}, \frac{J_{R1}}{r_{R1}^2}, \cdots, \frac{J_{Rn}}{r_{Rn}^2}\right] \tag{7}$$

$$[K_b] = \mathrm{diag}[K_{su}, K_{Nu}, K_{cu}, K_{ru}, 0, \cdots\cdots, 0] \tag{8}$$

$$[K_m] = \begin{bmatrix} \frac{L_s}{r_s \cdot 2\pi}\sum_{n=1}^{n} K_{sn} & & & & & & \\ & \frac{L_N}{r_N \cdot 2\pi}\sum_{n=1}^{n} K_{Nn} & & & & & \\ & & 0 & & & & \\ & & & \sum_{n=1}^{n} K_{rn} & -\sum_{n=1}^{n} K_{rn} & & \\ \frac{L_s}{r_s \cdot 2\pi} K_{s1} & -\frac{L_N}{r_N \cdot 2\pi} K_{N1} & 0 & -K_{r1} & K_{r1} & & \\ \vdots & \vdots & \vdots & \vdots & \vdots & \vdots & \vdots \\ \frac{L_s}{r_s \cdot 2\pi} K_{sn} & -\frac{L_N}{r_N \cdot 2\pi} K_{Nn} & 0 & -K_{rn} & 0 & \cdots & K_{rn} \end{bmatrix} \tag{9}$$

3 Dynamic characteristics analysis

3.1 Calculation of thread contact stiffness

The thread contact stiffness of PRSM, that is the thread meshing stiffness, refers to the ratio of the normal contact load and the contact deformation of the screw/roller or nut/roller. Since the PRSM is subjected to axial loads at run time, it is necessary to project the contact deformation of the thread to the axial direction. The normal deformation of the thread in contact area under the normal contact load is obtained by Hertzian contact theory [14]:

$$\delta = \delta^* \left[\frac{3F_n}{2\sum\rho}\left(\frac{(1-\mu_R^2)}{E_R} + \frac{(1-\mu_x^2)}{E_x} \right) \right]^{2/3} \frac{\sum\rho}{2} \tag{10}$$

where F_n is the normal contact load between two contact threads, and it can be obtained from the axial load F_a; $\Sigma\rho$ is the sum function of curvature on two contact surfaces; δ^* relates to the curvature difference function $F(\rho)$, which can be obtained by reference [15]; μ_R, E_R are the Poisson's ratio and Young's modulus of the roller; μ_x, E_x (x=s or N) are the Poisson's ratio and Young's modulus of the screw or nut, respectively.

The deformation of the thread in contact area along the axial direction is shown as

$$\delta_{xR-a} = \delta \cdot cos\beta_R \cdot cos\alpha \tag{11}$$

where β_R is the helix angle of the roller; α is the flank angle.

Therefore, the axial contact stiffness of the two components, especially the roller and screw or the roller and nut, can be expressed as:

$$K_{xRn} = \frac{F_a}{\delta_{xp-a}} \tag{12}$$

3.2 Calculation results of parameters

The structure parameters and material properties of the PRSM are presented in Table 1 and Table 2.

The calculation results of related parameters of the dynamic equations are shown in Table 3. The calculation method of the spur gear mesh stiffness can refer to [16].

3.3 Modal analysis

To simplify the calculation, the rollers are assumed equally spaced, the axial load is evenly distributed among threads of PRSM. Each roller and the ring gear mesh stiffness is equal, which is taken as the mean value of the internal mesh stiffness. Moreover, the mass and rotational inertia of the rollers are supposed the same with each other.

Table 1. Structure parameters of PRSM.

Parameters/(Units)	Screw	Roller	Nut
Nominal radius (mm)	12	4	20
Pitch (mm)	2	2	2
Number of starts	5	1	5
Helix angle (°)	7.6	4.5	4.5
Flank angle (°)	45	45	45
Lead(mm)	10	2	10
Roller profile radius(mm)	/	4.596	/
Gear			
Number of teeth	/	16	80
Modulus	/	0.5	0.5
Addendum coefficient	/	1	1
Pressure angle (°)	/	20	20
Helix angle (°)	/	0	0

Table 2. Material properties of PRSM.

Parameters	Material	Young's modulus (Pa)	Poisson's ratio
Screw	GCr15	2.12×10^{11}	0.29
Roller	GCr15	2.12×10^{11}	0.29
Nut	GCr15	2.12×10^{11}	0.29

Solving the problems of the system's natural characteristics can be achieved by obtaining the system's eigenvalues, and the natural frequencies and main modes of the system can be acquired by Eq. (13):

$$\omega_i^2 [M]\varphi_i = [K_b + K_m]\varphi_i \tag{13}$$

where $[M]$ is the mass matrix; $[K_b+K_m]$ is the stiffness matrix; ω_i is the i-th natural frequency; φ_i is the vibration mode corresponding to the i-th natural frequency, which can be expressed as:

$$\varphi_i = [\varphi_s^i, \varphi_N^i, \varphi_c^i, \varphi_r^i, \varphi_1^i, \cdots\cdots, \varphi_n^i]^T \tag{14}$$

Based on the parameters given in Table 3, the natural properties are studied by the present model. The natural frequencies and their multiplicities associated with each

Table 3. The calculation results of PRSM's parameters.

Parameters (Units)	Screw	Roller	Nut	Carrier	Ring gear
Rotational inertia ($kg \cdot m^2$)	1.832×10^{-4}	5.248×10^{-7}	8.68×10^{-4}	5.888×10^{-6}	4.48×10^{-5}
Torsional support stiffness (N·m/rad)	2.475×10^{4}	/	1.522×10^{6}	6.956×10^{6}	3.861×10^{6}
Mass (kg)	1.272	0.0328	2.170	0.023	0.112
Thread contact Stiffness ($N \cdot m^{-1}$)	2.078×10^{8}	/	2.198×10^{8}	/	/
Spur gear meshing stiffness ($N \cdot m^{-1}$)	/	/	/	/	2.464×10^{8}

mode are computed and listed in Table 4. The results show that the values of the natural frequencies gradually increase as the number of rollers increases except for the first and second-order natural frequencies and the number of multiplicity increases with the raise of roller number.

Table 4. Natural frequencies of PRSM with different number of the rollers (Hz).

Number of rollers	5	7	8	10	12
Multiplicity l=1	581	522	499	460	429
	971	835	786	709	650
	1019	1203	1285	1435	1571
	1657	1960	2095	2343	2566
	21666	24103	25233	27353	29320
Multiplicity l=n-1	13794	13794	13794	13794	13794

As shown in Table 5, Defining the number of the rollers is seven, the vibration modes of the PRSM can be obtained by Eq. (12), and the results indicate that:

1) The five natural frequencies in condition that the multiplicity is equal to one correspond to five main modes and each mode only represent the torsional vibration characteristic on the parts, and all rollers are subjected to torsional vibration along their own axes in the other four vibration modes except the second-order mode, where the vibration mode of each roller is the same. So these four modes can be defined as rotational mode.

2) The natural frequencies with the occasion of the multiplicity equivalent to (n-1) correspond to different modes, but the motion characteristics of all components are similar, meaning the main parts such as screw, nut, carrier and ring gear do not vibrate while one roller would perform the torsional vibration in each of the modes. Thus, this mode can be defined as the roller mode.

Table 5. Vibration mode of 7-rollers.

	522	835	1203	1960	24103	13794	13794	13794	13794	13794	13794
φ_s^i	0	0	0	0.2481	0	0	0	0	0	0	0
φ_N^i	0	0	0.1291	0	0	0	0	0	0	0	0
φ_c^i	0	1	0	0	0	0	0	0	0	0	0
φ_{r}^i	-0.9986	0	-1	1	-1	0	0	0	0	0	0
φ_1^i	-1	0	-0.9984	0.9923	0.4817	0	0	0	0	0	0
φ_2^i	-1	0	-0.9984	0.9923	0.4817	0	0	0	1	0	0
φ_3^i	-1	0	-0.9984	0.9923	0.4817	0	0	0	0	1	0
φ_4^i	-1	0	-0.9984	0.9923	0.4817	0	0	0	0	0	1
φ_5^i	-1	0	-0.9984	0.9923	0.4817	1	0	0	0	0	0
φ_6^i	-1	0	-0.9984	0.9923	0.4817	0	1	0	0	0	0
φ_7^i	-1	0	-0.9984	0.9923	0.4817	0	0	1	0	0	0

4 Conclusions

In this paper, the natural characteristics of planetary roller screw mechanism considering torsional degree of freedom are invesitigated: the mesh stiffness of thread and spur gear is calculated respectively, the relative displacement between roller and ring gear along the line of action is deduced; the natural frequencies and vibration modes are computed.

From the numerical results, it can be concluded:

1) The planetary roller screw mechanism has two typical modes of vibration: rotational mode and roller mode.

2) The values of the natural frequencies gradually increase as the number of rollers increases except for the first and second-order natural frequencies and the number of multiplicity increases with the raise of roller number.

3) The research on natural characteristic of planetary roller screw mechanism provides a theoretical basis for designers to predict the dynamics performance of transmission system during the early stages of design, where the resonance point and the parameter sensitive points of the transmission system can be avoided according to the dynamics analysis of the system's natural frequency.

4) Previous studies on the dynamics of planetary gear transmission, such as the establishment of a translation-torsion coupling dynamic model and the steady-state dynamic response can be considered to extend into the dynamics analysis of planetary roller screw mechanism.

Acknowledgments

This research was supported by the National Natural Science Foundation of China (Grant No. 51875458), Key Project of National Natural Science Foundation of China (Grant No. 51535009), the Fundamental Research Funds for the Central Universities (Grant No. 31020190504003) and the 111 Project (Grant No. B13044). The authors would like to express their appreciation to the agencies.

References

1. W. J. Zhang, G. Liu, R. T. Tong, S. J. Ma, Load distribution of planetary roller screw mechanism and its improvement approach, *Mech Eng Sci*, **230** (18), 3304-3318 (2016)
2. A. V. Zhdanova, V. V. Morozova, Theoretical study of the load distribution on the threads for roller screw mechanisms of a friction type, *International Conference on Industrial Engineering*, ICIE 2016, *Procedia Engineering*, **150**, 992-999 (2016)
3. J. Rys, F. Lisowski, The computational modal of the load distribution between elements in planetary roller screw, *J Theor App Mech-Pol*, **82**, 699-705 (2014)
4. S. J. Ma, G. Liu, R. T. Tong, A new study on the parameter relationship of planetary roller screw, *Math Probl Eng*, Article ID340437, 29 pages (2012)
5. G. Auregan, V. Fridrici, P. Kapsa, F. Rodrigues, Experimental simulation of rolling-sliding contact for application to planetary roller screw mechanism, *Wear*, **332-333**,1176-1184 (2015)
6. F. Abevi, A. Daidie, M. Chaussumier, M. Sartor, Static load distribution and axial stiffness in a planetary roller screw mechanism, *J Mech Design*, **138** (1), 012301-1~012301-11 (2016)
7. X. C. Zhang, G. Liu, S. J. Ma, R. T. Tong, H. Luo, Study on axial contact deformation of planetary roller screw, *Appl Mech Mater*, **155-156**, 779-783 (2012)
8. M. H. Jones, S. A. Velinsky, Kinematics of roller migration in the planetary roller screw mechanism, *J Mech Design*, **134** (6), 061006-1~0.61106-6 (2012)
9. M. H. Jones, S. A. Velinsky, Contact kinematics in the roller screw mechanism, *J Mech Design* ,**135** (5), 051003-1~051003-10 (2013)
10. Y. Q. Liu, J. S. Wang, H. G. Cheng, Y. P. Sun, Kinematics analysis of the roller screw based on the accuracy of meshing point calculation, *Math. Probl Eng*, Article ID303972, 10 pages (2015)
11. M. H. Jones, S. A. Velinsky, T. A. Lasky, Dynamics of the planetary roller screw mechanism, *J Mech Rob*, **8** (1), 014503-1~014503-6 (2016)
12. S. J. Ma, L. P. Wu, X. J. Fu, Y. J. Li, G. Liu, Modelling of static contact with friction of threaded surfaces in a planetary roller screw mechanism, *Mech Mach Theory*, **139**, 212-236 (2019)
13. X. J. Fu, G. Liu, R. T. Tong, S. J. Ma, Teik C. Lim, A nonlinear six degrees of freedom dynamic model of planetary roller screw mechanism, *Mech Mach Theory*, **119**, 22-36 (2018)
14. K. L. Johnson, Contact mechanics, *Cambridge University Press*, 84-104 (1985)
15. T. A. Harris, M. N. Kotzalas, Rolling bearing analysis, *Fifth Edition, New York: Taylor & Francis Group, LLC*, 104-120 (2007)
16. National Gear Standardization Technical Committee, Compilation of parts and related standards-gear and gear drive, *First Edition China Standard Press*, ISBN 9787506666435 186-187 (2012)

Study of Regenerative Braking System and Brake Force using Pulse Width Module

Warunchit Chueprasert*, and *Danai* Phaoharuhansa

Department of Mechanical Engineering, King Mongkut's University of Technology Thonburi, Bangkok, Thailand

Abstract. This paper aims to develop regenerative braking using pulse width module (PWM) control. It concerns frequency of regenerative braking period which is used to control regenerative braking. It is usually restricted by the electric power generation of AC synchronous axial motor. It is measured by motor bench tester. It indicates electronic brake torque and regenerative current. The resistance torque and regenerative current characteristics are expressed as second order polynomial equation. The results present the comparison between pulse signal and full period signal. The maximum deceleration is 10 Hz signal at 2.14 m/s^2, which is not exceed deceleration of brake comfort. The harvest energy at 10 Hz PWM control is 1.40 Wh. It is closely to use full period regenerative brake, but 10 Hz PWM control has comfortable than other frequencies and efficiency is 54.3 percent.

1 Introduction

Today, electric vehicle (EV) is usually powered by electric motor such as brushless DC (BLDC) motor and AC synchronous motor, which are control by inverter. The inverter convert energy from battery to powered motor so that the performance and the driving distance will depend on the battery capacity. The motor can be either actuator or generator. When vehicle is moved by inertia force of vehicle, the motor will be rotated to generate electricity which is charging to battery. It is call known as regenerative braking system (RSB) that it can slow the car [1-3] and it harvests the energy and then it charges to the battery for increasing the driving distance [4,5]. Powertrain system will be bi-directional power flow. RBS performs by using bi-directional because power able to supply power and charge in one system [6,7]. The energy harvest of RBS has some limitations such as speed and regenerative braking current [8,9]. The current charge to the battery so that it obtains as charging current. The maximum charging current depends on battery type [10,11].

Generally, electric vehicle consists of 2 types of braking systems such as mechanical and electrical brake. This research will study electric braking systems. The papers [12,13] studied the development of various control systems for braking force during regenerative energy and electronic brake directly affects to the vehicle movement so that vehicle controller unit (VCU) is required to determine the period of RBS [4].

VCU requires some control systems to perform RBS. There must be a system to control the operation of the brake system. Anti-lock Braking System (ABS) of EV performs that mechanical brake and electronic brake send the brake pulse according to the speed. Normally, ABS will work in case of emergency brake. To encourage brake performance of electric vehicle, mechanical brake system is accompanied with electronic brake force. Thus, anti-lock braking system for electric vehicle using the both brake systems has been presented in the paper [14,15]. The suitable deceleration of vehicle [16] is studied that maximum deceleration on dry road is 8 m/s^2 and uncomfortable of deceleration is between 5-8 m/s^2. The both conditions should be applied to emergency brake. The comfort deceleration should not over then 3 m/s^2.

Electronic brake has been developed for electric vehicle. It has another advantage such regenerative energy. The brake force of battery electric vehicle is enough to stop the vehicle in non-emergency situation. Therefore, this paper developed the regenerative braking control system considering deceleration of vehicle, brake force and regenerative.

Fig. 1. Free body diagram of tricycle vehicle.

* Corresponding author: warunchit.chu@mail.kmutt.ac.th

https://doi.org/10.1051/matecconf/202030601004

2 Vehicle system

2.1 Equation of motion

According to longitudinal motion of vehicle [17], the equation of motion can be presented as follows

$$m\ddot{x} = F_T - F_{TR} \tag{1}$$

$$F_{TR} = F_r + F_a + F_g \tag{2}$$

The total resistance (F_{TR}) consists of rolling resistance force (F_r), aerodynamic drag force (F_a), gradient force (F_g). When the vehicle is moving on the road, the tire has been deformed by torque of wheel and road's friction. It causes of the rolling resistance force. It is computed that

$$F_r = F_{r,f} + F_{r,r} \tag{3}$$

$$F_{r,n} = \mu_n m_n g \tag{4}$$

where $F_{r,n}$ denotes the rolling resistance force at wheel n, μ_n is the rolling resistance coefficient at wheel n, m_n is the vehicle mass at wheel n, g is gravitational acceleration constant.

The aerodynamic force (F_a) is occurred, when an object moves through the air. It resists the motion of the object. It depends on many factors, such as front area, shape and speed the vehicle. It is given as

$$F_a = \frac{1}{2}\rho_a C_d A V^2 \tag{5}$$

where ρ_a is the density of air, C_d is aerodynamic drag coefficient, A is frontal area of vehicle, V is speed of vehicle.

$$F_T = \begin{cases} \dfrac{\tau_m}{r_w} & ; \textit{acceleration} \\ -\dfrac{\tau_r}{r_w} & ; \textit{brake} \end{cases} \tag{6}$$

where τ_m denotes torque of motor, τ_r denotes resistance torque of motor.

Considering traction force (F_T), it can produce thrust force or resist force, which depends on control strategies. The motor will produce either driving torque (τ_m) or resistance torque (τ_r).

Table 1. Vehicle Specification.

Specification		Unit
Vehicle		
Wheelbase	2	m
Track width	1.5	m
Weight	600	kg
Gear ratio	5	
Wheel radial	0.3	m
AC synchronous axial motors		
Rated power	3.5	kW
Peak power	10	kW

2.2 Electric powertrain system

The powertrain system obtains a bi-directional power flow as shown in Fig. 2, which consists of 3 main components as listed in Fig. 3 such as battery pack, motor controller and AC synchronous axial motor. The electric current can be supplied to drive the motor and the regenerative braking current from motor is recharged to battery as shown in Fig. 2. The vehicle control unit will decide the task of motor such as actuator or generator.

The battery pack in Fig. 3(a) is compounded of 18650 Lithium ion battery cells. It is designed as 36 series and 7 parallel cells. The specifications of battery pack and cell is presented in Table II. It is designed using non-soldering technique and the pack consists of acrylic cover plates, copper taps and Li-ion cells.

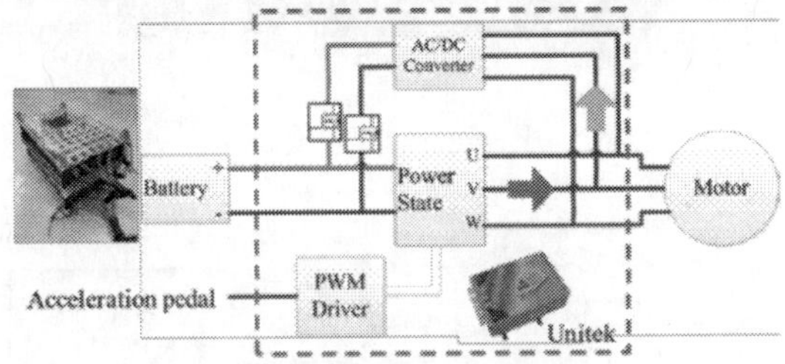

Fig. 2. Bi-directional power flow system of tricycle vehicle.

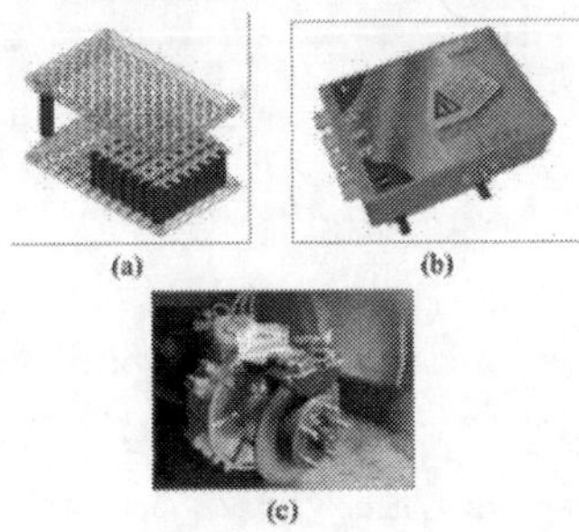

Fig. 3. Components in powertrain system (a) Battery pack
(b) UNITEK Bamocar D3 motor controller
(c) AC synchronous axial motor.

3 Regenerative Braking Control System

The control system consists of component shown in Fig. 4(c). The control system starts from Teensy 3.6 measure data from IMU sensor and generated pulse signal. Pulse signal is pulse width modulation (PWM). NI Myrio is center control unit that it processes signal from any sensor, switch and pulse signal from Teensy to control the enabled period of regenerative braking then operate relay which is connected with UNITEK. The data is record by data logger HIOKI LR8431. It records data from current sensor that measure current between battery and UNITEK shown in Fig. 4(a), hall effect sensor to measure the vehicle speed shown in Fig. 4(b) and regenerative braking signal from relay. The control structure has been presented in Fig. 5.

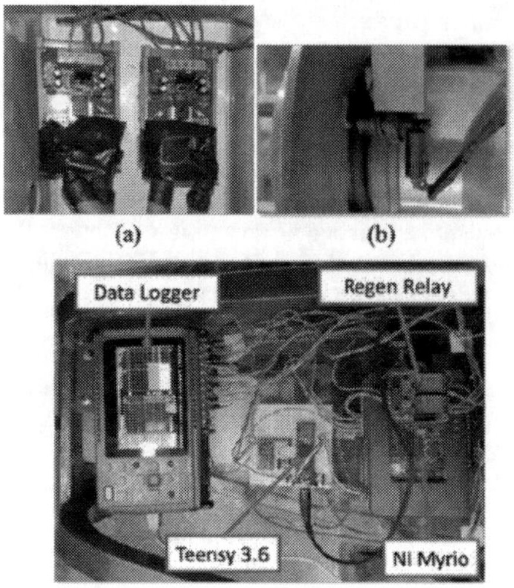

Fig. 4. Components in control system (a) Current sensor (b) Hall effect sensor (c) Control unit and data logger.

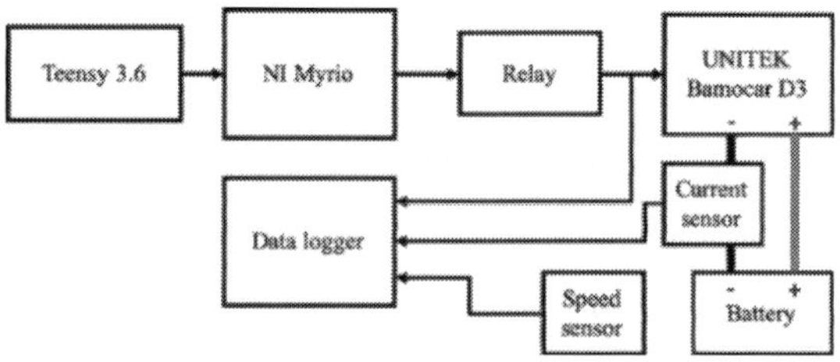

Fig. 5. Structure of control system.

During regenerative braking system, motor will perform as generator where motor coil is rotated, it induces electric back electromotive force (back EMF) and generates resistance torque and transforms the voltage level and current in order to charge battery.

The current and the resistance torque depend on the motor speed. The current must not exceed maximum charging current of battery pack ($I_{ch,max}$) because the overcurrent may harm the battery. The maximum regenerative current can be derived by

$$I_{ch,max} = N_p \, i_{ch,max} \tag{7}$$

where, $I_{ch,max}$ denotes maximum regenerative current, which charges to battery pack. N_p is number of parallel cells, $i_{ch,max}$ is maximum charging current per cell.

The resistance torque should not decelerate the vehicle rather than 3 m/s^2 for non-emergency brake because it may disturb passenger's comfort.

Therefore, RBS will be enabled when braking paddle is pushed and the charging current is less than maximum charging current, which can be derived by Eq. (7). The battery specification is shown in Table. II. The maximum charging current equals 28 A for the battery pack.

Table 2. Battery Specification.

Battery Cell	Value	Unit
Nominal voltage of battery pack	129.6	V
Battery pack's Capacity	21	Ah
Number of series (N_s)	36	
Number of parallel (N_p)	7	
Max. voltage of each cell	4.2	V
Nominal voltage per cell	3.6	V
Max. discharge current per cell	20	A
Max. charging current per cell	4	A

4 Electric power generation

Generally, MYWAY is motor bench tester, system of which has shown in Fig. 6. Test motor performed at desire speed, and then load motor generate resistance torque. The test motor will be controlled to increase torque to stabilize the desired speed.

To measure regenerative of AC synchronous axial motors in Fig. 3(c), load motor performs as actuator. Then, regenerative braking system is enabled. The electricity is generated, and it supplies to the tester. Charging current of RBS and resistance torque of test motor are measured at 250, 750, 1000, 1250, and 1500 rpm as illustrated by Fig. 7 and Fig. 8.

As experiment results in Fig. 7 and Fig. 8, the torque and the current characteristics during regenerative braking are similar that the characteristics can be expressed as second order polynomial equation as shown in Eq. (8) and (9). It represents torque of motor during brake as shown in Eq. (6). The solid lines in the figures exhibit the current and the resistance torque when RBS is enabled. The dashed lines illustrate the current and the resistance torque when RBS is disabled.

During regenerative braking system is disabled, the motor is rotated by inertia force of vehicle. The torque and current characteristics are linear. They are slightly increased corresponding with speed. Thus, they seem constant value as the dashed lines in Fig. 7 and 8.

$$I_{ch} = 0.0005N^2 + 0.028N + 17.434 \tag{8}$$

$$\tau_r = 0.0013N^2 + 0.2213N + 27.327 \tag{9}$$

where, I_{ch} is regenerative current of axial motor. τ_r is resistance torque of motor. N denotes rotation speed of motor.

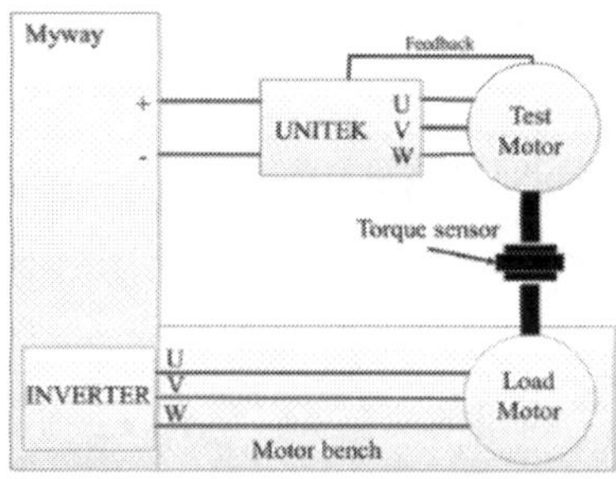

Fig. 6. Motor bench test system.

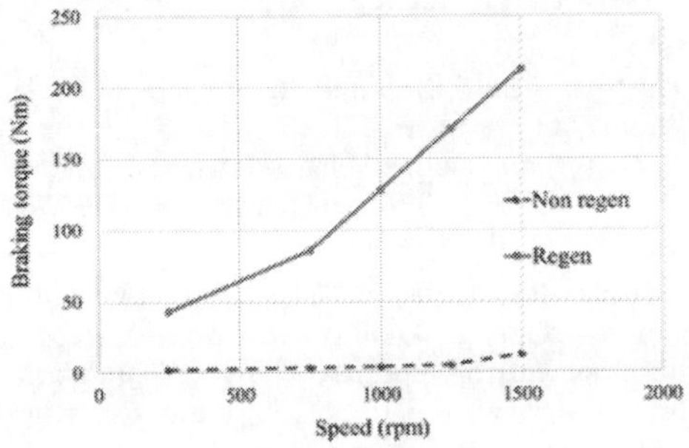

Fig. 7. Torque vs speed between using regen and non regen.

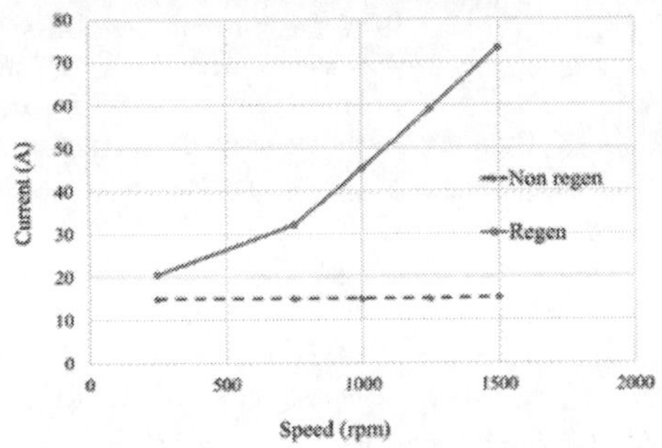

Fig. 8. Current vs speed between using regen and non regen.

5 Experiment Result

The experiment makes a car run straight on a flat road. Then record the data of wheel speed, current, and operating signal of regenerative braking system (RBS). The vehicle accelerates from initial speed to constant speed at 20 km/h, and then the RBS will be started till the vehicle stop. The data are used to compute deceleration and harvest energy. The regenerative braking control signal consist of pulse frequency and full period. Pulse signal form seems as pulse width modulation (PWM) and relates to anti-lock brake system (ABS). The experiment will vary frequency from 1 Hz, 5 Hz, 10 Hz to control regenerative braking system. Full period signal always enables RBS as long as the brake switch is on.

The results of regenerative braking system using 1 Hz, 5 Hz, 10 Hz and full period as shown in Fig. 9. The deceleration of full period, 10 Hz, 5 Hz and 1 Hz are 2.70, 2.14, 2.13 and 1.99 m/s^2, subsequently. The deceleration of 10 Hz signal is closest to full period signal. Harvest energy (E_h) are computed as 1.91, 1.40, 1.28 and 1.55Wh, respectively. Harvest energy at 1 Hz frequency is greater than other frequencies. The energy is only accumulated during the enabled period.

$$\eta = \frac{3600\,E_h}{E_k} \times 100 \tag{10}$$

Where η is efficiency of regenerative braking system. E_k denotes the kinetic energy of vehicle at constant speed, it equal 9,274.08 J.

The efficiency of regenerative braking system can be derived by Eq. (10), it equal 74.1, 54.3, 49.7 and 60.2 percent, respectively.

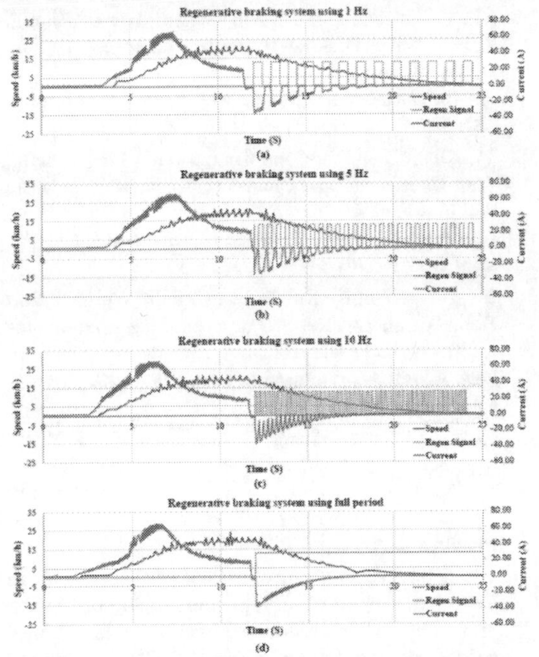

Fig. 9. Experiment result (a) RBS at 1 Hz (b) RBS at 5 Hz (c) RBS at 10 Hz (d) RBS at Full period enable signal.

6 Conclusion

As results, the torque and the charging current characteristics are expressed as second order polynomial equation, when regenerative braking system is disabled, the torque and current characteristics are constant.

Regenerative braking system using full period enable signal has deceleration and energy greater than other pulse frequencies, but brake feel is uncomfortable condition. The harvest energy at 1 Hz enable frequency gets more efficiency, but braking motion has sway motion.

At 10 Hz enable frequency, the vehicle is obtained maximum deceleration at 2.14 m/s^2, which is greater than other frequencies and feeling while braking is comfortable than other frequencies. The harvest energy at 10 Hz enable frequency is between 1 Hz and 5 Hz enable frequencies.

Therefore, the maximum charging current is 28 A and maximum charging current at 10 Hz frequency is 37.51 A that it exceeds the limitation of battery pack in the first period when regenerative braking is enabled. It may damage the battery.

References

1. Rahul Ganpati Chougale and C. R. Lakade, Regenerative braking system of electric vehicle driven by brushless DC motor using Fuzzy logic,

IEEE International Conference on Power, Control, Signals and Instrumentation Engineering, Chennai, India, September 21-22 (2017)

2. Xiaohong Nian, Fei Peng, and Hang Zhang, Regenerative Braking System of Electric Vehicle Driven by Brushless DC Motor, IEEE Transactions on Industrial Electronics, **61**, Issue 10, October (2014)
3. Yu-Chan Chen, Yu-Chen Chang, Jiang-Feng Cheng, Wen-Cheng Yu, and Chun-Liang Lin, Regenerative braking-driving control system, *13th IEEE Conference on Industrial Electronics and Applications (ICIEA)*, June (2018)
4. Ye Tao, Xiaohua Xie, Haiyan Zhao, Wei Xu, and Hong Chen, A Regenerative braking system for electric vehicle with four in-wheel motors based on Fuzzy logic, *Chinese Control Conference, Dalian, China*, July 26-28 (2017)
5. Chen Lv, Junzhi Zhang, Yutong Li, and Ye Yuan, Mechanism analysis and evaluation methodology of regenerative braking contribution to energy efficiency improvement, Energy Conversion and Management, **92**, pp. 469-482, March (2015)
6. Wanzhong Zhao, GangWu, ChunyanWang, Leiyan Yue, and Yufang Li, Energy transfer and utilization efficiency of regenerative braking with hybrid energy storage system, Journal of Power Sources, **427**, pp. 174-183, July (2019)
7. Jin-san Kim, Feel-soon Kang, Sun-phil Kim, and Sung-Jun Park, Bidirectional DC-to-DC converter for motor control unit of electric vehicle, *41st Annual Conference of the IEEE Industrial Electronics Society (IECON 2015)*, Nov. 9-12 (2015)
8. Shinn-Ming Sue, Yi-Shuo Huang, Jhih-Sian Syu, and Chen-Yu Sun, A bi-directional power flow IPM-BLDC motor drive for electrical scooters, *IEEE Conference on Industrial Electronics and Applications*, Taichung, Taiwan, June 15-17 (2010)
9. A. Joseph Godfrey and V. Sankaranarayanan, A new electric braking system with energy regeneration for a BLDC motor driven electric vehicle, an International Journal Engineering Science and Technology, **21**, pp. 704-713, August (2018)
10. Zhengyu Chua, XuningFeng, Minggao Ouyanga, ZuofuWang,LanguangLu, JianqiuLi, and XuebingHana, Optimal charge current of lithium ion battery, Energy Procedia, **142**, pp. 1867-1873, December (2017)
11. Upender Rao Koleti, ChengZhang, Romeo Malik, Truong Quang Dinh, and JamesMarco, The development of optimal charging strategies for lithium-ion batteries to prevent the onset of lithium plating at low ambient temperatures, Journal of Energy Storage, **24**, August (2019)
12. T. Sakamoto, K. Hirukawa, and T. Ohmae, Cooperative control of full electric braking system with independently driven four wheels, *9th IEEE International Workshop on Advanced Motion Control*, Istanbul, Turkey, March 27-29 (2006)
13. Gaurav A. Chandak and A. A. Bhole, A review on regenerative braking in electric vehicle, *Innovations in Power and Advanced Computing Technologies*, Vellore, India, April 21-22 (2017)
14. Chun-Liang Lin and Weng-Ching Lin, ABS control design for two-wheel drive electric vehicles, *IEEE International Conference on Mechanic Automation and Control Engineering*, Hohhot, China, July 15-17 (2011)
15. Sagar Maliye, Pragyanpriyanka Satapathy, Sudeendra Kumar, and Kamalakanta Mahapatra, Regenerative and Anti-Lock braking system in electric vehicles, *IEEE International Conference on Advanced Communications, Control and Computing Technologies*, May 8-10 (2014)
16. Zong Chang-fu, Yang Sheng-nan, and Zheng Hong-yu, A control strategy of electronic braking system based on brake comfort, *2011 International conference on transportation, mechanical, And electrical engineering*, Changchun, China, December 16-18 (2011)
17. R. Rajamani, *Vehicle dynamics and control*, 1st edition, Springer-Verlag US, 2006. ISBN 978-0-387-28823-9.

A new impact dynamics model of a clearance joint considering the adhesive effects in space environment

Ruiting Tong[1,*], *Zefen* Quan[2], *Qi* Wan[1], *Xiaojun* Fu[1], and *Geng* Liu[1]

[1]Shaanxi Engineering Laboratory for Transmissions and Controls, Northwestern Polytechnical University, 710072 Xi'an, P. R. China
[2]Shanghai Aircraft Design and Research Institute, 200436 Shanghai, P. R. China

Abstract. In space environment, there is severe adhesive effect, and the impact contact process shows different phenomenon due to the adhesive forces. In this paper, a new impact dynamics model of a clearance joint is developed considering the adhesive effects. The Hertz contact model, L-J potential and Dugdale model are combined, and an adhesive force term is included in the original hybrid contact force model. Taking a clearance joint as an example, the theory of adhesive forces during the impact contact process is given, and the influence of the adhesive forces is investigated. Negative forces are discovered at the beginning of the contact, and the contact forces of the modified model are higher than the hybrid model. Besides, the indentation depth is also higher than the hybrid model. The energy conservation principle is employed to explain the phenomena, and the contact forces will be underestimated if the adhesive effects are ignored. This work could be contributed to estimating the contact forces and friction forces more reasonablely in the space environment.

1 Introduction

In space environment, the microgravity and high vacuum are the typical characteristics. Under the microgravity, there are irregular impacts all the time, and the high vacuum induces severe adhesive effects due to clean contact surfaces. To realize movement or power transmission, the motion of the mechanical mechanism is necessary, and the clearance is evitable. For a clearance joint, even a small clearance can produce an obvious influence on the dynamic characteristics, wear and lifetime of mechanical systems [1-2], and the impact will greatly influence the friction properties [3].

Many works are done on modelling the impact contact problems, and the clearance joints are always taken as examples. Two classical models, Hertz model and Persson model [4-5] are widely used to investigate the contact behaviors of the clearance joint. Considering the energy loss, Lankarani and Nikravesh [6] improved the Hertz contact model by incorporating a hysteresis damping function, and a Lankarani-Nikravesh(L-N) model was developed. For a cylindrical clearance joint, Liu et al. [7-8] developed an approximate model, and it was more precise than the Hertz model and Persson model. A hybrid contact force model was developed by Bai and Zhao [9], in which the L-N model and an improved elastic foundation model were combined, and the model showed a wide applicable scope. Ma et al. [10] also presented a hybrid contact force model by extending the L-N contact force model and the elastic foundation model, and the hybrid model was validated by comparing with experimental results. Alves et al. [11] compared various contact force models, and pointed out that the dynamic behaviors of contact bodies greatly depended on the contact force model. A nonlinear model for a clearance joint was introduced by Wang et al. [12], and the impact velocities, material properties and geometry dimensions of the journal and bearing were considered, which made it get a wide suitable range.

The adhesive effects are not considered in the references above due to that they are so slight in ambient environment and could be neglected. In the space environment, adhesion forces are more important between clean metal surfaces than the ambient environment, and the friction coefficient in vacuum environment could be six times higher than the case of ambient environment [13]. Therefore, the adhesive effects should be considered in the impact dynamics model in space environment.

Several classical theories could consider the influence of the adhesion force in contact force models. Bradley [14] investigated the adhesive contact between two contact rigid spheres, and the expression of the adhesion was given. Derjaguin-Muller-Toporov (DMT) theory [15] was developed based on the adhesive contact model between a ball and a plane, and a relationship between the adhesion force and the radius of the ball was described. The Johnson-Kendall-Roberts (JKR) theory of Johnson et al. [16] presented the formula of the adhesive contact force between two elastic spherical solid surfaces. Tabor [17] defined a Tabor parameter, and the competition between the DMT theory and JKR theory could be solved. By using the Dugdale model, Maugis [18] developed a Maugis-Dugdale (M-D) theory that was

* Corresponding author:tongruiting@nwpu.edu.cn

a JKR-DMT transition, and the intermediate region between the JKR theory and DMT theory was discovered. Extending the M-D theory, Baney and Hui [19] developed a cohesive zone model to describe the adhesion between long cylinders in contact. Based on these theories, more research in this area were carried out, while the adhesion theories above were not used in impact contact force models.

In this paper, taking a clearance joint as an example, based on the hybrid contact force model [9], a modified contact force model is developed combining the adhesive effects in space environment. The theory of adhesive forces during the impact process is given, and an example is presented in the final. The results of these two models are compared, and the adhesive effects are discussed.

2 Model description

2.1 Hybrid contact force model [9]

Bai and Zhao [9] developed a hybrid contact force model, and the contact force is

$$F_{Nhyb} = K_n \delta^n + D_{mod} \dot{\delta} \tag{1}$$

where K_n is a nonlinear stiffness coefficient, δ is the deformation, the exponent n is equal to 1.5, D_{mod} is the modified damping coefficient and it is related to the nonlinear stiffness coefficient, K_n, and $\dot{\delta}$ is the relative deformation velocity. Based on the Persson contact model, the relationship between the normal force P and the deformation δ is

$$P = \frac{1}{2}\pi\delta E^* \sqrt{\frac{\delta}{2(\Delta R + \delta)}} \tag{2}$$

where E^* is the equivalent elastic modulus, and ΔR is difference between the radii of the bearing and journal.

$$\frac{1}{E^*} = \frac{1-\mu_B^2}{E_B} + \frac{1-\mu_J^2}{E_J} \tag{3}$$

$$\Delta R = R_B - R_J \tag{4}$$

where E_B and E_J are the elastic modulus, μ_B and μ_J are the Poisson's ratio, and R_B and R_J are the radii of the bearing and journal, respectively.

The nonlinear stiffness coefficient can be obtained as

$$K_n = \frac{dP}{d\delta} = \frac{d\left(\frac{1}{2}\pi\delta E^* \sqrt{\frac{\delta}{2(\Delta R + \delta)}}\right)}{d\delta} = \frac{1}{8}\pi E^* \sqrt{\frac{2\delta\left[3(R_B - R_J) + 2\delta\right]^2}{(R_B - R_J + \delta)^3}} \tag{5}$$

The modified damping coefficient can be obtained as

$$D_{mod} = \frac{3K_n\left(1-c_e^2\right)e^{2(1-c_e)}\delta^n}{4\dot{\delta}_0} \tag{6}$$

where c_e is the coefficient of restitution, and $\dot{\delta}_0$ is the initial impact velocity, which are used to represent impact.

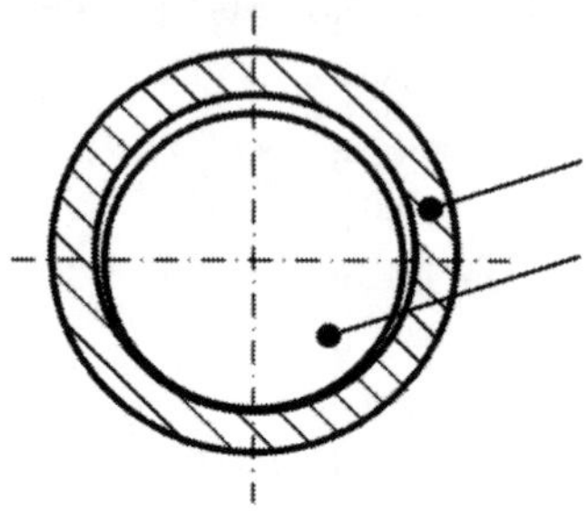

Fig. 1. A simplified clearance joint

2.2 Adhesive force model

If the adhesive effects are considered, there will be adhesive pressure p_A on the interface in addition to the Hertzian pressure p_H. The Hertzian pressure acts on the region $|r| < a$, and the adhesive pressure acts on the region $|r| < c$, as shown in Fig. 2.

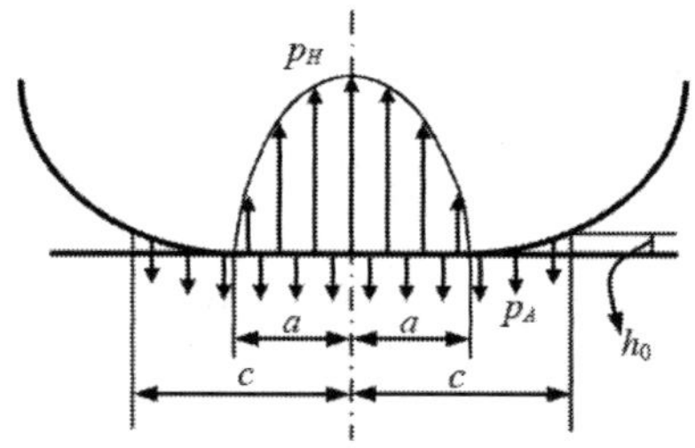

Fig. 2. Adhesion at the interface of the contact bodies [20]

Generally, the adhesive pressure p_A is assumed as a function of separation z between the two contact surfaces, and it follows the law derived from the Lennard-Jones (L-J) potential

$$p_A(z) = \frac{8\omega}{3z_0}\left[\left(\frac{z}{z_0}\right)^{-3} - \left(\frac{z}{z_0}\right)^{-9}\right] \tag{7}$$

where ω is the adhesion energy, and z_0 is the equilibrium separation of the surfaces.

For a two-dimensional adhesive contact between a cylinder and a plane, Baney and Hui [19] simplified the L-J force, as shown in Fig. 3. The adhesive force out of the contact area was assumed as a constant $\sigma_0 \cdot A$ (σ_0 is a constant called theoretical stress, and $\sigma_0 = \omega / h_0$, $h_0 = 0.97z_0$, A is the contact area of the region $a \le |r| \le c$), up to a maximum separation h_0, and it falls to zero if the separation is greater than h_0.

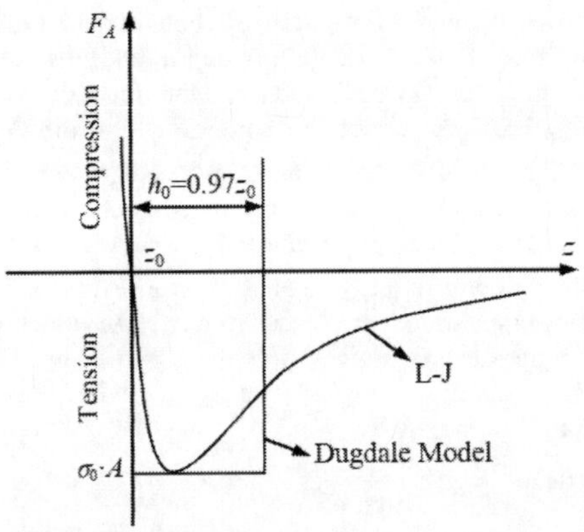

Fig. 3. Adhesive forces derived from the L-J potential and the Dugdale model in fracture mechanics

The pressure caused by the adhesive force for cylinders can be expressed as

$$p_A(r) = \begin{cases} \sigma_0 & a \le |r| \le c \\ \dfrac{\sigma_0}{\pi}\tan^{-1}\sqrt{\dfrac{c^2-a^2}{a^2-r^2}} & |r| < a \end{cases} \quad (8)$$

When $|r| < a$, the compression plays the dominant role, so the pressure caused by the adhesive force in the range of $|r| < a$ is ignored in this paper.

Considering the clearance joint in Fig. 1, when the impact occurs, the deformation is shown in Fig. 4. We define that $\widehat{DA} = a$ is the Hertzian contact half-width, and $\widehat{AC} = c - a$ is the region where adhesive forces are considered. Then the adhesive region can be rewritten as $\widehat{AC} = R_B \cdot \theta$.

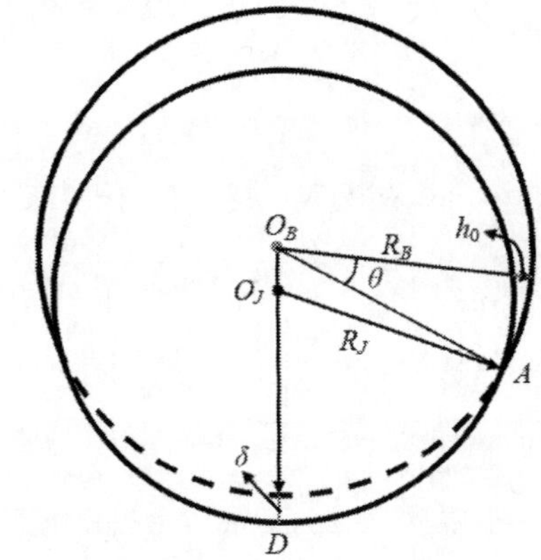

Fig. 4. Deformation in impact contact between the journal and bearing

Based on the definition and the simplication, we can obtain the adhesive pressure

$$p_A = \sigma_0 \quad a \le |r| \le c \quad (9)$$

and the adhesive force in a two-dimensional problem can be written as

$$F_A = p_A \cdot \widehat{AC} = \sigma_0 \cdot \widehat{AC} = \sigma_0 \cdot R_B \cdot \theta \quad a \le |r| \le c \quad (10)$$

2.3 Modified contact force model

Coupling the hybrid contact force model [9] and the adhesive force model in this paper, we can obtain a modified contact force model as follows

$$F_{N mod} = K_n \delta^n + D_{mod} \dot{\delta} - F_A \quad (11)$$

In equation (11), the first two terms are the hybrid contact force, and the parameters are the same as Bai and Zhao's model [9].

3 Example

Taking a clearance joint as an example, the materials of the journal and bearing are the same, and the parameters used in the simulation are listed in Table 1. Using the software Matlab, both the hybrid contact force model and modified contact force model are simulated, and the results are compared in Fig. 5-Fig. 7.

Table 1. Parameters of the example.

Parameters	**Value**
Material	Iron(BCC Fe)
Elastic modulus E/Pa	2.11×10^{11}
Poisson's ratio μ	0.3
Radius of the journal R_J/m	9.5×10^{-3}
Radius of the bearing R_B/m	10.0×10^{-3}
Adhesion energy ω/J·m^{-2}	3.07[21]
Equilibrium separation z_0/m	1.58×10^{-10}
Mass of the journal m/kg	0.08
Initial impact velocity $\dot{\delta}_0$/m·s^{-1}	0.3
Coefficient of restitution c_e	0.8

Fig. 5 shows the relationship between the time and the contact forces. The negative contact forces at the beginning of the impact contact indicates that there are obvious adhesive effects. When the journal just contacts the bearing, the adhesive forces act on the journal, which means that the journal is accelerated, and the contact forces of the modified model are higher than the hybrid model. In addition, due to that the energy done by the adhesive forces acts on the journal, so the impact contact time is longer for the modified model compared with the hybrid model. At the end of the impact process, the contact forces also show negative values to show the adhesive effects.

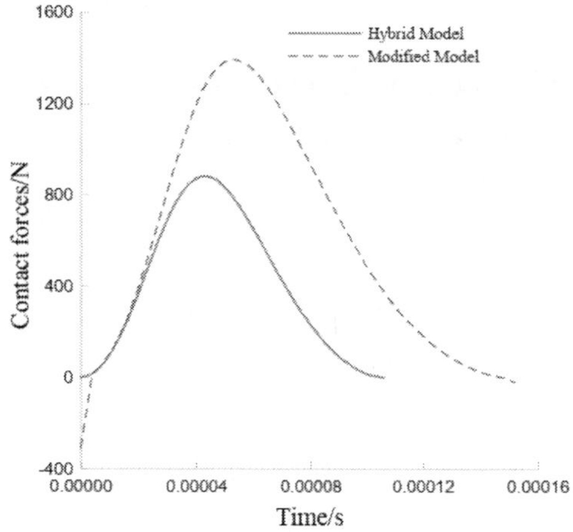

Fig. 5. The relationship between the time and the contact forces

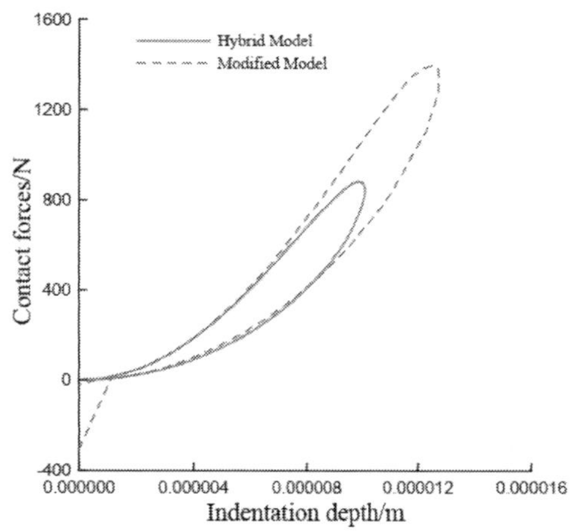

Fig. 6. The relationship between the indentation depth and the contact forces

The impact-restitution curves are given in Fig. 6, and the indentation depth and contact forces obtained from the modified model are higher than the hybrid model. From the view of energy conservation principle, the kinetic energy of the journal will be converted to the mechanical energy during the impact process. In the modified model, the total energy includes the work done by the adhesive forces, so the indentation depth and the contact forces are higher than the hybrid model. Furthermore, the energy conversion process should be the reason for the phenomenon in Fig. 7.

4 Conclusions

Considering the adhesive effects in the space environment, a modified contact force model is developed based on the Hertz contact model, L-J potential and Dugdale model. A clearance joint is used as an example, and the relationship between the time and contact forces, the indentation depth and the contact forces and the time and indentation depth are investigated. Some conclusions are drawn as follows:

(1) If the adhesive effects are considered, there will be negative contact forces, and the contact forces of the modified model are higher than the hybrid model.

(2) The kinetic energy and the work done by the adhesive forces are converted to the mechanical energy in the impact process, so the indentation depth of the modified model is higher than the hybrid model.

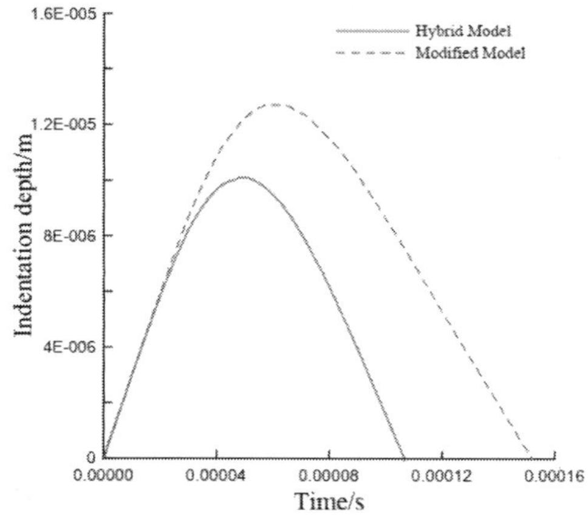

Fig. 7. The relationship between the time and the indentation depth

The authors would like to thank the National Natural Science Foundation of China (51675429), Key Project of National Natural Science Foundation of China (51535009), the Fundamental Research Funds for the Central Universities (31020190503004) and the 111 Project (B13044) for their financial support.

References

1. S. Erkaya, J. Mech. Sci. Technol. **26**, 1419-1430 (2012)
2. X.P. Wang, G. Liu, J. Mech. Sci. Technol. **29**, 4113-4120 (2015)
3. R.T. Tong, G. Liu, Microgravity Sci. Tec. **31**, 85-94 (2019)
4. M. Ciavarella, P. Decuzzi, Int. J. Solids Struct. **38**, 4507-4523 (2001)
5. M. Ciavarella, P. Decuzzi, Int. J. Solids Struct. **38**, 4523-4533 (2001)
6. H.M. Lankarani, P.E. Nikravesh, J. Mech. Des. **112**, 369-376 (1990)
7. C.S. Liu, K. Zhang, R. Yang, Acta Mech. Sinica **21**, 451-458 (2005)
8. C.S. Liu, K. Zhang, R. Yang, Mech. Mach. Theory **42**, 183-197 (2007)

9. Z.F. Bai, Y. Zhao, Int. J. Nonlin. Mech. **48**, 15-36 (2013)
10. J. Ma, L. Qian, G. Chen, M. Li, Mech. Mach. Theory **94**, 148-164 (2015)
11. J. Alves, N. Peixinho, M. Tavares da Silva, P. Flores, H.M. Lankarani, Mech. Mach. Theory **85**, 172-188 (2015)
12. X.P. Wang, G. Liu, S.J. Ma, J. Mech. Sci. Technol. **30**, 1537-1545 (2016)
13. D.F. Moore, *Principles and Applications of Tribology*. (Pergamon Press, Oxford, 1975)
14. R.S. Bradley, Philos. Mag. **13**, 853-862 (1932)
15. B.V. Derjaguin, V.M. Muller, Y.P. Toporov, J. Colloid Interface Sci. **53**, 314-326 (1975)
16. K.L. Johnson, K. Kendall, A.D. Roberts, Proc. R. Soc. London, Ser. A **324**, 301-313 (1971)
17. D. Tabor, J. Colloid Interface Sci. **58**, 2-13 (1977)
18. D. Maugis, J. Colloid Interface Sci. **150**, 243-269 (1992)
19. J.M. Baney, C.-Y. Hui, J. Adhes. Sci. Technol. **11**, 393-406 (1997)
20. T.X. Liu, G. Liu, Q. Xie, Q.J. Wang, ASME J. Tribol. **128**, 40-48 (2006)
21. A.M. Guellil, J.B. Adams, J. Mater. Res. **7**, 639-652 (1992)

Numerical simulations of blood cell flow in non-Newtonian fluid

Kazuhiro Shitara[1], and *Toru* Hyakutake[2,*]

[1]Graduate School of Engineering, Yokohama National University, 240-8501 Yokohama, Japan
[2] Faculty of Engineering, Yokohama National University, 240-8501 Yokohama, Japan

Abstract. We investigated how non-Newtonian viscosity behavior affects the flow characteristics of blood cells. Our findings offer insight about how shear thinning affects the dispersion of liposome-encapsulated hemoglobin and red blood cells in blood. The lattice Boltzmann method was used for fluid calculations, and the rheological properties of the non-Newtonian fluid were modeled with power-law relationships. The deformable three-dimensional red blood cell model was applied. First, we investigated the effects of shear thinning on the flow behavior of single blood cell. Simulation results indicate that shear thinning promotes the axial concentration of red blood cells. Next, varied the hematocrit to see how mutual interference between blood cells affects flow. At low hematocrit, shear thinning clearly promotes the axial concentration of red blood cells. As the hematocrit increases, in contrast, mutual interference has a greater effect, which counteracts shear thinning so the red blood cell distribution resembles the distribution within a Newtonian fluid.

1 Introduction

Red blood cells (RBCs) play an essential role in supplying oxygen to all tissues in the body. When the diameters of RBCs and vessels are on the same order of magnitude, the rheological properties of RBCs greatly influence blood flow. RBCs are biconcave and disk-shaped, and they deform passively under external force. This deformability has some effect on RBCs' tendency to concentrate along the axis of flow paths, which helps them to flow smoothly through narrow microvessels.

Several recent numerical studies have focused on the behavior of deformable RBCs in microvascular flows [1-9]. The motion of RBCs is described well with existing numerical techniques. However, most studies have focused on the flow of RBCs in the Newtonian fluid of blood plasma. Currently, liposome-encapsulated hemoglobin (LEH) is being promoted vigorously in Japan as synthetic alternative to RBCs [10]. Typical LEH particles are 200–250 nm in diameter, which is 1/30 smaller than a human RBC. A solution of plasma and LEH shows non-Newtonian behavior in some conditions [11]. To better understand this behavior, we have simulated the flow of surrounding plasma-LEH solution on the flow of erythrocytes in microvessels. The simulation results offer valuable insights into the mechanisms by which the flow of RBCs delivers oxygen through microvascular networks.

2 Numerical method

We constructed a three-dimensional microvascular model and investigated the flow behavior of RBCs in the Newtonian and non-Newtonian fluid. Figure 1 shows schematics of the microvascular model we employed, in which the vessel diameter (D) is 20 μm and the vessel length (L_x) is 40 μm. The mean velocity in the vessel U_m was set to 1.5×10^{-3} m/s. Three hematocrit levels were simulated, at 3.0 %, 6.0 % and 15 %. Blood flow was simulated using the lattice Boltzmann method (LBM) [12] in conjunction with the immersed boundary method (IBM) [13] to account for fluid-membrane interactions between the flow field and deformable RBCs. Guo's method was adopted to account for spatiotemporal changes in forces on the flow and to model unsteady flows accurately [14]. At the upstream and downstream boundaries, periodic boundary conditions with a pressure difference were imposed. The no-slip boundary condition was applied to the boundary nodes to represent that the flow velocity at the wall matches the velocity of the wall.

To model RBCs, the unstressed structure was given the biconcave shape proposed by Evans and Fung [15]. Each RBC membrane was modeled with 1,272 triangular elements. Membrane forces due to strain, bending,

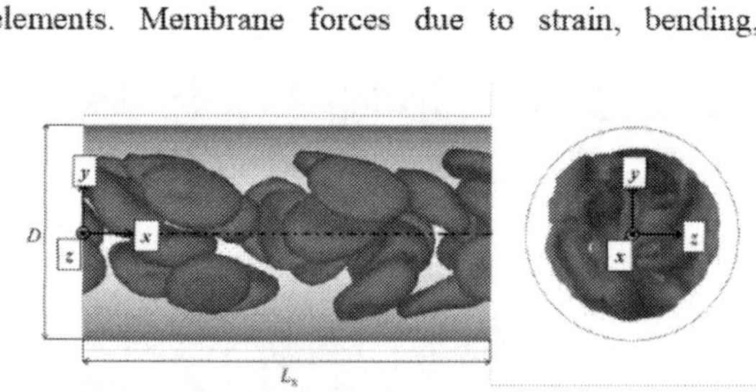

Fig. 1. Numerical model

* Corresponding author: hyaku@ynu.ac.jp

 https://doi.org/10.1051/matecconf/202030601006

surface, and volume constraints were computed from the corresponding energy terms using the principle of virtual work [16]. Reference values were used to determine the parameters of the RBC membrane [17]. Viscosity inside the RBCs was assumed to be five times greater than the viscosity of plasma without LEH.

The surrounding fluid was modeled as both plasma without LEH (Newtonian) and plasma with LEH (non-Newtonian). For non-Newtonian fluid, we applied the following power-law model to describe the fluid viscosity,

$$\mu = \mu_1 \gamma^{n-1} \qquad (1)$$

where μ is the apparent viscous coefficient, γ is the shear rate, and n is an index that distinguishes non-Newtonian and Newtonian fluids. μ_1 means the viscous coefficient at n = 1.0 (Newtonian fluid). Figure 2 plots the relationship between shear rate and the viscosity of the surrounding fluid that follows from Eq. 1. For the non-Newtonian fluid, we assumed n = 1 and μ_1 = 1.2 mPa·s. Figure 3 shows the distribution of viscosity in the microvessel. The viscosity of the Newtonian fluid was constant in the vessel. For the non-Newtonian fluid, in contrast, the high-viscosity fluid was concentrated around the vessel axis since the flow velocity was low near the axis. This high-viscosity region was approximately five times more viscous than the fluid near the vessel walls.

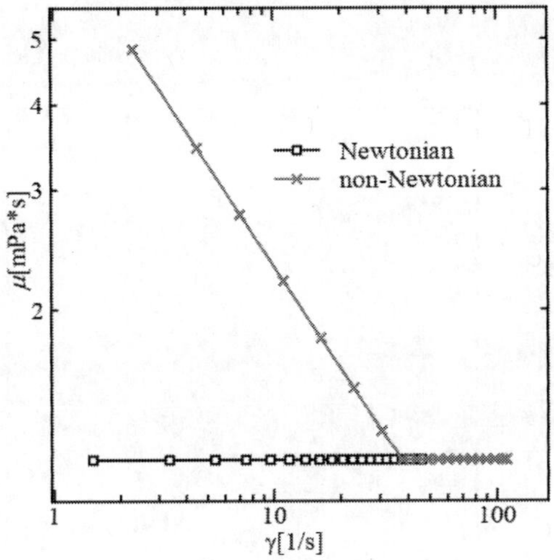

Fig. 2. Relationship between shear rate and viscosity

3 Results and discussion

3.1 Single RBC

First, we calculated the flow of single RBC in Newtonian and non-Newtonian fluids. Figure 4 shows the trajectories of the center of gravity of the simulated RBCs. We considered two initial positions with y_0 set to 2.0 and 5.0 μm. The position of y_0 = 5.0 μm is the midpoint between the vessel centerline (y = 0 μm) and wall (y = 10 μm). As time passes, the y-coordinates of the RBCs changed. Shear thinning in the non-Newtonian fluid tended to promote the axial concentration of RBCs. In a non-Newtonian fluid, viscosity is higher at the center of the vessel. In this simulation, since the mean velocity was constant, the capillary number (Ca = $\mu U_m/DE_s$) increases toward the vessel center, where E_s is the strain modulus of the RBC membrane. A higher capillary number causes larger deformation of the RBC. The associated variation in the viscosity in the vessel encourages the RBCs to cluster around the vessel center.

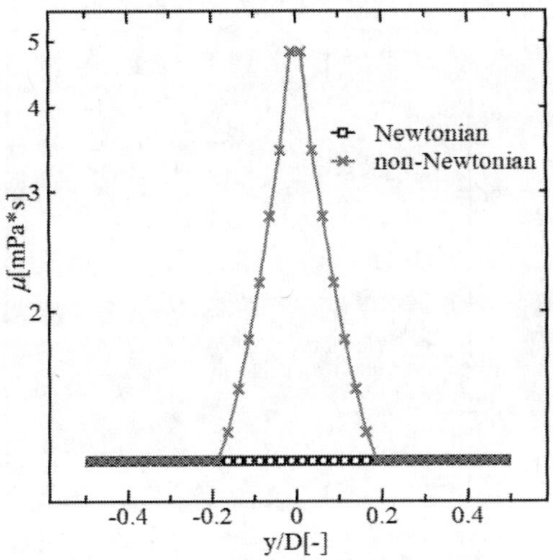

Fig. 3. Viscous distribution in the microvessel

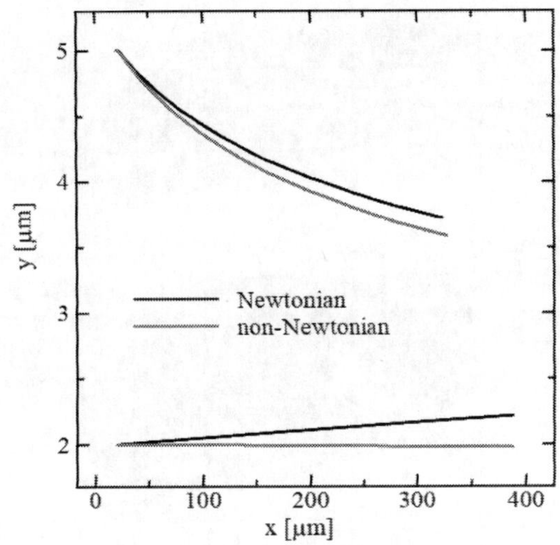

Fig. 4. Trajectories of single RBC

3.2 Multiple RBCs

Next, we calculated the flow of multiple RBCs. Figure 5 plots the RBC flow for the three hematocrit values we simulated. Figure 6 shows the time average of the RBC distribution in the vessel. The horizontal axis gives the distance from the vessel centerline, and the vertical axis indicates the area ratio of RBC passing through the downstream cross-section. These results show that a cell-free layer formed due to the axial concentration of RBCs. This cell-free layer became thicker as the hematocrit decreased. In the case of Hct = 3.0 %, the RBC distributions in Newtonian and non-Newtonian fluids are very different. With small hematocrit values, shear thinning dominates the axial concentration of RBCs. As the hematocrit increases, mutual interference between RBCs has more of an effect, so the distribution of RBCs is less different between Newtonian and non-Newtonian fluids. When Hct = 6.1 %, the RBCs were more clustered near r = 4 μm in non-Newtonian fluid than in Newtonian fluid. To investigate this phenomenon in more detail, we defined the orientation angle, which is the angle between x-direction and the longitudinal direction of the RBC. We recorded the orientation angles

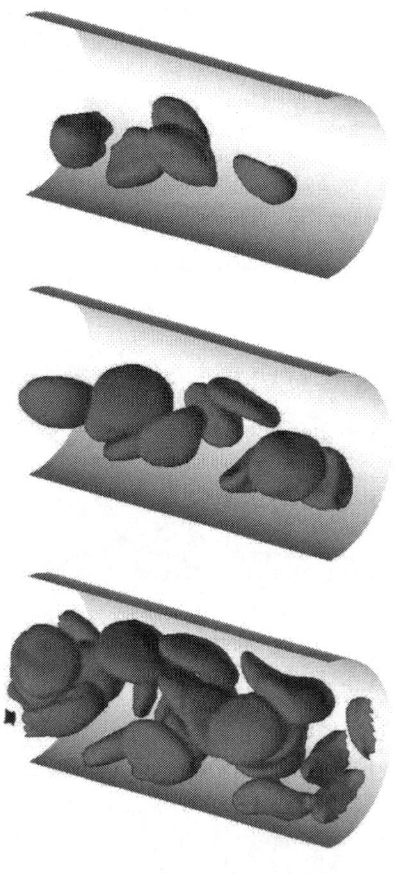

Fig. 5. Image of RBC flow for various hematocrit (upper : Hct = 3.0 %, middle : Hct = 6.1 %, lower : Hct = 15 %)

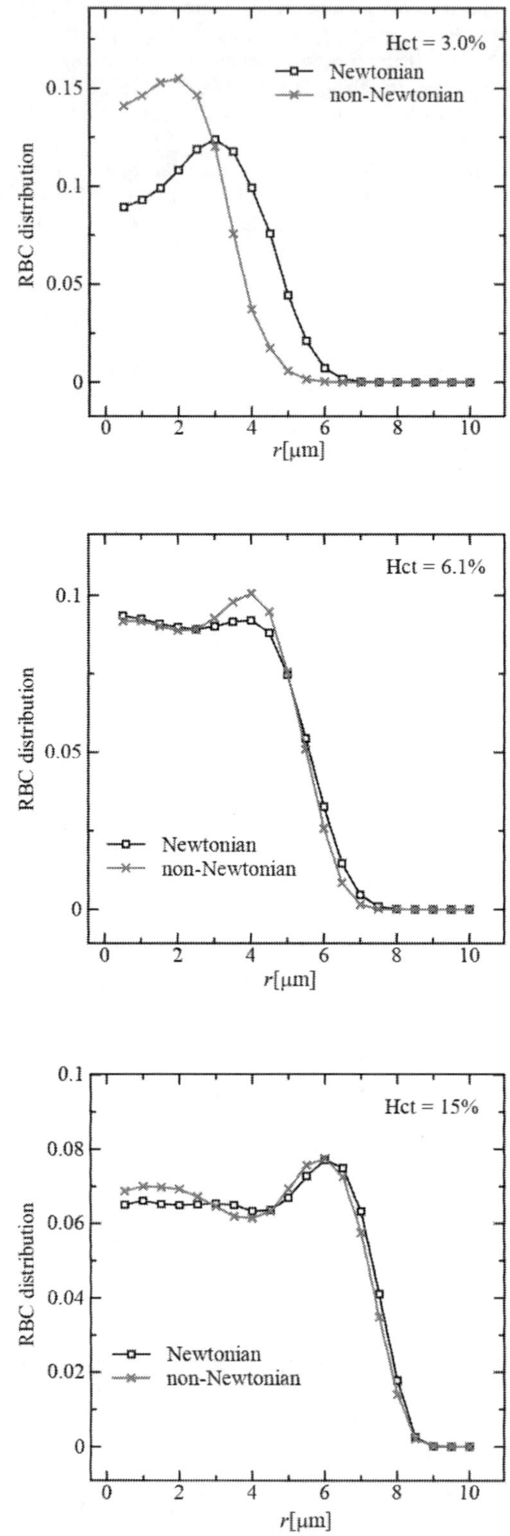

Fig. 6. RBC distributions for various hematocrit (upper : Hct = 3.0 %, middle : Hct = 6.1 %, lower : Hct = 15 %)

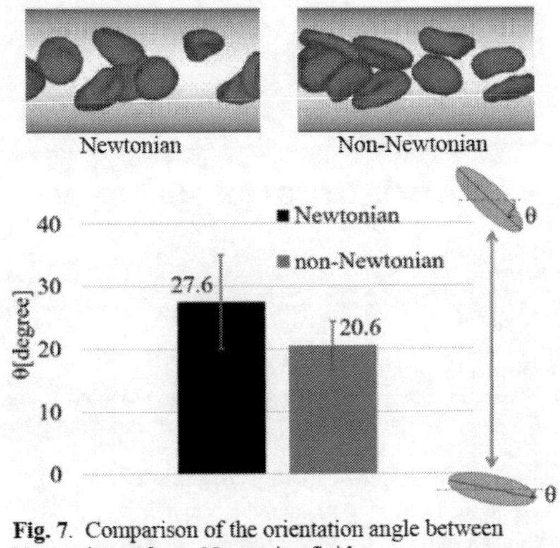

Fig. 7. Comparison of the orientation angle between Newtonian and non-Newtonian fluids

of the RBCs using image analysis. In Fig. 7, the RBC flow images and the orientation angles are compared between Newtonian and non-Newtonian fluids at Hct = 6.1%. The angles for non-Newtonian fluid tended to be larger than those in Newtonian fluids. This explains the concentration of RBCs around r = 4 μm. On the other hand, with Hct = 15 %, the difference in the orientation angles almost disappeared.

4 Conclusion

In this study, to clarify the flow behavior of RBCs in blood with added LEH, the effects of shear thinning on the flow in blood vessels were investigated using three-dimensional microvasucular models. First, we investigated the flow behavior of single blood cell in Newtonian and non-Newtonian fluids. The simulation results indicate that shear thinning encourages RBCs to concentrate around the vessel axis. Next, we investigated the flow behavior of multiple blood cells by varying the hematocrit. These simulation results indicate that shear thinning has more of an effect at low hematocrit levels. The orientation angle of the RBCs was finally investigated. The difference in the orientation angles in Newtonian and non-Newtonian fluids explains the variation in RBC distribution within microvessels. As the hematocrit increases further, the interactions between the RBCs have more of an effect, so the discrepancy in the distribution RBCs in Newtonian and non-Newtonian fluids nearly disappeared.

References

1. Boryczko, K., Dzwinel, W., Yuen, D.A., Dynamical clustering of red blood cells in capillary vessels, J. Mol. Model., **9** (2003), 16–33.
2. Dupin, M.M., Halliday, I., Care, C.M., Alboul, L., Munn, L.L., Modeling the flow of dense suspensions of deformable particles in three dimensions, Phys. Rev. E, Stat. Nonlin. Soft Matter Phys., **75** (2007), 066707.
3. Zhang, J., Johnson, P.C., Popel, A.S., Effects of erythrocyte deformability and aggregation on the cell free layer and apparent viscosity of microscopic blood flows, Microvasc. Res., **77** (2009), 265–272.
4. Sugiyama, K., Ii, S., Takeuchi, S., Takagi, S., Matsumoto, Y., Full Eulerian simulations of biconcave neo-Hookean particles in a Poiseuille flow, Comput. Mech., **46** (2010), 147–157.
5. Zhao, H., Isfahani, A.H.G., Olson, L.N., Freund, J.B., A spectral boundary integral method for flowing blood cells, J. Comput. Phys., **229** (2010), 3726–3744.
6. Tsubota, K., Wada, S., Effect of the natural state of an elastic cellular membrane on tank-treading and tumbling motions of a single red blood cell, Phys. Rev. E, Stat. Nonlin. Soft Matter Phys., **81** (2010), 011910.
7. Imai, Y., Kondo, H., Ishikawa, T., Teck Lim, C., Yamaguchi, T., Modeling of hemodynamics arising from malaria infection, J. Biomech., **43** (2010), 1386–1393.
8. Secomb, T.W., Mechanics and computational simulation of blood flow in microvessels, Med. Eng. Phys., **33** (2011), 800–804.
9. Hyakutake, T., Nagai, S., "Numerical simulation of red blood cell distributions in three-dimensional microvascular bifurcations", Microvasc. Res., **97** (2015), 115–123.
10. Sakai, H., Sou, K., Horinouchi, H., Kobayashi, K., Tsuchida, E., Review of hemoglobin-vesicles as artificial oxygen carriers, Artif. Organs, **33** (2009), 139–145.
11. Sakai, H., Sato, A., Okuda, N., Takeoka, S., Maeda, N., Tsuchida, E., "Peculiar flow patterns of RBCs suspended in viscous fluids and perfused through a narrow tube (25μm)", Am. J. Physiol.-Heart Circul. Physiol., **297** (2009), 583–589.
12. Succi, S., The lattice Boltzmann equation: for fluid dynamics and beyond, Oxford: Oxford University Press (2001).
13. Peskin, C.S., Numerical analysis of blood flow in the heart, J. Comput. Phys., **25** (1977), 220–252.
14. Guo, Z., Zheng, C., Shi, B., Discrete lattice effects on the forcing term in the lattice Boltzmann method, Phys. Rev. E, Stat. Nonlin. Soft Matter Phys., **65** (2002), 046308.
15. Evans, E.A., Fung, Y.C., Improved measurements of the erythrocyte geometry, Microvasc. Res., **4** (1972), 335–347.
16. Krüger, T., Computer simulation study of collective phenomena in dense suspensions of red blood cells under shear, Berlin: Springer (2012).
17. Oulaid, O., Saad, A.K.W., Aires, P.S., Zhang, J., Effects of shear rate and suspending viscosity on deformation and frequency of red blood cells tank-treading in shear flows, Comp. Methods Biomech. Biomed. Engin., **19** (2016), 648–662.

Numerical Simulation of Slurry flow in horizontal pipe based on CFD-DEM

Jinchao Fan[1], *Benchun* Yao[1], *Yi* Hao[2], *Shimin* Zhang[1] *, and *Xiaoxiao* Zhu[1]

[1]China University of Petroleum-Beijing, No.18 Fuxue Road, Changping District Beijing, China
[2]China COMAC Shanghai Aircraft Design and Research Institute, 5188 jinke road, zhangjiang, pudong new area, Shanghai, China

Abstract. In this paper, we propose a novel pipeline cleaning method utilizing slurry. The reason why slurry can be used for pipeline cleaning is that the collisions between the particles and the contaminant in the pipe wall can enhance the cleaning effect. A slurry with polypropylene particles embedded in water is used to cleaning a horizontal pipe is conducted to realize this method. Because the flow characteristics of the slurry is crucial for the cleaning process, it is valuable to conduct a simulation and investigate the influence of several different factors including the convey velocity and the particle size. A 3D CFD-DEM model has been established. The indicators including pressure loss, particle accumulation level at the top of the pipe are choses to characterize the slurry flow and the influence of convey velocity and particle size has been investigated accordingly. In addition, an effective method is proposed to determine the critical convey velocity for each size of the particle.

1 Introduction

Pipeline is an important tool for industrial production. It plays a very important role in the transportation of goods including food, water, and oil&gas etc [1]. Contaminant will generate inside the pipe wall, especially in the long-distance pipeline, which will not only cause pollution to the goods but also increase the flow resistance. In severe cases, it will even leads to suspension accident[2]. Therefore, the pipeline needs to be cleaned regularly. The traditional pipeline cleaning method can be divided into chemical cleaning and physical cleaning. Chemical cleaning needs the use of the pharmacy which may lead the pollution of the environment. Physical cleaning mainly includes waterjet cleaning, cleaning PIG, and mechanical cleaning etc. The waterjet cleaning needs complex equipment leading the cleaning cost to be very high. The cleaning PIG cannot get through the pipe which has a complex geometry. The mechanical cleaning may damage the pipe wall. In short, each physical cleaning method has its own obvious disadvantage. The slurry which carries particles has a good cleaning effect on the contaminant due to the continuous collision of the particles with the dirt inside the pipe during the flow.

In the process of pipeline cleaning using slurry, the flow characteristics is an important factor which affects the cleaning result a lot. The flow characteristics of the slurry could be roughly divided into three stages, namely stationary bed, sliding bed and suspension[3]. In order to enhance the cleaning, the flow is expected to reach the state of suspension as far as possible to ensure effective collision between particles and contaminant inside the pipe. However, it results in a very high flow velocity and great pressure loss as well as an increasing cost of the cleaning process. Therefore, a research on the flow characteristics of the slurry is a necessity to make the slurry-based pipeline cleaning possible. Specifically, the role played by the parameters such as particle density and diameter, flow velocity should be investigated.

In order to conduct an elementary research, water is used as the carrier because it is cheap, easy available and environment friendly. Because suspension slurry is preferred, the polypropylene particle is chosen as its density is closs to the carrier.

In the research of slurry flow characteristics, most slurries are sedimentary. The particles are heavier than the water[4-5]. In recent years, the flow characteristics of a slurry composed of light particles has gradually attracted attention. It mainly includes research on the ice slurry transport[6], pipeline transport of agricultural slurry such as wheat straw[7]. In the research of the slurry composed of polypropylene particles, Denis Edelin[8] used experimental methods to study the flow characteristics of the slurry. In his experiment the particle diameters are less than 1 mm, however, the particles which have a diameter more than 1mm are not studied.

In this paper, the computational fluid dynamics and discrete element method coupling calculation method (CFD-DEM) is used to simulate the flow characteristics of slurry formed by polypropylene particles with diameters of 1mm and above. The influence of flow velocity and particle diameter on the flow characteristics of slurry has been study, and according to the simulation results, the critical flow velocity of the slurry which is suitable for pipeline cleaning has been obtained.

* Corresponding author: zsm7481976@163.com

2 CFD-DEM model

2.1 Governing equations for carrier phase

In the CFD-DEM model, the carrier is considered as the continuous phase, and the governing equations of the carrier are sloved by the CFD. As mentioned above, water is used as the carrier, and it could be properly treated as Newtonian fluid and incompressible in our case. So, the governing equations for the carrier should be continuity equation and momentum equation, namely,

$$\frac{\partial(\rho_f)}{\partial t}+\nabla(\rho_f V_f)=0, \quad (1)$$

and

$$\frac{\partial(\rho_f V_f)}{\partial t}+\nabla(\rho_f V_f V_f)=-\nabla p+\nabla(\tau_f+\tau_t)+\rho_f g+F_{fp}, \quad (2)$$

where ρ_f is the density of the fluid, V_f is the average velocity of the fluid at the particle location, p is the pressure of the fluid at the particle location, F_{fp} is the fluid-particle interaction source term, τ_f is the viscous stress tensor, τ_t is the turbulent stress tensor. The turbulence is modeled by the standards $k-\varepsilon$ model[9].

2.2 Governing equations for particle phase

In this paper, the particle is considered as the discrete element and the translational motion of the particles is described by the second law of Newton[10]

$$m_p\frac{dV_p}{dt}=F_f+F_{fp}+F_C, \quad (3)$$

where m_p is the mass of the particle, V_p is the velocity of the particle, F_f is the floating weight of the particle, F_{fp} is the liquid-particle interaction source term, F_C is the contact force of the particle.

2.2.1 The fluid-solid interaction

$$F_{fp}=F_S+F_M+F_D+F_A, \quad (4)$$

where F_S ie the Saffman lift force, F_M is the Magnus lift force, F_D is the drag force, and F_A is the add mass force.

The Saffman lift force F_S is modeled as [9,11]

$$F_S=1.615d^2\sqrt{\frac{\rho_f\mu_f}{\|\omega_f\|}}(V_f-V_p)\times\omega_f f(\mathrm{Re}_p,\mathrm{Re}_s), \quad (5)$$

$$f(\mathrm{Re}_p,\mathrm{Re}_s)=\begin{cases}0.0524\sqrt{\alpha\mathrm{Re}_p} & \mathrm{Re}_p\geq 40\\ (1-0.3314\sqrt{\alpha})e^{-0.1\mathrm{Re}_p}+0.3314\sqrt{\alpha} & \mathrm{Re}_p<40\end{cases}$$

where d is the particle diameters, μ_f is the viscosity of the fluid, ω_f is the angular velocity of the fluid, the particle Reynolds number is defined as $\mathrm{Re}_p=d\rho_f V_p/\mu_f$, and the rotational Reynolds number is $\mathrm{Re}_s=d\rho_f\omega_f/\mu_f$, and α is the ratio of rotational Reynolds number to particle Reynolds number $\alpha=0.5\mathrm{Re}_s/\mathrm{Re}_p=d\omega_f/V_p$.

The Magnus lift force F_M is modeled as[9,12]

$$F_M=\left(\frac{\pi}{8}\right)\rho_f d^3\omega_r\times V_r, \quad (6)$$

where ω_r is the relative angular velocity between the particle and the fluid, V_r is the relative velocity between the particle and the fluid.

The drag force F_D is modeled as [9,13]

$$F_D=\frac{1}{8}\pi d^2 C_d\rho_f\|V_f+V_f'-V_p\|(V_f+V_f'-V_p)C_f^{-\left(3.7-0.65e^{\left(1.5-\log_{10}(\mathrm{Re}_p)\right)^2}\right)}, \quad (7)$$

where V_f is the average velocity of the fluid at the particle location, V_f' is the speed fluctuation of the fluid at the particle location, V_p is the velocity of the particle, C_d is the drag coefficient, C_f is the volume fraction of the fluid, Re_p is particle Reynolds number.

The add mass force F_A is modeled as [9,14]

$$F_A=-0.5\rho_f V_p\frac{d(V_p-V_f)}{dt}. \quad (8)$$

2.2.2 Particle-particle and particle-wall interaction

The interaction between two particles can be modeled by the tangential force and normal force [15],as illustrated in Fig.1

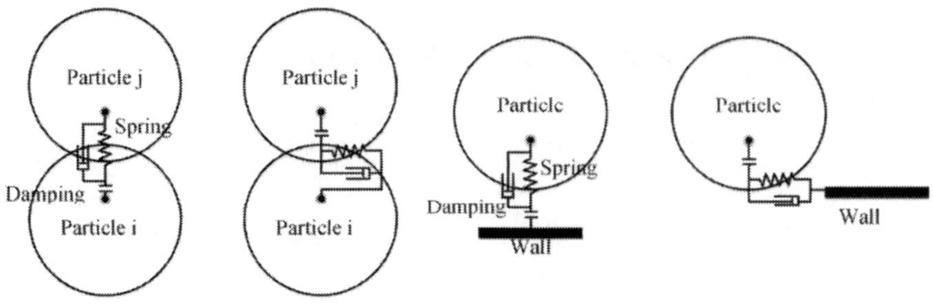

Fig. 1 The contact force of the particle[15]

The normal force $\vec{F}_{nij}$ is modeled as

$$\vec{F}_{nij} = \left(-k_n\beta^{\frac{3}{2}} - \eta_n \vec{G}\vec{n}\right)\vec{n}\cdot \quad (9)$$

The tangential force $\vec{F}_{tij}$ is modeled as

$$\vec{F}_{tij} = \begin{cases} -k_t\delta - \eta_t \vec{G}_{ct} & \left|\vec{F}_{tij}\right| \le \mu\left|\vec{F}_{nij}\right| \\ \mu\left|\vec{F}_{nij}\right|\vec{t} & \left|\vec{F}_{tij}\right| > \mu\left|\vec{F}_{nij}\right| \end{cases}, \quad (10)$$

where the k,η is the spring stiffness and the damping coefficient, $\vec{G}$ is the relative velocity, β is the normal displace of the contact point, δ is the tangential displace of the contact point, μ is the friction coefficient, the tangential unit vector is modeled as

$$\vec{t} = \frac{\vec{G}_{ct}}{\left|\vec{G}_{ct}\right|} \quad (11)$$

In this paper, all of the particles are of the same material and diameter, so according to the Hertz contact theory the normal spring stiffness is modeled as

$$k_n = \frac{\sqrt{2rE}}{3\left(1-v^2\right)}, \quad (12)$$

and the tangential spring stiffness is modeled according to the Mindlin-Dersiewicz contact theory

$$k_t = \frac{2\sqrt{2G}}{\left(2-v\right)}\alpha^{1/2}, \quad (13)$$

where E is the elastic modulus of the particle, G is the shear modulus of the particle, and V is the poisson's ratio of the particle.

The damping coefficient of the normal and tangential are calculated by:

$$\eta_n = 2\sqrt{m_p k_n}\text{, and } \eta_t = 2\sqrt{m_p k_t}. \quad (14)$$

3 Equations and mathematics

In this paper, a 3D CFD-DEM model has been built to research the flow characteristics of the slurry which is composed of the Polypropylene particle. A 3D computation model is constructed with diameter of 50 mm and length of 1000 mm. The model is discreted by hexahedron mesh, and the number of the whole grid is 21543. The fluid phase equation was calculated using the ANSYS FLUENT© 15.0 and the particle motion equation was calculated using DEM software. In the process of the simulation, the particles are randomly generated in the inlet. The particle information in the DEM software is listed in Table 1.

Table 1 The set up in the DEM software

Material	particle	pipeline
density	910kg/m^3	7850kg/m^3
Shear modulus	3.158×10^{8}pa	8.077×10^{11}pa
Poisson's ratio	0.4103	0.3

4 Result and discussion

In this section, the flow characteristics of the slurry is analyzed according to the simulation. The influence of the convey velocity and particle size on the flow characteristics have been analyzed, and the flow characteristics of the slurry is represented by the concentration distribution in the outlet.

Firstly, the slurry with particles of 1mm in diameter is used to investigate the influence of the convey velocity on the flow characteristics. Then particles of five diameters of 1 mm, 1.5 mm, 2 mm, 2.5 mm, and 3 mm were selected for the research of the influence of particle size on the flow characteristic.

In the groups of simulation, the particle generation rate at the inlet keeps the same, and the convey velocity has been chosen as 1m/s, 1.5m/s, 2m/s, 2.5m/s, 3m/s, 3.5m/s, 4m/s respectively.

4.1 The influence of convey velocity

The convey velocity is an important factor affecting the slurry flow. An increase of convey velocity will lead to increase of the output pressure of the pumping equipment which will result in more energy consumption. In this paper, we choose particle of 1mm in diameter to research the influence of convey velocity on flow characteristics as well as the pressure loss.

Fig. 2 shows the relationship between pressure loss and convey velocity. It can be observed that the pressure loss increase with convey velocity. In addition, pressure loss seems to increase faster with the convey velocity increases. Therefore, one can conclude that while ensuring the cleaning effect it is necessary to reduce the convey velocity as soon as possible in order to reduce the capacity of the pumping equipment.

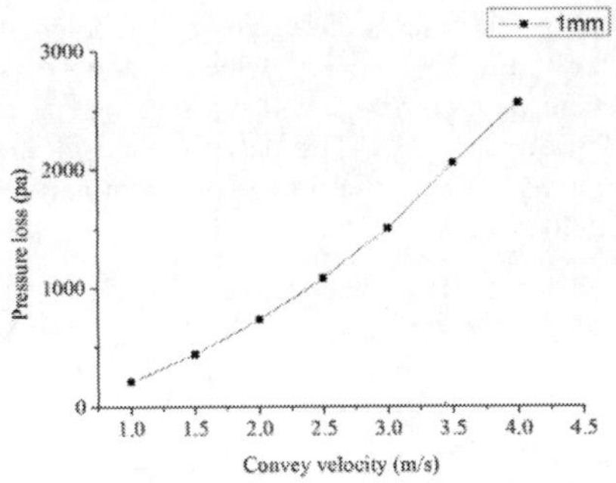

Fig. 2 The pressure loss of the slurry

The particle concentration in the outlet is an apparent feature which can be used to characterize the flow status. Fig. 3 shows the particle concentration contour in the outlet. It can be observed from Fig. 3 that slurry flow could be grouped to two categories. When the convey velocity is less than 2 m/s, most of the particles accumulate at the top, and only few of the particles are suspended in the lower part of the pipeline. It implies that the particle concentration is denser at the top than anywhere else. In other cases, the dense zone shifts to pipeline center as convey velocity increases, and it means that the particle and the fluid gradually well mixed.

In order to intuitively display the level of the particle accumulation at the top for different convey velocity, a scale factor 'C' defined as the ratio of particle concentration above the pipe center at the 0.8R position to the average concentration of the pipe outlet, namely to characterize the particle concentration.

$$C = \frac{i_{0.8R}}{i_{out}} , \qquad (15)$$

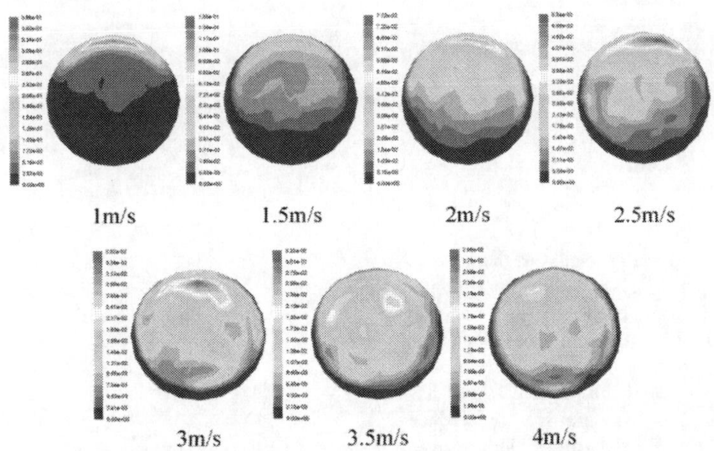

Fig. 3 Pipeline outlet particle concentration distribution contour

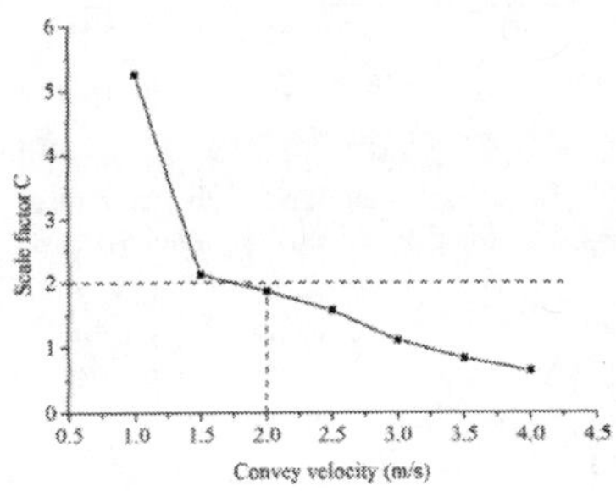

Fig. 4 The relationship of characterized concentration with convey velocity

Fig. 4 shows the curve of the scale factor C as a function of the slurry convey velocity. The scale factor shows a significant decrease with increasing convey velocity, and this indicates that the level of particle accumulation at the top is gradually weakened. This is consistent with the phenomenon in Fig 3. However, it can be observed from Fig. 2 that as the slurry convey velocity increases, the pressure loss is also significantly increased, which will lead to higher requirements for pumping equipment. A benchmark that C~2 was determined so that the convey velocity is desirable for pipeline cleaning. The curve in Fig. 4 shows that as the velocity increases from the 1 m/s to the 1.5 m/s, C decreases significantly, and this suggests that the mixing state of the slurry has been significantly improved. When the velocity is increased above the 2 m/s, C is less than 2.

Therefore, for the slurry with a diameter of 1mm the velocity of 2m/s could be as the critical velocity for pipeline cleaning.

4.2 The influence of particle size

The particle size is another factor which might affect the pipeline cleaning. At the same velocity, the increasment of the particle size can make the particle achieve higher kinetic energy, but the flow characteristic will also be changed. In this paper, particles of five different particles including diameters of 1 mm, 1.5 mm, 2 mm, 2.5 mm, 3 mm are chosen to determine how the particle size affects the slurry flow. In the simulations, the mass generation rate of different size particles are kept the same, and the flow characteristic of the five different diameter particles at different velocity are obtained. The particle generation rate at the inlet is listed in Tab. 2.

Table. 2 The number generated rate at the inlet

diameter (mm)	1	1.5	2	2.5	3
Generation rate (/s)	135000	40000	16875	8640	5000

It can be seen from Sec. 4.1 that the particles tend to accumulate at the top of the pipe when convey velocity is low. In order to research how the particle size affects the accumulation level at the low convey velocity, the flow characteristic of the slurry which composed of different particle diameter was firstly researched at the velocity of 1m/s and 1.5m/s. The particle concentration distribution curves at the outlet for the different particles were shown in the Fig. 5.

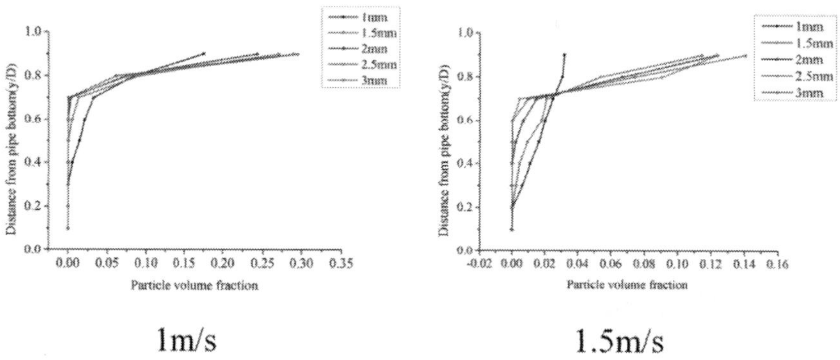

Fig. 5 Concentration profile of outlet at velocity of 1 m/s and 1.5 m/s

It can be observed from the Fig. 5 that in the case of the velocity is 1m/s, all particles show the phenomenon of accumulation at the top, but the level of the accumulation has a significantly enhance with the particle size increased. When the velocity increases to 1.5m/s the accumulation level of the particle which has the diameter of 1mm has a significantly reduce compared to the other size of the particles. This proves that the 1mm diameter particles can be easier mixed than the larger particles.

In order to directly show how the particle size affects the flow characteristic of the slurry, in this section the scale factor 'C' of all particles at different velocity are obtained from the simulation results, and the curves are shown in the Fig. 6.

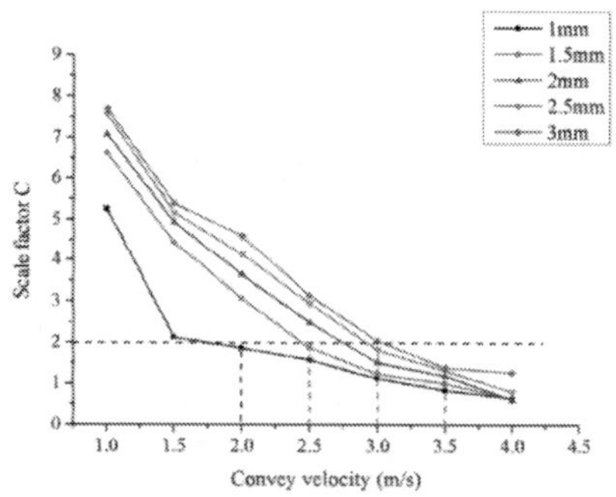

Fig. 6 Curve of the scale factor of slurry with different diameter particles as a function of convey velocity

It can be observed from the Fig. 6 that as the velocity increases the scale factor of all different particles has a significant reduce, at the same velocity the particles which have a low diameter have a lower scale factor, this proves that it can be easier mixed. According to the judging criteria proposed in the section 4.1, the critical velocity for pipeline cleaning of all different particles were obtained, and the results were shown in the Tab. 3.

The results in the Tab. 3 show that increasing the particle size can make the critical velocity have a significant increase, and this leads a higher requirement of the equipment. Therefore, when using the slurry for pipeline cleaning, it is necessary to reduce the diameter of the particles as much as possible while ensuring the cleaning effect. In this paper, the cleaning effect was not mentioned, the research on this issue will be completed in the subsequent work and refer to the results of this paper.

Table. 3 Critical velocity of slurry of different diameters

diameter（mm）	1	1.5	2	2.5	3
velocity（m/s）	2	2.5	3	3	3.5

5 Conclusion

1 The convey velocity of the slurry is an important factor of the flow characteristic. As the velocity increases, in the one hand the level of the particle accumulation at the top gradually decreases, in the other hand the pressure loss gradually increases. According to the standard proposed in this paper, for the particles with diameter of 1mm the critical velocity for pipeline cleaning is 2m/s.
2 The size of the particle in the slurry is another important factor of the slurry flow characteristic. In the case of the same velocity, the larger-diameter particle exhibits a more serious tendency to accumulate at the top of the pipe and the critical convey velocity which can be used for the pipeline cleaning is higher. Therefore when using the slurry for pipeline cleaning the particle size should be reduced as much as possible while ensuring the cleaning effect.

References

1. Zhu X, Zhang S , Li X , et al. Numerical simulation of contact force on bi-directional pig in gas pipeline: At the early stage of pigging[J]. Journal of Natural Gas Science and Engineering, 2015, 23:127-138.
2. Zhu X , Wang W , Zhang S , et al. Experimental Research on the Frictional Resistance of Fluid-Driven Pipeline Robot with Small Size in Gas Pipeline[J]. Tribology Letters, 2017, 65(2):49.
3. Doron P , Barnea D . Flow pattern maps for solid-liquid flow in pipes[J]. International Journal of Multiphase Flow, 1996, 22(2):273-283..
4. Wilson K C, Addie G R, Sellgren A, et al. Slurry Transport Using Centrifugal Pumps[M]// Slurry transport using centrifugal pumps. 2006.
5. Newitt D M . Hydraulic conveying of solids in horizontal pipes[J]. Trans. Instn chem. Engrs, 1955..
6. Mellari S. Experimental investigations of ice slurry flows in horizontal pipe based on monopropylene glycol[J]. International Journal of Refrigeration, 2016, 65: 27-41..
7. Vaezi M , Katta A K , Kumar A . Investigation into the mechanisms of pipeline transport of slurries of wheat straw and corn stover to supply a bio-refinery[J]. Biosystems Engineering, 2014, 118(1):52–67..
8. Edelin D, Czujko P-C, Castelain C, et al. Experimental determination of the energy optimum for the transport of floating particles in pipes[J]. Experimental Thermal and Fluid Science, 2015, 68: 634-643.
9. Uzi A, Levy A. Flow characteristics of coarse particles in horizontal hydraulic conveying[J]. Powder Technology, 2018, 326: 302-321.
10. Brosh T, Kalman H, Levy A, et al. DEM–CFD simulation of particle comminution in jet-mill[J]. Powder Technology, 2014, 257: 104-112.
11. Mei R . An approximate expression for the shear lift force on a spherical particle at finite reynolds number[J]. International Journal of Multiphase Flow, 1992, 18(1):145-147.
12. Rubinow S I, Keller J B. The transverse force on a spinning sphere moving in a viscous fluid[J]. Journal of Fluid Mechanics, 1961, 11(3):447-459.
13. Felice R D . The voidage function for fluid-particle interaction systems[J]. International Journal of Multiphase Flow, 1994, 20(1):153-159.
14. Maxey M R. The motion of small spherical particles in a cellular flow field[J]. Physics of Fluids (00319171), 1987, 30(7):1915-1928.
15. Tsuji Y, Tanaka T, Ishida T. Lagrangian Numerical Simulation of Plug Flow of Cohesionless Particles in a Horizonal Pipe[J]. Powder Technology, 1992, 71(3):239-250.

MATEC Web of Conferences **306**, 02001 (2020) https://doi.org/10.1051/mateconf/202030602001
ICMME 2019

Analytical Calculation of Annular Gap Throttling of Damping Valve

Yijie Chen, [1,*], *Xiaodong* Gao[1], *Xiaoling* Han[1], *Ning* Zhao[1], and *Yiqiang* Wan[1]

[1] China North Vehicle Research Institute, Beijing, China

Abstract. For the throttling characteristics of the suspension system damping valve, an analytical method for gap throttling based on turbulent flow state was proposed for the first time. The boundary layer theory was used to give the expression of the three-layer velocity distribution of the fluid under turbulent flow state, and the analytical formula of the gap flow of the parallel plate was derived by the integral theorem. Compared with the numerical solution of CFD software, the error was very small, which verified the correctness of the equation. Based on the structural characteristics of the damping valve, the turbulent flow formula of the concentric annular gap was derived, and the influence of different physical parameters on the differential pressure and flow of the gap was analyzed, which lay the foundation for improving the design precision of the damping valve.

1 Introduction

In the prior art, the gap throttling can be analysed mainly by the following methods: (1) The analytic calculation of the gap flow using the Hydraulic Engineering Handbook is simple and well known, and has been widely used in engineering design. The viscosity of the oil in the damping device is large, the throttle area is relatively small, but due to the significant increase in the over-flow, the oil is mainly in a turbulent state when it is deformed by the damping valve, which leads to a large calculation error. (2) The differential solution of fluid motion N-S second-order nonlinear non-homogeneous partial differential equations is solved, but the calculation process is too complicated, and only the numerical solution in specific cases can be obtained, which is not suitable for engineering applications; (3) Using fluid dynamics software modeling and post-processing analysis, which can get more accurate numerical solution of relevant parameters, however, it is impossible to draw specific expression, and it is difficult to play important role in the design of the damping valve.[1-2]

Based on the above situation, this paper uses the theory of fluid mechanics to deduct, verify and systematically study the analytical formula of turbulent flow gap of damping valve in different assembly modes.

2 Parallel plate slot throttling analytical calculation

Fig 1 is a schematic illustration of fluid passing through a parallel plate. p_1 is the inlet end pressure, p_2 is the outlet end pressure, δ is the plate spacing, l is the plate length, V_f is the fluid movement speed, and the pressure difference between the two ends of the plate is $\Delta p = p_1 - p_2$, and the plate thickness in the z direction is set to the unit length.

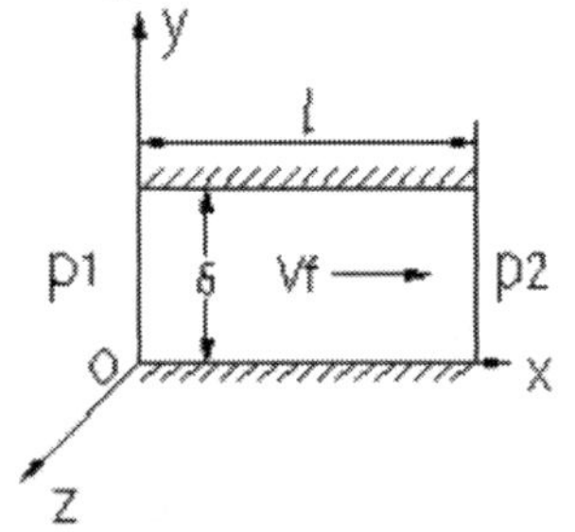

Fig.1. Sketch of parallel-plate.

The formula of the turbulent flow rate of the parallel plate gap is derived by the fluid velocity boundary layer theory.

(1) Velocity distribution law

As the distance of the fluid in the turbulent state changes from the wall surface, the degree of influence of the viscous shear stress and the turbulent additional shear stress on its flow also changes. From the wall surface to the center of the gap, as y increases, the effect of the viscous shear stress gradually decreases, however, the additional shear stress of turbulent flow has an increasing influence on the velocity flow. According to the turbulent flow-average momentum equation, the viscosity vorticity theory and the Prandtl mixing length theory, the sum of the two is a constant and equal to the wall shear stress[3-4]:

* Corresponding author: chenyijie1206@163.com

$$\mu\frac{d\overline{u}}{dy}+\rho l_{\mathrm{p}}^{2}\left(\frac{d\overline{u}}{dy}\right)^{2}=\tau_{\mathrm{w}} \qquad (1)$$

Where $\overline{u}$ is the time-averaged velocity distribution, τ_{w} is the wall shear stress, l_{p} is the Prandtl mixing length, ρ is the fluid density, and μ is the hydrodynamic viscosity.

The wall turbulence velocity boundary layer is generally divided into an inner layer and an outer layer, and the inner layer includes an adhesive bottom layer, a transition layer and a logarithmic layer; The outer layer includes a trailing layer and a sticky top layer. In a closed gap flow, the influence of the outer layer may not be considered. Fig 2 is the experimental curve of the average velocity distribution in the boundary layer[5]:

$$\overline{u}^{+}=\frac{\overline{u}}{V^{*}},\quad y^{+}=\frac{y}{l^{*}},\quad l^{*}=\frac{v}{V^{*}},\quad V^{*}=\sqrt{\frac{\tau_{w}}{\rho}} \qquad (2)$$

Where $\overline{u}^{+}$ is a dimensionless velocity, y^{+} is a dimensionless to wall distance, l^{*} is the characteristic length, and V^{*} is the wall shear stress velocity.

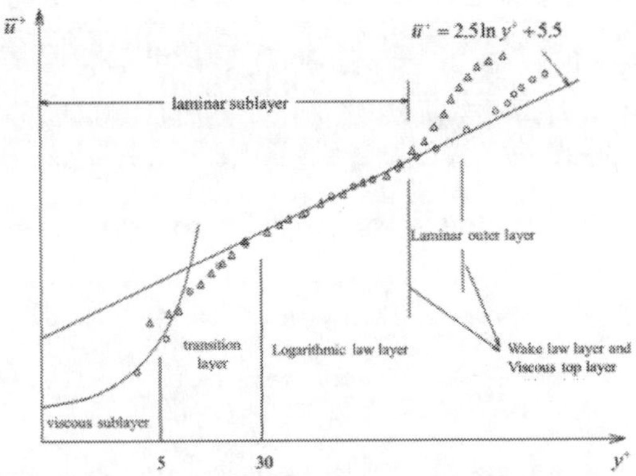

Fig. 2. Boundary layer velocity average distribution curve.

(2) Analytical formula of single-wide parallel plate turbulent flow

Integral solution for single-wide parallel plate turbulent flow analytical equation:

$$Q_{d}=2\left(\int_{0}^{5}\overline{u}_{c}^{+}dy^{+}+\int_{5}^{30}\overline{u}_{g}^{+}dy^{+}+\int_{30}^{\frac{\delta}{2}}\overline{u}_{d}^{+}dy^{+}\right) \qquad (3)$$

Where $\overline{u}_{c}^{+}$ is the viscous bottom layer velocity distribution, $\overline{u}_{g}^{+}$ is the transition layer velocity distribution, and $\overline{u}_{d}^{+}$ is the logarithmic layer velocity distribution.

The time-average velocity distribution and thickness expression of each layer in the above turbulent state are brought into the integral form and simplified:

$$Q_{d}=-127.79v+\left[3+2.5\ln\left(\frac{\delta}{2v}\sqrt{\frac{\tau_{w}}{\rho}}\right)\right]\delta\sqrt{\frac{\tau_{w}}{\rho}} \qquad (4)$$

The total friction generated between a single wide parallel plate and a fluid is:

$$F_{\mathrm{m}}=2l\tau_{\mathrm{w}} \qquad (5)$$

The pressure difference across the gap is:

$$F_{y}=\Delta p\delta \qquad (6)$$

According to the principle that the pressure difference is balanced with the frictional force on the plate, the wall shear stress expression can be obtained[6]:

$$\tau_{w}=\frac{\Delta p\delta}{2l} \qquad (7)$$

Bring into (3) and organize it:

$$Q_{\mathrm{d}}=-127.79v+\left[3+2.5\ln\left(\sqrt{\frac{\Delta p\delta^{3}}{8v^{2}l\rho}}\right)\right]\sqrt{\frac{\Delta p\delta^{3}}{2l\rho}} \qquad (8)$$

The above formula is the flow analysis formula of the parallel plate gap in the turbulent flow state, and based on this, the analytical equation of the concentric annular gap turbulent flow can be derived.

3 Concentric calculation of concentric annular gap throttling

When the throttle plate is mounted on the inside of the piston and subjected to a uniform load, the oil flows through the annular gap between the valve plate and the piston wall, and the pressure difference is generated; In the case of reasonable tolerance size design, it can be approximated as concentric annular gap. Fig 3 is schematic view of the fluid passing through concentric annular gap, r_{n} is the inner radius of the annular gap, and the remaining parameters are defined as above.

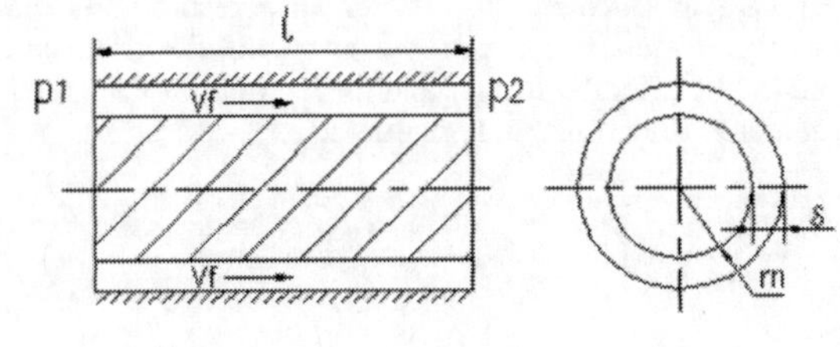

Fig. 3. Concentric annular gap schematic

After the concentric annular gap is unfolded, it is equivalent to a parallel plate slit having the width $B_{\mathrm{k}}=2\pi r_{\mathrm{n}}$, the circumferential integral of the formula (6) can be used to sort out the turbulent flow of the concentric annular gap:

$$Q_{\mathrm{t}}=2\pi r_{\mathrm{n}}\left\{-127.79v+\left[3+2.5\ln\left(\sqrt{\frac{\Delta p\delta^{3}}{8v^{2}l\rho}}\right)\right]\sqrt{\frac{\Delta p\delta^{3}}{2l\rho}}\right\} \qquad (9)$$

4 Analytical verification and analysis of crevice flow

In the prior art, since there is no known empirical formula for the turbulent flow rate of the gap, so the proposed analytical equation (6) is verified by the fluid dynamics software FLUENT which is widely used in engineering and has high calculation accuracy. First, use the pre-processing module GAMBIT to establish the physical model of the gap, and divide the rectangular grid, import the post-processing solver, given the velocity boundary condition, set the residual to 0.001 and select the k-epsilon turbulence model for numerical calculation.

The relevant parameters for the gap and oil are set as follows: gap width is $\delta = 0.004\text{m}$, long gap is $l = 0.1\text{m}$, Oil kinematic viscosity is $\nu = 1 \times 10^{-5}\text{m}^2/\text{s}$, density is $\rho = 890\text{kg/m}^3$. The flow velocity when taking the oil through the gap is $V_\text{f} = 15\text{m/s}, 20\text{m/s}, 25\text{m/s}, 30\text{m/s}$, Ensure Reynolds number $\text{Re} > 1100$, The pressure difference comparison results are shown in Tab1.

Tab1. Parallel plate gap pressure difference comparison.

$V_\text{f}/(\text{m/s})$	Δp_c / Mpa	Δp_w / Mpa	Δp_F / Mpa	γ / %
15	0.010	0.040	0.042	5.0
20	0.013	0.066	0.070	5.7
25	0.017	0.096	0.100	4.0
30	0.020	0.130	0.140	7.1

In the table, is the turbulent flow formula to calculate the differential pressure, is the FLUENT software to calculate the differential pressure, is the relative deviation between the turbulent flow formula and the FLUENT software, and the laminar flow formula calculates the differential pressure as:

$$\Delta p_\text{c} = \frac{12\mu l Q_\text{d}}{\delta^3} \tag{10}$$

It can be seen from the data in the table that the calculation results of the parallel plate gap turbulent flow formula and the numerical solution of the fluid dynamics software are relatively small, thus verifying the correctness of the above derivation process and the formula. However, the analytical solution calculated by the formula of the laminar laminar flow rate has a large error and the numerical value is obviously small, it is shown that the formula (10) is not suitable for the case of gap throttling under turbulent flow conditions, and it also shows that the self-derived turbulent flow formula has application background and practical value.

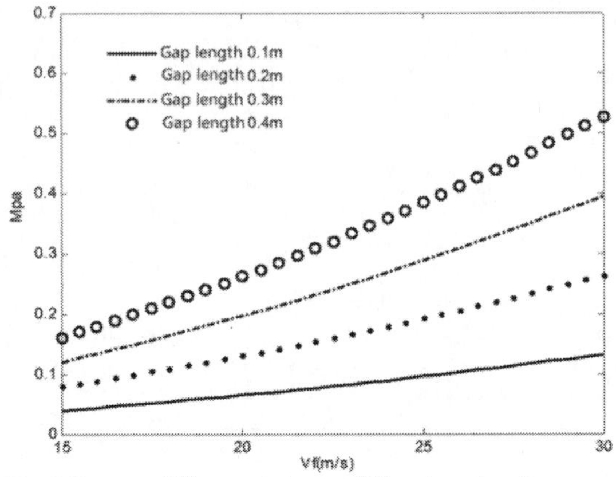

Fig.4. Pressure difference between different gap lengths.

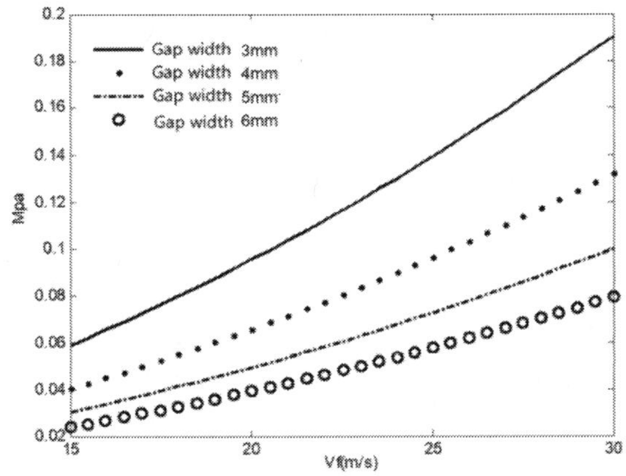

Fig.5. Pressure difference between different gap widths.

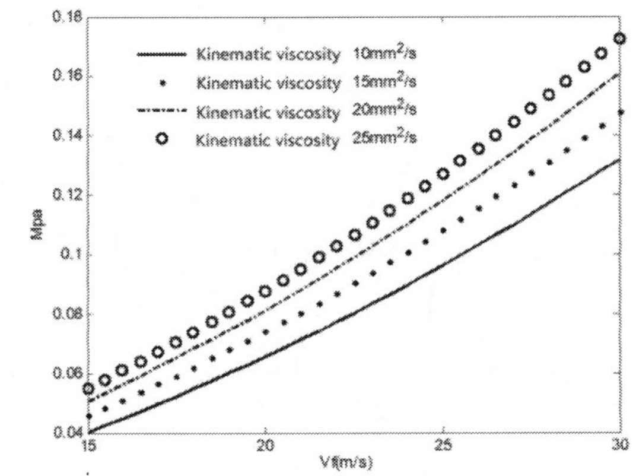

Fig.6. Pressure difference between different kinematic viscosities.

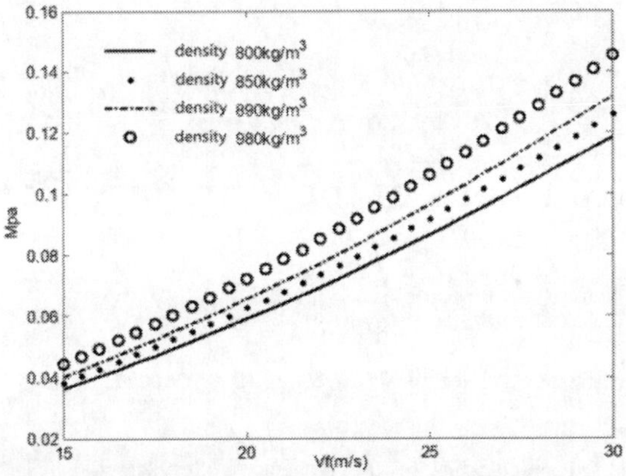

Fig.7. Pressure difference between different oil densities.

The turbulent flow formulas (6) and (10) are used to analyze the influence of the gap structure parameters and the oil-related characteristics on the differential pressure when the oil-related characteristics change within a certain range. Set the oil flow rate to $V_f = 15\text{m/s} : 30\text{m/s}$,The calculation results are shown in Figure 4–7,it can be seen that as the length of the gap increases, the contact area between the oil and the wall becomes larger, and the total friction generated also increases; As the viscosity of the oil moves, the attraction between the molecules of the oil and the friction with the wall gradually increase; As the density of the oil increases, the probability of collision during molecular motion increases, and more energy can be consumed. Due to the above factors, the pressure difference of the oil at the same overcurrent speed shows an upward trend. In addition, as the gap width increases, the throttling effect gradually decreases, and the pressure difference of the oil at the same moving speed is significantly reduced.

5 Conclusions

For the damper valve form of the slit throttling, the analytic formula of the turbulent flow under the annular gap structure is deduced and systematically studied by the theory of fluid mechanics. The work carried out is as follows:

Using the fluid dynamics boundary layer theory, the closed gap flow in the turbulent flow state is divided into three layers of the adhesive bottom layer, the transition layer and the logarithmic layer in order from the wall to the far side. The expression of the time-averaged velocity distribution of each fluid in each layer of the fully developed section is given, and the analytic formula of single-wide parallel plate turbulent flow is derived by integral. Compared with the numerical solution of the pressure difference of FLUENT at different speeds, the correctness of the formula is verified. After that, the influence of relevant parameters on the pressure difference at both ends of the gap is further analyzed, which provides reference for the design of the damping valve. Based on the formula of single-width parallel plate turbulent flow, the analytical formula of layer turbulent flow in concentric annular gap is derived.

References

1. Zuo L, Zhang PS. Energy Harvesting, Ride Comfort, and Road Handling of Regenerative Vehicle Suspensions[J]. Journal of Vibration & Acoustics, 2013, 135 (1): 48-65.
2. Wang R, Chen Z, Xu H, et al. Modelling and validation of a regenerative shock absorber system[C]. International Conference on Automation and Computing, 2014: 32-37.
3. Maravandi A, Moallem M. Regenerative Shock Absorber Using a Two-Leg Motion Conversion Mechanism[J]. IEEE/ASME Transactions on Mechatronics, 2015, 20 (6):2853-2861.
4. Roshan YM, Maravandi A, Moallem M. Power Electronics Control of an Energy Regenerative Mechatronic Damper[J]. IEEE Transactions on Industrial Electronics,2015, 62 (5): 3052-3060.
5. Sabzehgar R, A Maravandi, M Moallem. Energy Regenerative Suspension Using an Algebraic Screw Linkage Mechanism[J]. IEEE/ASME Transactions on Mechatronics, 2014, 19 (4): 1251-1259.

Experimental Research on Static Stiffness of the Planetary Roller Screw Mechanism

Shangjun Ma*, *Guanyu* Wu, *Jianxin* Zhang, and *Geng* Liu

Shaanxi Engineering Laboratory for Transmissions and Controls, Northwestern Polytechnical University, Xi'an, 710072, China

Abstract. This paper investigates the static stiffness characteristics of the planetary roller screw mechanism (PRSM). Firstly, an analytical model of elastic deformation in the PRSM considering contact deformation, combined deformation of screw shaft and thread deformations is established. Then, an accuracy coefficient is introduced to take into account the effect of machining error and assembly error on elastic deformation. Secondly, the stiffness tests of the PRSM are performed under different loads and nut positions. Lastly, the correctness of the analytical results is verified via the comparison with experimental results. The results show that the analytical results are in good agreement with the experimental results, and the stiffness variations under different loads and nut positions are obtained.

1 Introduction

Planetary roller screw (PRSM) is a mechanical transmission device, which can realize transformation of linear motion and rotary motion. As shown in Fig. 1, a PRSM primarily consists of a nut, screw and group of rollers, and it provides more contact points than conventional ball screw mechanisms (BSM) at a specific lead length. The advantages of the PRSM are high stiffness, high load capacity and compact structure, etc. Therefore, the PRSM has been increasingly used in engineering for motion and position control.

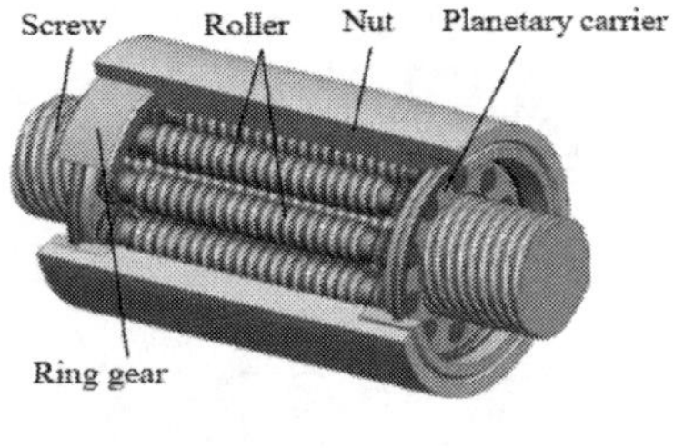

Fig. 1. The structures of the PRSM.

Earlier works on stiffness characteristics of the PRSM have included research on a calculation method for the static stiffness of PRSM [1-2]. In recent years, Jones [3] used a direct stiffness method to construct a stiffness model of the roller screw mechanism. In addition to predicting the overall stiffness of the mechanism, the distributions of the load across the threads of the individual bodies are also calculated. Yang et al. [4] used the roller as an object and investigated the load distribution and static rigidity of the PRSM. Morozov et al. [5] proposed a method for increasing the load capacity and rigidity of the PRSM by adjusting the screw surfaces. Abevi et al. [6] proposed an original approach based on the discrete model to calculate the static load distribution and the axial stiffness of the inverted PRSM. However, there are few experimental validations in the above models to support its engineering application. Also, the effects of machining error and assembly error on elastic deformation are not addressed.

Therefore, this paper provides an experimental research of the axial stiffness of the PRSM. Firstly, in section 2, the elastic deformation is studied, to obtain more accurate analytical results, an accuracy factor is introduced to revise the total deformation. Then, in section 3, the elastic deformation tests and axial stiffness calculations under different loads and different nut positions are further performed. Finally, the correctness of the analytical results is verified via the comparison with experimental results, and the variation trends between the loads and the stiffnesses are revealed.

2 Elastic deformations of the PRSM

2.1. Definition of axial stiffness

The axial stiffness of the PRSM can be defined as the ratio of axial load to deformation, which can be expressed as:

$$k = \triangle F / \triangle l \tag{1}$$

where k is axial stiffness of the PRSM, $\triangle F$ is axial load and $\triangle l$ is deformation under load $\triangle F$.

2.2. Elastic deformations of the PRSM

* Corresponding author: mashangjun@nwpu.edu.cn

In this paper, three kinds of elastic deformation such as contact deformation, combined deformation of screw shaft and thread deformations are considered.

2.2.1 Contact deformation

The contact deformation calculated by Hertzian contact theory is easily derived as [7]:

$$\delta_H = (9P_N^2/16E^2R_e)^{1/3}F_2(e) \quad (2)$$

where R_e is the equivalent radius, P_N is the normal load, E is the equivalent elastic modulus. $F_2(e)$ can be expressed as:

$$F_2(e) = 2/\pi(b/a)^{1/2}[F_1(e)]^{1/3}K(e) \quad (3)$$

$$F_1(e) = (4/\pi e^2)^{1/3}(b/a)^{1/2}\{[a^2/b^2K(e)\text{-}E(e)][K(e)\text{-}E(e)]\}^{1/6} \quad (4)$$

where e is the eccentricity of ellipse, and a and b are the semimajor axis and the semiminor axis of the contact ellipse, respectively. $K(e)$ and $E(e)$ are elliptic integrals of the first kind and second kind, respectively.

2.2.2 Combined deformation of screw shaft

The tension or pressure deformation, torsional deformation and bending deformation of screw shaft are considered in this section. The combined deformation is represented as [8]:

$$\delta_S = (4/E\pi d_s^2+8L_s^2/\pi^3 d_s^4\eta G)Fx+y\tan\alpha \quad (5)$$

where F is the working load, x is the working position of the nut and d_s is the nominal diameter of screw thread. G is shear modulus, L_s is the lead of the screw and η is the transmission efficiency. α is helix angle of the screw and y is the deflection caused by the weights of the screw shaft.

In Eq. (3), the first item in brackets on the right of the equal sign is tension or pressure deformation, the second item in brackets is torsional deformation and the last item is bending deformation.

2.2.3 Thread deformations

The thread deformations in the axial direction include deformations caused by bending σ_1, shear σ_2, thread root lean σ_3, thread root shear σ_4 and radial shrinkage (or radial expansion) σ_5.

The detailed calculation method can be found in references [9]. The total deformation of the screw thread σ_S, roller thread σ_R and nut thread σ_N can be written as:

$$\sigma_S = \sigma_1+\sigma_2+\sigma_3+\sigma_4+\sigma_{5S} \quad (6)$$

$$\sigma_R = \sigma_1+\sigma_2+\sigma_3+\sigma_4+\sigma_{5S} \quad (7)$$

$$\sigma_N = \sigma_1+\sigma_2+\sigma_3+\sigma_4+\sigma_{5N} \quad (8)$$

The total deformation of the thread in the axial direction is denoted as δ_T, resulting in:

$$\delta_T = \sigma_S+2\sigma_R+\sigma_N \quad (9)$$

The total deformation of the PRSM in the axial direction can be expressed as:

$$\delta_{PRSM} = \delta_H+\delta_T+\delta_S \quad (10)$$

2.3. Accuracy coefficient

In order to consider the effects of machining error and assembly error on elastic deformation on the load bearing region, an accuracy factor is introduced to revise the total deformation. According to the comparison research of stiffness between PRSM and BSM [2], the value of accuracy coefficient of the PRSM can refer to the value of BSM [10], as shown in Table 1.

Table 1. Accuracy coefficient of the PRSM.

Accuracy level	1	2	3	4	5
f_{ar}'	1.667	1.724	1.818	1.887	2

The total deformation of the PRSM considering accuracy coefficient in the axial direction can be rewritten as:

$$\delta'_{PRSM} = f_{ar}'(\delta_H+\delta_T)+\delta_S \quad (11)$$

3 Stiffness experiment

3.1. Experimental system

The test rig is shown in Fig. 1. In stiffness testing, two physical quantities need to be collected: axial load and axial displacement (axial elastic deformation). According to the definition of stiffness, the measuring principle of axial stiffness is that the screw is fixed and the rotation of the nut is limited, the axial load is applied on nut, and the displacement change of the nut relative to the screw under the action of axially loaded is measured. Therefore, the clamping sets are used to fix the screw and the anti-rotation device is designed to limit the rotation of nut. At the load end, the hydraulic cylinder is used as the power source to achieve axial loading, the linear grating ruler is used to detect the displacement of the nut relative to the screw.

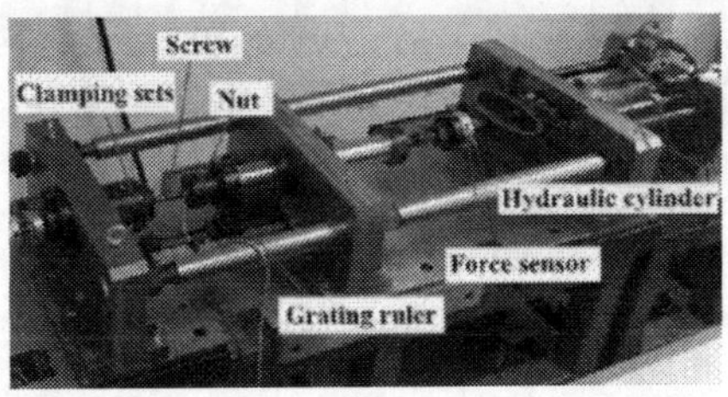

Fig. 2. The test rig of axial stiffness.

3.2. Experimental results

Force-deformation curves are obtained under different nut positions (The distance l_{Nut} between the nut and the fixed end of the screw is used to denote the nut position), as shown in Figs. 3-5. The accuracy coefficient is 1.818.

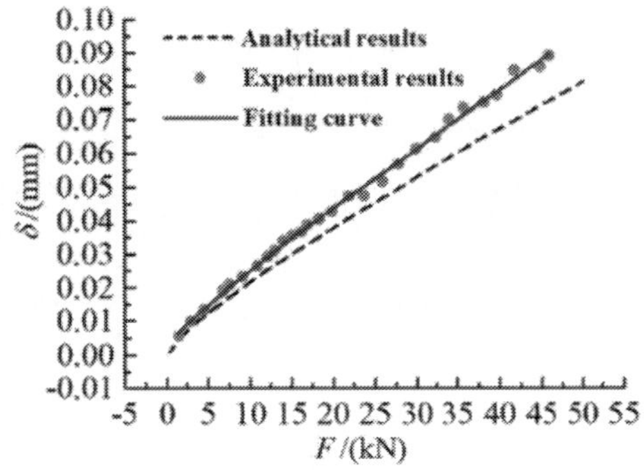

Fig. 3. l_{Nut}=60mm.

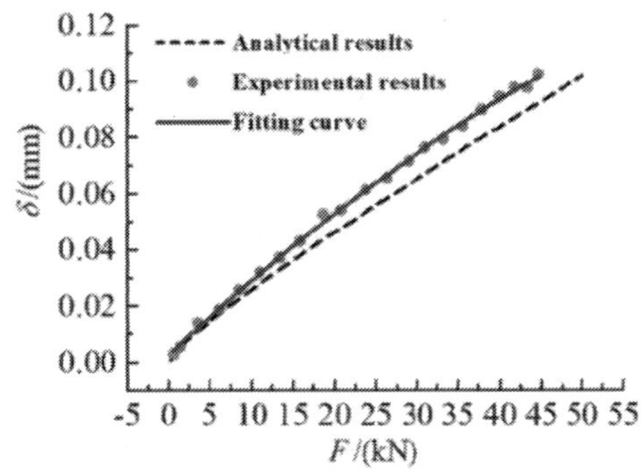

Fig. 4. l_{Nut}=92mm.

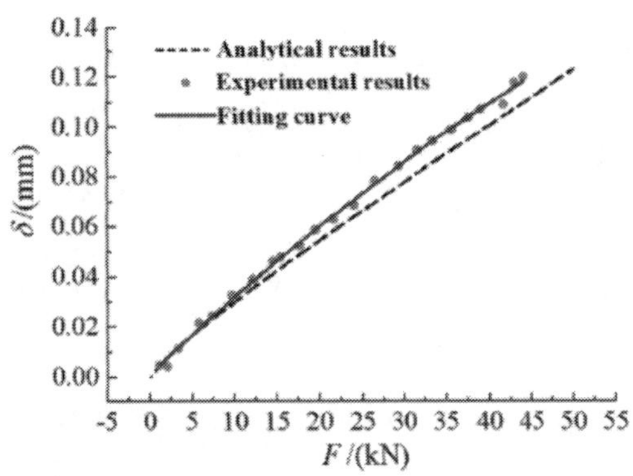

Fig. 5. l_{Nut}=125mm.

Figs. 3-5 show that the fitting curves of axial deformation are similar to that of theoretical curves. When the load is less than 10 kN, the difference is small. However, when the load is greater than 10 kN, the different increases with the increase of the load. The larger the distance l_{Nut}, the larger the deformation, that is, the axial stiffness of the PRSM decreases with the increase of the distance l_{Nut}.

The possible reasons are as follows: 1) the deformation measured by experiment also contains other deformations such as the structural deformation of the test rig, the clearances of the sensors and the clearances of the components of PRSM [11-12]; 2) the accuracy coefficient is an empirical value, which requires a lot of experiments to obtain. In addition, the selection of accuracy coefficient in this paper refers to the BSM, and the axial stiffness of PRSM may be more sensitive to machining errors.

According to the data in the Figs. 3-5 and Eq. (1), the experimental and analytical results of the axial stiffness of PRSM under different loads and different nut positions can be calculated respectively, as shown in Tables 2-4.

Table 2. Comparison of experimental and analytical results of axial stiffness (l_{Nut}=60mm)

Stiffness \ Load	10 kN	20 kN	30 kN	40 kN	46 kN
Experimental results (10^5N/mm)	4.6394	5.2817	5.6634	5.9345	6.0659
Analytical results (10^5N/mm)	4.0236	4.6424	4.9286	5.1285	5.2115
Relative error	13.27%	12.10%	12.97%	13.58%	14.09%

Table 3. Comparison of experimental and analytical results of axial stiffness (l_{Nut}=92mm).

Stiffness \ Load	10 kN	20 kN	30 kN	40 kN	46 kN
Experimental results (10^5N/mm)	3.9063	4.3478	4.6012	4.7790	4.8352
Analytical results (10^5N/mm)	3.4087	3.8499	4.0553	4.2298	4.3518
Relative error	12.74%	11.45%	11.86%	11.49%	9.99%

Table 4. Comparison of experimental and analytical results of axial stiffness (l_{Nut}=125mm).

Stiffness \ Load	10 kN	20 kN	30 kN	40 kN	46 kN

Load / Stiffness	10 kN	20 kN	30 kN	40 kN	46 kN
Experimental results (10^5N/mm)	3.3520	3.6749	3.8557	3.9795	4.0196
Analytical results (10^5N/mm)	2.9933	3.3650	3.4810	3.6477	3.6681
Relative error	10.70%	8.43%	9.72%	8.34%	8.74%

As shown in the tables above, the relative error between the experimental and analytical results is between 8% and 15%, and the relative error of the axial stiffness decreases with the increase of the distance between the nut and the fixed end of the screw. This is because the deformation on the load bearing region is affected by the machining accuracy. With the increase of the distance between the nut and the fixed end of the screw, the proportion of the deformation on the load bearing region in the overall deformation decreases continuously, which further leads to the decrease of the relative error. Therefore, to improve the static stiffness characteristic of the PRSM, the thread thickness should be increased appropriately, and the auxiliary support should also be increased when the nut stroke is longer.

4 Conclusions

This study develops an analytical model to investigate the static stiffness characteristics of a PRSM based on the Hertz contact theory. The experimental research under different loads and nut positions was performed to verify the correctness of the analytical model. The relative error between the experimental and analytical results is between 8% and 15%, which is acceptable. The results can be used to guide the parameter design and help improve the transmission quality. Planned future research will focus on a large number of experiments to obtain more accurate accuracy coefficients and identifying the influencing factors of deformation in the test rig.

The research is supported by National Natural Science Foundation of China (Grant No. 51875458), Key Project of National Natural Science Foundation of China (Grant No. 51535009) and the Fundamental Research Funds for the Central Universities (Grant No. 31020190504003).

References

1. J. Otsuka, T. Osawa, S. Fukada, A study on the planetary roller screw, comparison of static stiffness and vibration characteristics with those of the ball screw, Bull. Jpn. Soc. Precis. Eng. **23**, 3 217-223 (1989)
2. Q.Z. Jin, J.J. Yang, J.L. Sun, The comparative research on the static stiffness of ball screw and planetary roller screw, J. Mech. Sci. Technol. **18**, 2 230-232 (1999)
3. M.H. Jones, S.A. Velinsky, Stiffness of the roller screw mechanism by the direct method, Mech. Based Des. Struct. Mach. **42**, 1 17-34 (2014)
4. J.J. Yang, Z.X. Wei, J.S. Zhu, W. Du, Calculation of load distribution of planetary roller screws and static rigidity, Huazhong Univ. Sci. Technol. **39**, 4 1-4 (2011) (in chinese)
5. V.V. Morozov, V.I. Panyukhin, A.V. Zhdanov, Increasing the load capacity and rigidity of roller-screw mechanisms by adjusting the screw surfaces, Vestn. Mashinostr. **2** 15-21 (2016)
6. F. Abevi, A. Daidie, M. Chaussumier, M. Sartor. Static load distribution and axial stiffness in a planetary roller screw mechanism, J. Mech. Des. **138** 1 012301-8 (2011)
7. K.L. Johnson, *Contact Mechanics* (Cambridge University Press, ISBN: 0521347963, 1987)
8. S.J. Ma, W. Cai, L.P. Wu, G. Liu, P. Cheng, Modelling of transmission accuracy of a planetary roller screw mechanism considering errors and elastic deformations, Mech. Mach. Theory. **134** 151-168 (2019)
9. W.J. Zhang, G. Liu, R.T. Tong, S.J. Ma. Load distribution of planetary roller screw mechanism and its improvement approach, Proc. Inst. Mech. Eng. Part C-J. Eng. Mech. Eng. Sci. **203-210** 1989-1996 (2015)
10. Ball screw mechanism, Part four: Axial static stiffness, (Standards Press of China, GB/T17587.4-2008, Beijing, 2008)
11. S.J. Ma, L.P. Wu, X.J. Fu, Y.J. Li, G. Liu. Modelling of static contact with friction of threaded surfaces in a planetary roller screw mechanism, Mech. Mach. Theory. **139** 212-236 (2019)
12. S.J. Ma, L.P. Wu, G. Liu, X.J. Fu. Local contact characteristics of threaded surfaces in a planetary roller screw mechanism, Mech. Based Des. Struct. Mach, Online Doi: 10.1080/15397734. 2019. 16159 44 (2019)

Vertical Dynamic Analysis of Ballastless Tracks on Train-Track-Bridge System

Haoran Xie[1,2,*], *Bin* Yan[1,2], *Jie* Huang[1,2]

[1]School of Civil Engineering, Central South University, Changsha, China
[2]National Engineering Laboratory for High Speed Railway Construction, Changsha 410075, China

Abstract. In order to investigate the vertical dynamic response characteristics of train-track-bridge system on CWR (Continunously Welded Rail) under dynamic load of train on HSR (High-Speed Railway) bridge. Based on the principle of vehicle train-track-bridge coupling dynamics, taking the 32m simply supported bridge of a section of Zhengzhou-Xuzhou Passenger Dedicated Line as an example, the finite element software ANSYS and the dynamic analysis software SIMPACK are used for co-simulation, and bridge model of the steel spring floating slab track and the CRTSIII ballastless track (China Railway Track System) considering the shock absorbing steel spring, the limit barricade and the contact characteristics of track structure layers are established. On this basis, in order to study the dynamic response laws of the design of ballastless track structure parameters to the system when the train crosses the bridge and provide the basis for the design and construction, by studying the influence of the speed of train on the bridge, the damage of fasteners and the parameters of track structure on the train-track-bridge system, the displacement of rail, vertical vibration acceleration and wheel-rail force response performance are analyzed. Studies have shown that: At the train speed of 40 km/h, the displacement and acceleration of the rail and track slab in the CRTSIII ballastless track are smaller than the floating slab track structure, but the floating slab track structure has better vibration reduction performance for bridges. The acceleration of rail, track slab and bridge increases obviously with the increase of train speed, the rail structure has the largest increasement. Reducing the stiffness of fasteners could decrease the vertical acceleration response of the steel spring floating slab track system, the ability to absorb shock can be enhanceed by reducing the stiffness of the fastener appropriately. Increasing the density of the floating slab can increase the vertical acceleration of the floating slab and the bridge, thereby decreasing the vibration amplitude of the system.

1 Introduction

In recent decades, the world railway ballastless track structure is mainly characterized by HSR (High-Speed Railway) in Germany and Japan. At present, with the rapid development of Chinese railways, in order to eliminate the dynamic impact of traditional seamed rails, the CWR (Continunously Welded Rail) is widely used. For protecting the environment, reduceing the occupation of cultivated land and preventing foundation settlement, as well as the need to cross complex terrain and existing lines, CWR is commonly laid on bridges rather than the subgrade [1]. For example, the Beijing-Shanghai HSR subgrade section is only 18.5%, and the rest of the lines are arranged on the bridges. Such as the CRTSI double block ballastless track, the CRTSII longitudinal ballastless track, the CRTSIII ballastless track and the steel spring floating slab track [2]. Among them, the CRTSIII ballastless track on the bridge adopts the unit block structure where there is an intermediate isolation layer between the baseplate and the self-compacting concrete (SCC) layer, and using the limit barricade to transmit the longitudinal and lateral forces of the track [3]; As for the steel spring floating slab track, it floats the concrete slab with a certain quality and stiffness on the steel spring isolator, 30mm away from the top surface of the base cushion, which forms the mass-spring-foundation isolation system [4]. Domestic and foreign scholars have carried out extensive and profound researchs on the mechanics and deformation performance of ballastless track under the action of trains, including foundations and situations such as subgrade [5], simple-supported bridge and steel-concrete bridge [6], seismic response [7], mortar layer separation [8-9], track slab structure voiding [10], damage induced by temperature stress [11] and the CRTS series track types [12-13]. The vibration and impact of the system can be greatly reduced due to CWR on the bridge using ballastless track. In view of the special structure of CRTSIII ballastless track and steel spring floating slab track, the vibration mechanism and time-history characteristics of the train-track-bridge system are not very clear. In addition, there are a little of studies on the influence of train speed and fastener stiffness on the dynamic performance of the system and the co-simulation of 3D entities.

* Corresponding author: m15955099594@163.com

 https://doi.org/10.1051/matecconf/202030602003

In order to explore the dynamic response of the ballastless track on CWR bridge, based on the train-track-bridge coupling dynamics principle, and take a 32m simple box girder bridge of Zheng-Xu Passenger Dedicated Line as an example. Based on the method of ANSYS+SIMPACK co-simulation, the modal analysis is carried out to investigate the dynamic response of the system under the parameters of the structure itself, the fastener damage and the train speed, at the same time, the dynamic characteristics such as vertical displacement, vibration acceleration and wheel-rail force are discussed.

2 Spatial simulation model of train-track-bridge system

2.1 Vibration model of the system

The infinite-length "Eular beam" supported by the elastic point is adopted, and the three-dimension stiffness and damping coefficient spring are used to simulate the constraint of the WJ-8C fastener on the rail structure. The track slab and baseplate are regarded as double-layer elastic composite beam [14] as the reference mass, as shown in Figure 1.

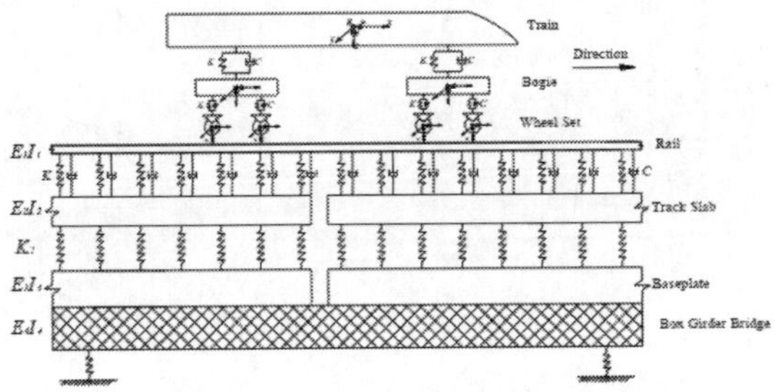

(a) Vibration model

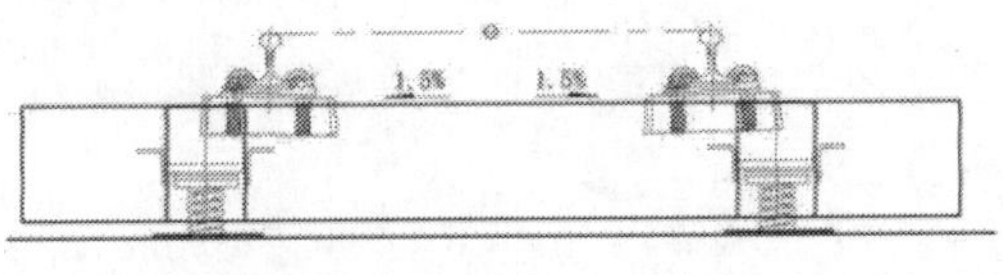

(b) Steel spring floating slab track

Fig. 1. Schematic diagram of system structure

2.2 Vehicle dynamics model

Based on the CRH3 type train in China, a multi-body dynamics spatial simulation model is established, which is mainly composed of a rigid body, a spring and a damper, including a train body, a bogie, four wheelsets, a suspension system and a secondary suspension system, with a total of 35 DOF. Among them, the weight of the wheel set is 1200kg, 3200kg for the frame, and 48,000kg for the train body. Besides, the track irregularity adopts the American grade six spectrum.

2.3 FEM of track-bridge system

Simply supported box girder is supported by fixed + active mode, the finite element model of steel spring floating slab track and CRTSIII ballastless track is established, respectively. (Fig. 2)

Structural parameters are shown in the table 1 below:

Table 1. Structural parameters

Material Values	CRTSIII Ballastless Track			
	Track slab	SCC	Baseplate	Bridge
Elastic modulus/GPa	35.5	32.5	32.5	33.5
Poisson ratio	0.2	0.2	0.2	0.2
Density/ kg·m^{-3}	2500	2500	2500	2500
Moment of inertia/m^4	-	-	-	10.76
Material Values	Steel Spring Floating Slab Track			
	Floating slab		Bridge	
Elastic modulus/GPa	32.5		32.5	
Poisson ratio	0.2		0.2	
Density/ kg·m^{-3}	3000		2500	
Moment of inertia/m^4	-		10.83	

Among them, Solid45 element is used to simulate the structure of bridge and track structures, and Target170 and Contact174 element are used in ANSYS to simulate the surface-surface contact between the track layers.

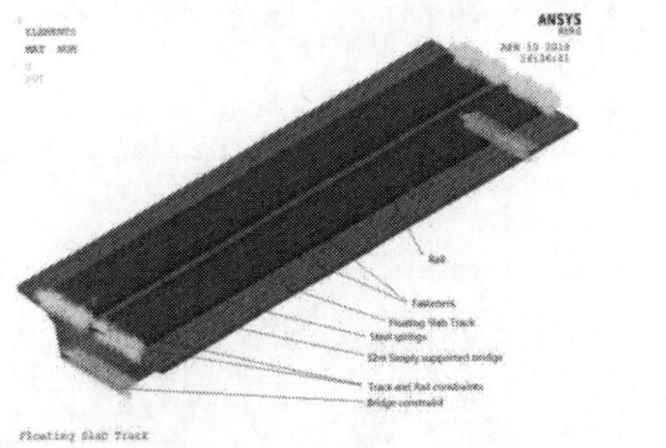

(a) FEM of floating slab track

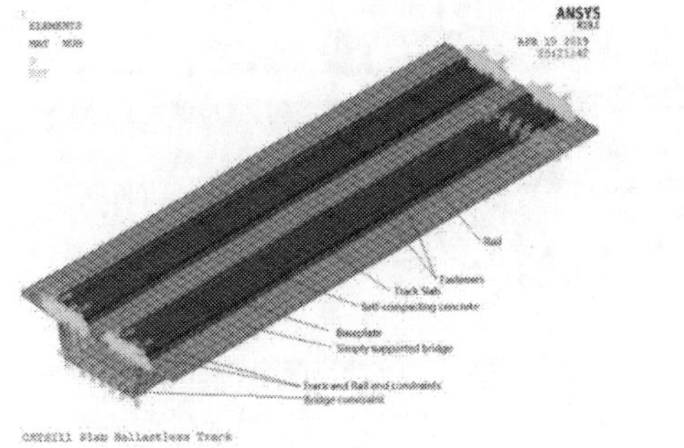

(b) FEM of CRTSIII slab ballastless track

Fig. 2. FEM of ballastless track on bridge

 https://doi.org/10.1051/matecconf/202030602003

Based on the two type ballastless track models show in above, the multi-body dynamics simulation model of train-track-bridge is established. (Fig. 3)

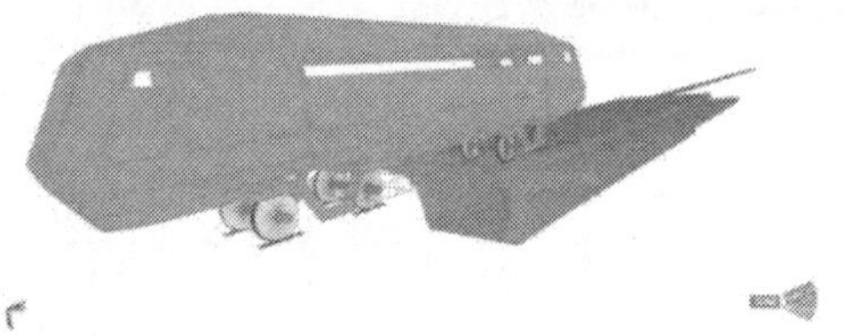

(a) System coupling model

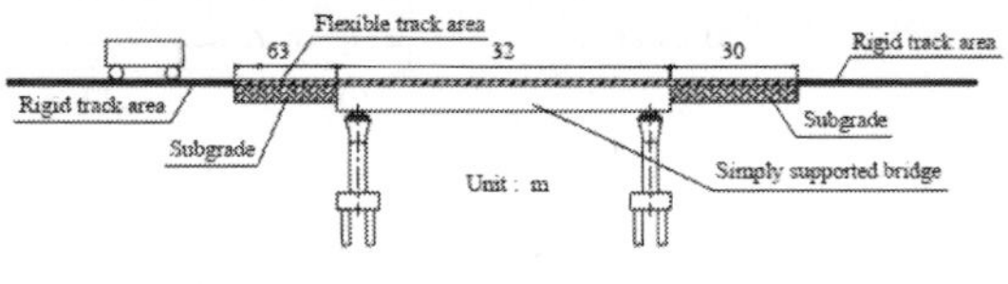

(b) Location layout

Fig. 3. Schematic diagram of train-track-bridge coupling model

3 Characteristics of dynamic response of different track types on bridge

3.1 Dynamic response of CRTSIII ballastless track

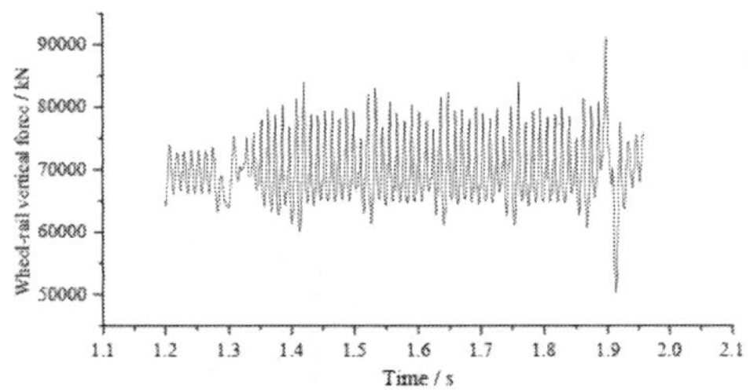

(a) Wheel-rail vertical force

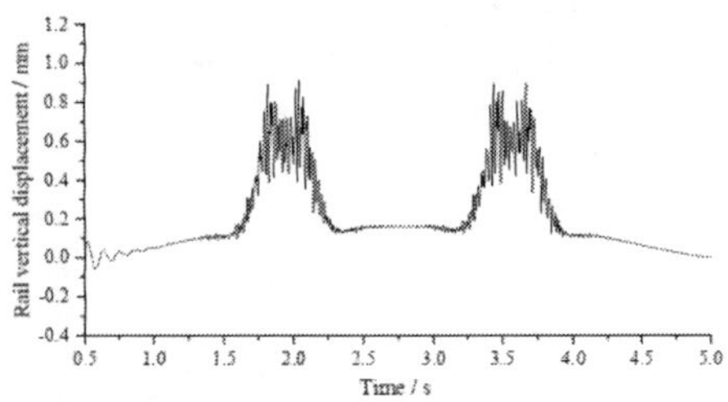

(b) Vertical displacement of rail

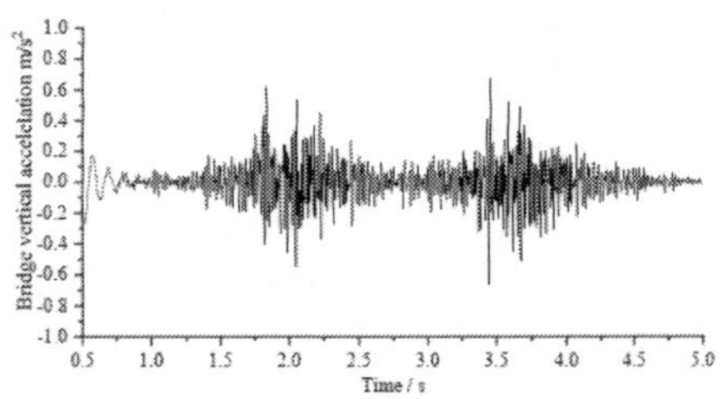

(c) Vertical acceleration of bridge

Fig. 4. Dynamic Response of CRTSIII

In Figure 4, the dynamic response under the train dynamic action of the CRTSIII ballastless track is given, and it can be clearly seen in Fig (a) that when the train passes through the track slab seam, the wheel-rail interaction surges, and the vertical force of them increases sharply by 10.4%. Attention should be paid to the dynamic response law of the train body at the 0.07 m slab seam of adjacent track slab. By observing the time-history curve of the displacement of the steel spring floating slab, it can be found that the curve tendency of the displacement of floating slab is very similar to that of rail structure, and it's going to be m-shaped. The maximum vertical displacement of the rail is 0.93mm, and the maximum acceleration of the bridge is $0.71 m/s^2$, both of which increase significantly at both ends of the bridge, and the dynamic response in the mid-span region is relatively small.

3.2 Dynamic response of steel spring floating slab track

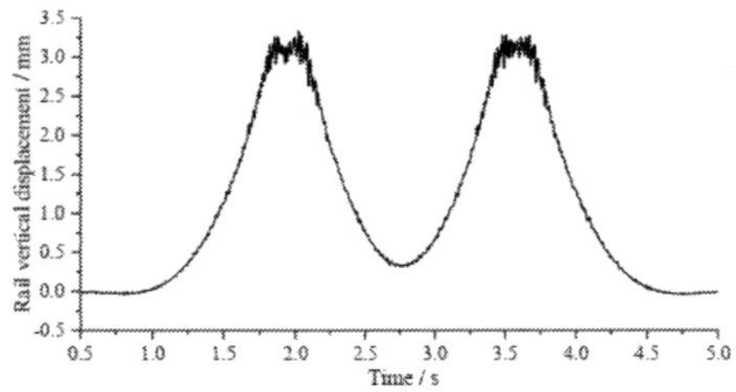

(a) Vertical displacement of rail

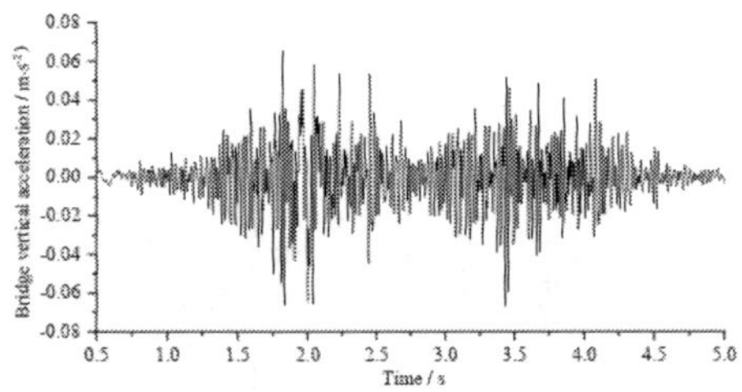

(b) Vertical acceleration of bridge

Fig. 5. Dynamic response of steel spring floating slab track

Figure 5 shows the system dynamic response of a 40 km/h train speed of the floating slab track.

By comparing the dynamic response of the CRTSIII ballastless track structure and the steel spring floating slab track structure, it can be found that the vertical displacement of the rail structure of the floating slab track in Fig (a) is greater than that of the the III type track structure. The maximum rail vertical displacement of the floating slab track is 3.33 mm, and 0.93 mm for CRTSIII ballastless track. It can be found from the comparison of rail vertical displacement time-history curve that the stiffness of floating slab track structure is less than that of the CRTS track structure due to the installation of steel spring shock absorber. In Fig (b), the tendency of curves of bridge vertical acceleration is similar, but the floating slab track structure of bridge are much smaller than that of CRTSIII track. In conclusion, the vibration damping performance of floating slab track structure is better.

4 Influencing factors on dynamic response of steel spring floating slab track

Combined with the actual train operating conditions, for the ballastless track structure of shock-absorbing steel spring floating slab, the factors such as train speed, fastener stiffness, floating slab density and other parameters have a great impact on the track structure. Therefore, it is necessary to select these three factors to analyze the dynamic response of train-track-bridge system and provide basis for design and construction.

4.1 Effect of train speed on vertical dynamic response

When analyzing the influence of speed on vertical dynamic response, the operating speeds are 20km/h, 40km/h, 60km/h, and 80km/h, respectively. And the other parameters of the model remain unchanged. The calculation results are shown in Table 2.

Table 2. Vertical dynamic response of steel spring floating slab track system on bridge under different speed

Item	A	B	C	D
Speed (km/h)	20	40	60	80
Rail displacement (mm)	3.25	3.33	3.45	3.57
Slab displacement (mm)	2.73	2.77	2.75	2.76
Bridge displacement (mm)	0.24	0.25	0.24	0.24
Rail acceleration (m/s^2)	169	370	507	584
Slab acceleration (m/s^2)	3.56	6.12	7.40	9.55
Bridge acceleration (m/s^2)	0.042	0.065	0.074	0.089

With the increase of the train speed, compared with the floating slab and bridge structure, the vertical displacement of the rail increases more obviously, 9.8%. In addition, the acceleration of the rail, the floating slab and the bridge is obviously increased. Among them, the increase of the rail structure is up to 245.6%, and 168.3% for floating slab, 112% for bridge. The dynamic effect of the wheel-rail system is significantly increased, and the wheel load shedding rate and the derailment coefficient are slightly increased. Therefore, in order to reduce the vibration of the special section, the appropriate speed limit of the subway train can be carried out in the vibration-damping section.

4.2 Influence of fastener stiffness on vertical dynamic response

The fastener stiffness has great significance to the dynamic response of the train-track-bridge system. The stiffness variation in the actual environment will significantly affect the train operation safety, resulting in the change of the geometric shape of the track structure. The fastener stiffness is 1.5e7N/m, 3e7N /m, 4.5e7N/m, and 6e7N/m, respectively. The corresponding laws of system dynamics are given in Table 3.

Table 3. Vertical dynamic response under different fastener stiffness

Item	A	B	C	D
Stiffness (N/m)	1.5e7	3e7	4.5e7	6.0e7
Rail displacement (mm)	3.77	3.33	3.15	3.11
Slab displacement (mm)	2.72	2.77	2.75	2.75
Bridge displacement (mm)	0.24	0.25	0.24	0.24
Rail acceleration (m/s^2)	239	292	367	444
Slab acceleration (m/s^2)	4.94	6.12	7.97	9.42
Bridge acceleration (m/s^2)	0.041	0.065	0.071	0.078

As can be seen from table 3, with the fastener stiffness increases, the vertical displacement of the rail decreases significantly, 17.51%. However, the vertical displacement of the floating slab and the bridge does not change significantly. Therefore, the rail is more sensitive to the change of the fastener stiffness. In order to maintain the smoothness of the line, it should be ensured that the fastener stiffness can not be too small. Besides, the acceleration of the rails, floating slab and bridge increases to varying degrees. Therefore, reducing the rigidity of the fastener can reduce the acceleration of the of the steel spring floating slab track structure on bridge, thereby achieving the effect of shock absorption. However, the reduction in the stiffness of the fastener increases the vertical displacement of the rail. In summary, the fastener stiffness should be appropriately reduced under the premise of ensuring the geometric position of the track. This is consistent with the design principle of low stiffness fasteners for rail transit on the bridge.

4.3 Effect of steel spring floating slab density on dynamic response

In order to study the influence of the density of the steel spring floating slab on the vertical dynamic response of the track structure on the bridge, the density of the floating slab is 2.6e3kg/m^3, 2.8e3kg/m^3, 3.0e3kg/m^3 and 3.2e3kg/m^3, respectively. Other calculation parameters remain unchanged. The results are shown in Table 4.

Table 4. Vertical dynamic response under different floating slab density

Item	A	B	C	D
Density (kg/m^3)	2.6e3	2.8e3	3.0e3	3.2e3
Rail displacement (mm)	3.33	3.33	3.34	3.33
Slab displacement (mm)	2.75	2.76	2.78	2.76
Bridge displacement (mm)	0.23	0.23	0.25	0.23
Rail acceleration (m/s^2)	377	357	330	314
Slab acceleration (m/s^2)	6.25	5.99	5.86	5.53
Bridge acceleration (m/s^2)	0.073	0.060	0.052	0.042

Increasing the density of the slab can reduce the vertical acceleration of the rail, the floating slab and the bridge, and the decline ratios were 16.7%, 11.5% and 42.5%, respectively, but has little effect on the vertical displacement of them. Therefore, the affect of vibration damping can be increased by increasing the density of the floating slab.

5 Conclusions

Based on ANSYS+SIMPACK, the CRTSIII ballastless track and steel spring floating slab track on the 32m simply supported box girder bridge are established considering the shock absorbing steel spring, the limit barricade and the contact characteristics of track structure layers.

The difference of dynamic response with different track types when the train passes has been studied. Taking the floating slab track system on the bridge as an example, the influence of train speed, fastener stiffness and the density of the floating slab on the dynamic characteristics of the train-track-bridge system is analyzed. The main conclusions obtained include:

(1) At the running speed of 40km/h, the displacement and acceleration of the rail and track slab in the CRTSIII ballastless track structure are smaller than that of the floating slab track structure, but the bridge acceleration is obviously larger, so the ability to absorb shock of the floating slab track on the bridge more excellent.

(2) The acceleration of rail, floating slab and bridge increases with the increase of train speed. The increase of rail structure is up to 245.56%, and the dynamic effect of wheel-rail system is significantly increased.

(3) Reducing the stiffness of the fastener will reduce the vertical acceleration of the rail, track slab and bridge. Under the premise of ensuring the geometric shape of the track, the stiffness of the fasteners should be appropriately reduced to achieve the effect of vibration reduction.

(4) Increasing the density of the floating slab has little effect on the vertical displacement of the system, but it can reduce the vertical acceleration of the floating slab and the bridge. The vibration amplitude of the system can be reduced by increasing the density of the floating slab.

The authors would like to thank the National Key R&D Program of China (2017YFB1201204), National Natural Science Foundational of China (51608542) and Natural Science Foundation of Hunan Province(2017JJ3387).

References

1. B. Yan, G.L. Dai, N. Hu. *Recent development of design and construction of short span high-speed railway bridges in China*[J]. Engineering Structures. **100**: 707-717, (2015)
2. *Code for Design of High Speed Railway*: TB 10621-2014[S], (2014)
3. J.J. Wang, M. Wang, W.B. Liu, Y. Zhao. *Technology Of CRTS III Ballastless Track System*[J]. Chinese Railways, (08):11-15, (2017)
4. M.H. Li, M. Ma, W.N. Liu, et al. *Influence of static preload on vibration reduction effect of floating slab tracks*[J].Journal of Vibration and Control, **25**(6):1148-1163, (2019)
5. Jiang, G. Hong, et al. *"Geometry mapping and additional stresses of ballastless track structure caused by subgrade differential settlement under self-weight loads in high-speed railways."* Transportation Geotechnics **18**: 103-110, (2019)
6. Podworna, Monika.*Dynamic response of steel-concrete composite bridges loaded by high-speed train*[J].Structural engineering and mechanics, **62**(2):179-196, (2017)
7. B. YAN, S. LIU, H. PU, G.L. DAI, X.P. CAI. *Elastic-plastic seismic response of CRTS II slab ballastless track system on high-speed railway bridges* [J]. SCIENCE CHINA Technological Sciences, **60**(6): 865-871, (2017)
8. Han J, Zhao G, Xiao X, et al. *Effect of cement asphalt mortar damage location on dynamic behavior of high-speed track*[J]. Advances in Mechanical Engineering, **10**(4): 1687814018770779, (2018)
9. J.J. Ren, et al. *"Criteria for repairing damages of CA mortar for prefabricated framework-type slab track."* Construction and Building Materials, **110**: 300-311, (2016)
10. J.J. Ren, X.B. Yan, G.H. Xu, K. Xu. *Effects of Contact Loss underneath Concrete Roadbed on Dynamic Performances of Slab Track-Subgrade System*[J]. Journal of Southwest Jiaotong University
11. Y. Tian, N. Zhang, H. Xia.*Temperature effect on service performance of high-speed railway concrete bridges*[J].Advances in structural engineering, **20**(6):865-883, (2017)
12. J.F. Wang, et al. *"Performance of Cement Asphalt Mortar in Ballastless Slab Track over High-Speed Railway under Extreme Climate Conditions."* International Journal of Geomechanics **19.5**: 04019037, (2019)
13. P. Wang,Q. Yi,C.Y. Zhao, et al.*Elastic wave propagation characteristics of periodic track structure in high-speed railway*[J].Journal of Vibration and Control, **25**(3):517-528, (2019)
14. L. Xu, W.M. Zhai. *Stochastic Model Used for Temporal-spatial Analysis of Vehicle-track Coupled Systems*[J].Journal of the China Railway Society, **40**(01):74-79, (2018)

MATEC Web of Conferences **306**, 02004 (2020) https://doi.org/10.1051/matecconf/202030602004
ICMME 2019

Modelling of Electro-mechanical Servo System Based on the Planetary Roller Screw Mechanism

Jianxin Zhang[1,2], *Chuanming* Du[1], *Shangjun* Ma[1,]* *and Geng* Liu[1,]*

[1]Shaanxi Engineering Laboratory for Transmissions and Controls, Northwestern Polytechnical University, Xi'an 710072, China
[2]Hubei Jiangshan Heavy Industry Co.,Ltd. , Xiangyang 441057, China

Abstract. Taking the electro-mechanical servo system as the research object, considering the contact stiffness, friction and clearance of the main components in the electro-mechanical servo system, the analysis model of the electro-mechanical servo system based on Planetary roller screw mechanism (PRSM) is established by using AMESim software. The results showed that the response speed of the system slowed down when the friction of PRSM was taken into account. The larger the clearance or the smaller the stiffness, the greater the fluctuation amplitude of the system response. After the controller was adjusted, the steady-state error of the system caused by the load force can be reduced quickly.

1 Introduction

The electro-mechanical actuator (EMA) is a core device of a fully electro-mechanical device. As an actuator for converting the rotational movement of the motor into linear motion, the planetary roller screw mechanism (PRSM) has the advantages of high bearing and high precision compared with the ball screw mechanism (BSM), it can be widely applied to the fields of industrial automation, ship and airplane braking [1].

The domestic and foreign scholars have made extensive research on the EMA, Liscouët et al. [2-3] proposed the EMA power transfer design criterion based on the simulation model. Ma et al. [4] established the finite element model of PRSM in EMA, and analyzed its dynamic characteristics, which provided a reference for studying the improvement of the overall performance of EMA. Huang et al. [5] established the EMA model to consider the non-linear factors such as clearance and friction, and set up the sub-system equations. Habibi et al. [6] analyzed the dynamic characteristics of EMA to control the rudder surface of aircraft. It showed that the dead zone caused by empty range and static friction was an important factor limiting the input and output characteristics of EMA, and the influence of dead zone on the system was reduced by using inner loop control strategy.

This paper considered the contact stiffness, friction and clearance of the main components in the electro-mechanical servo system, the analysis model of the electro-mechanical servo system based on the PRSM was established by using AMESim software.

2 Introduction to EMA System

2.1 The components of EMA system

The composition of EMA in the rudder loop system of aircraft is shown in Fig. 1. The controller accepts the instructions from the flight control computer and drives the motor to move forward and backward, then reduces the speed and increases the torque through the gear device, finally converts the rotating motion of the screw into the linear output and drives the rudder surface to realize the corresponding action. At the same time, the current signal in the motor, the speed signal of the output shaft of the motor and the position signal of the rudder surface are all involved in the closed-loop feedback control.

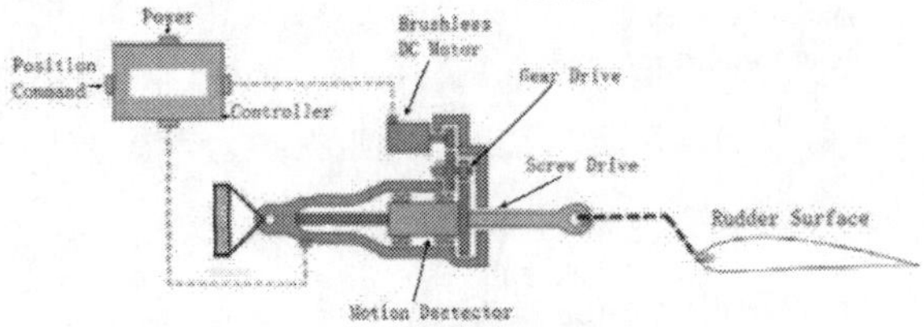

Fig. 1. Working principle of EMA in rudder loop system.

2.2 The principle of EMA System

The whole system model of EMA established in AMESim is showed in Fig. 2. The modelling of electro-mechanical system mainly includes the following parts: Region *a* includes sine or given instruction input signal. Region *b* and region *c* are controller and motor model respectively. Region *d* and region *e* are the mechanical transmission models of gear pair and PRSM

* Corresponding author: mashangjun@nwpu.edu.cn
npuliug@nwpu.edu.cn

or BSM respectively. Region f and region g are equivalent stiffness and damping. The region h is the rudder surface model.

3 Model of motor and controller

3.1 Motor model

The motor adopts 270V permanent magnet brushless DC motor, and the DC motor model in the AMESim drive library are selected, as shown in Fig. 3. The parameters are shown in Table 1.

The relationship between the armature current of the motor I_c and the input voltage V_c is described by a first-order transfer function G_e.

$$I_c = G_e(V_c - K_\omega \omega_m) \tag{1}$$

$$G_e = \frac{1/R_c}{L_c/R_c s+1} = \frac{1/R_c}{\tau_e s+1} \tag{2}$$

where w_m is the output angular speed of the motor, K_w is the back EMF coefficient, R_c is the winding resistance of the motor, L_c is the inductance of the motor winding and τ_e is the electrical time constant of the motor. The torque generated by the motor is expressed as follows:

$$T_m = K_c I_c \tag{3}$$

When the current is saturated, the following expression is obeyed:

$$T_{peak} = K_c I_{lim} \tag{4}$$

$$T_m = J_m \dot{\omega}_m + K_{mv}\omega_m + \frac{T_{load}}{i} \tag{5}$$

where J_m is the moment of inertia of the motor shaft, K_{mv} is the friction coefficient of the motor, T_{load} is the load torque and i is the product of the reducer ratio of the gear and the screw.

Table 1. Parameters of EMA system

Parameters	Symbols	Value	Units
Nut stiffness	S_N	108	N/m
Nut damping	D_N	8944	N/(m/s)
Nut mass	M_N	3	kg
Screw nominal diameter	ds	20	mm
Screw lead	Ls	2	mm
Screw contact stiffness	S_S	108	N/m
Screw contact damping	D_S	8944	N/(m/s)
Gear reducer ratio	R_G	3.33	_
Gear transmission efficiency	f_g	0.98	_
Actuator fixed stiffness	S_F	1.4×107	N/m
Actuator fixed damping	D_F	334	N/(m/s)
Actuator drive damping	D_D	3347	N/(m/s)
Actuator neutral length	L_a	500	mm
Actuator shell mass	M_A	20	kg
Actuator drive stiffness	S_D	1.4×107	N/m
Rudder surface inertia	A	20	kg*m2
Mass of steering surface	M_{SS}	2000	kg
Current loop feedback channel gain	G_C	1	_
Speed loop feedback channel gain	G_S	0.15	_

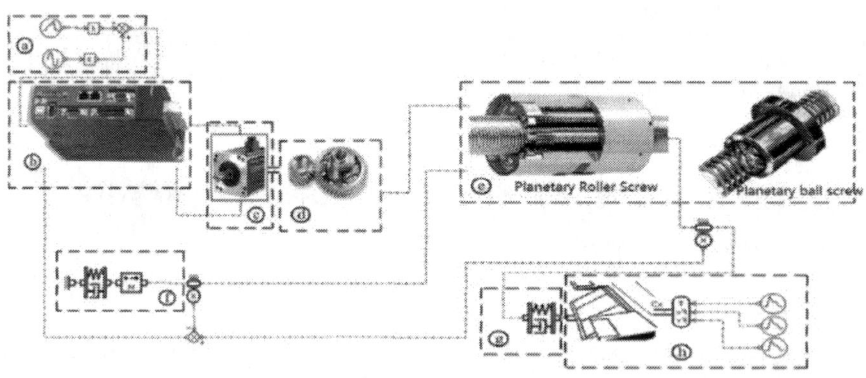

Fig. 2. EMA system model.

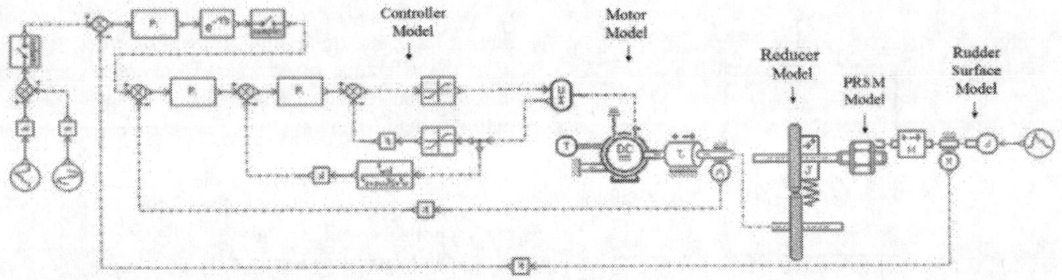

Fig. 3. AMESim simulation model.

3.2 Controller model

The loop of electro-mechanical action subsystem adopts three closed-loop control structures, the innermost loop is current loop, the outermost loop is position loop, and the middle loop is velocity loop, as shown in Fig. 3.

For the motor control system, the current loop is essentially the torque link, the introduction of the current loop in the control link has the following functions: By limiting the current of the motor bus, it can effectively protect the motor from damage due to excessive current in starting, braking, overload and blocking. Through proper control, the current loop can effectively reduce the electromechanical time constant of motor inner loop and broaden the bandwidth of speed loop, which lays a good foundation for the design of fast response speed loop controller.

The speed loop is a very important inner loop in the position servo system, which can provide system damping to improve the dynamic response characteristics of the control system, and adjust the frequency band of the speed loop can take into account the rapidity and anti-interference ability of the system. When the stiffness of motor and mechanical transmission structure is not large enough, the stiffness of the system can also be increased by the design of speed ring, but if the stiffness of speed ring is too large, which can easily cause mechanical vibration and lead to system jitter.

The position loop receives the position given instruction signal sent by the upper computer or the flight control computer, and measures the actual position of the steering gear through the position sensor, and controls the deviation signal between the position given and the position feedback, so that the actual output of the steering gear tracks the change of the instruction signal. The feedback loop of position loop has integral link, so it is only necessary to use proportional controller to improve the rapidity of the system and eliminate the steady-state error. The adjustment of the parameters of the proportional controller is finally determined by observing the steady-state error of the step response by the trial method.

In AMESim, the sampling frequency is set to 10Hz, and the load is added from 5s to 10000N. The signal input is 0.1 V, applied from 0.1 s, and given 0 when 0～0.1s. The output results were disturbed by injecting 10000N load force into the system at 5s, and recovered at 7s. as shown in Fig. 4.

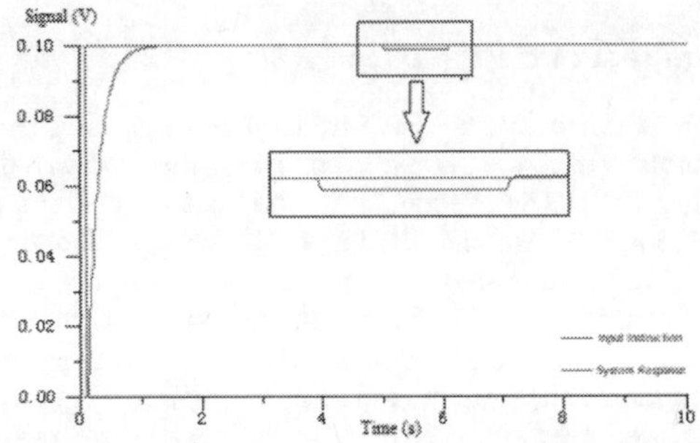

Fig. 4. Simulation results of motor and controller.

4 The reducer model

The reducer model is established in AMESim, as shown in Fig. 3. In this model, the contact stiffness, energy loss and clearance can be considered.

5 The rudder model

The hydrodynamic load on the rudder surface is mainly related to these factors: the area of the rudder surface, the flying altitude, the airspeed and the upwind angle of the rudder surface. During flight, the load torque of aerodynamics is the main load type of rudder surface. In AMESim, the rudder surface is equivalent to a concentrated mass block, as shown in Fig. 3.

6 The PRSM model

The contact stiffness, clearance, Coulomb friction coefficient and other parameters between the screw and the nut can be considered, the model is shown in Fig. 3.

6.1 Stiffness impact of the PRSM

The stiffness of the actuator is set to be 1×10^9 N/m, 1.5×10^9 N/m, 2×10^9 N/m respectively, as shown in Fig. 5. The simulation results show that when the system inputs the load force, the higher the stiffness value, the smaller the fluctuation amplitude, the more easily the steady state is achieved, and the smaller the steady-state error of the system.

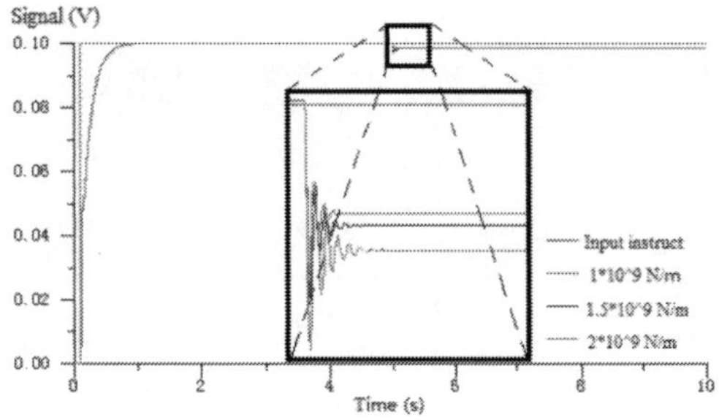

Fig. 5. Stiffness effect of PRSM.

6.2 Impact of clearance in PRSM

The clearance is derived from the machining and assembly error of the parts such as the PRSM, the bearing, etc. The clearance in the actuator is set as 0.001mm, 0.01mm and 0.1mm respectively, as shown in Fig. 6. The results show that in EMA, the clearance has a significant influence on the control accuracy and stability of the system. The greater clearance, the greater amplitude of the system response, the higher frequency, the greater steady-state error.

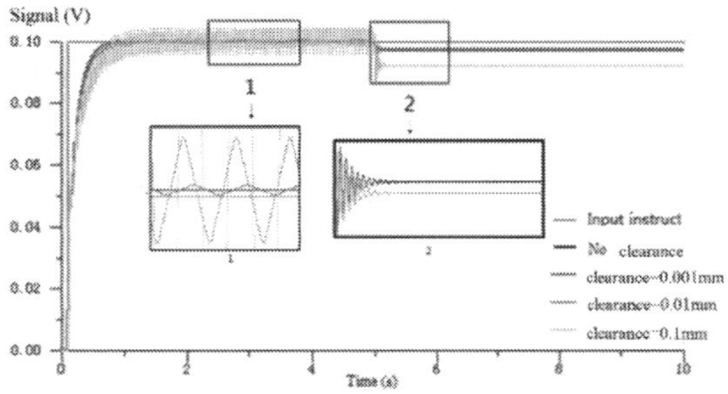

Fig. 6. Clearance impact of actuator.

6.3 Friction effects of PRSM

The formula of friction moment F_{fric} is:

$$F_{fric} = [F_C + F_S e^{-\frac{|\omega_m|}{\omega_s}} + |T_{load}|(c + d\,\mathrm{sgn}(\omega_m T_{load}))]\mathrm{sgn}(\omega_m) \quad (6)$$

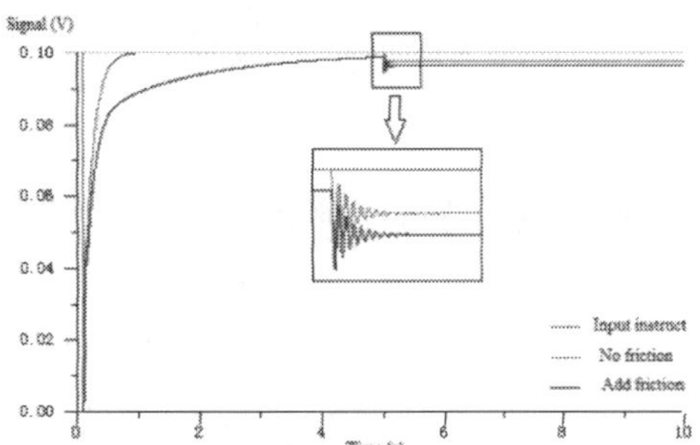

Fig 7. PRSM simulation results for friction effects

where F_c and F_s are the torque produced by Coulomb friction and Stribeck friction respectively. The output moment of the motor T_m is the sum of the friction moment T_{fric} and the output moment of the load end T_{load}, w_s is the reference speed, c is the average coefficient of the external force influence, and d is the quadrant determining coefficient.

The established AMESim model is shown in Fig. 3, and the simulation results are shown in Fig. 7. The results show that when considering the friction torque of PRSM, the response speed of the system slows down and the steady-state error is increased.

7 Conclusions

Through the simulation calculation, the conclusion is drawn as follows:

(1) when the controller is perfected, the steady-state error of the system caused by the load force can be reduced quickly.

(2) when the friction torque of PRSM is taken into account, the response speed of the system becomes slower and the steady-state error becomes larger.

(3) the larger clearance or the smaller stiffness value, the larger the fluctuation amplitude of the system response, the higher frequency and the greater steady-state error.

This research was supported by the National Natural Science Foundation of China (Grant No. 51875458), Key Project of National Natural Science Foundation of China (Grant No. 51535009) and the Fundamental Research Funds for the Central Universities (Grant No. 31020190504003).

References

1. S. J. Ma, G. Liu, R. T. Tong, Friction Heat Analysis of the Planetary Roller Screw Considering Load Distribution, *Mech Based Des of Struc*, **43,** 164-182 (2015)
2. J. Liscouët, M Budinger, J. C. Maré, S. Orieux, Modelling approach for the simulation-based preliminary design of power transmissions, *J Mech Mach Theory*, **46** (3), 276-289 (2011)
3. J. C. Maré, Dynamic loading systems for ground testing of high speed aerospace actuators, *Aircr Eng Aerosp Tec*, **78** (4), 275-282 (2006)
4. S. J. Ma, G. Liu, R. T. Tong, A new study on the parameter relationship of planetary roller screw, Math Probl Eng, Article ID340437, 29 pages (2012)
5. J. Huang, Y. Liu, X. H. Zhang, W. Guan, Z. Y. Song, Dynamic Modeling and Simulation of Electro-Mechanical Actuator Coupled with Nonlinear Factors, Proceedings of the 37th Chinese Control Conference, CCC. 2018, Wuhan. China, 3776-3780 (2018)
6. S. Habibi, J. Roach, G Luecke, Inner-loop control for electro-mechanical (EMA) flight surface actuation systems, *J Dyn Syst-T ASME*, **130** (051002), 1-13 (2008)

MATEC Web of Conferences **306**, 02005 (2020) https://doi.org/10.1051/matecconf/202030602005
ICMME 2019

A Referenced Cyber Physical System for Compressor Manufacturing

Jin Cao[1,*], *Junliang* Wang[2], and *Junqing* Lu[1]

[1]Shanghai Highly(group) Co., Ltd, Shanghai, China
[2]College of Mechanical Engineering, Donghua University, Shanghai, China

Abstract. Compressor is a typical high-end discrete product, with the shortening of product life cycle and the enhancement of the degree of product customization, the traditional compressor manufacturing system architecture cannot meet the requirements of comprehensive digital management of compressor from body scheme design to parts production line, logistics management, operation and maintenance monitoring and evaluation. This paper presents a compressor manufacturing system architecture based on digital twinning, and establishes an Internet platform for compressor industry oriented to remote coordination from three aspects of compressor design, production, operation and maintenance. The platform includes industrial Internet infrastructure layer, physical space entity model layer, virtual space multidimensional model layer, physical space and virtual space multidimensional model correlation and mapping layer, big data intelligent analysis decision-making layer, and digital twin application layer. Through the establishment of the compressor product design and simulation model of digital twin, compressor production process digital twin model, compressor fault diagnosis and remote operations digital twin model, implementation is based on the number of compressor collaboration in manufacturing industrial Internet platform twin system, leading the transformation and upgrading of intelligent manufacturing industry, compressor industry sustainable development ability and international competitiveness.

1 Introduction

Compressor is a typical high-end discrete product whose rotor manufacturing accuracy is as high as micron level. There are more than 1000 kinds of products and tens of thousands of parts. Therefore, compressors workshop is a typical high-end discrete manufacturing system [1]. In recent years, with the integration of new generation information technology, equipment and processes have been promoted from automation to intelligence. At present, the traditional compressor manufacturing system architecture can't meet the needs of comprehensive management from compressor design to parts production line, logistics management, operation and maintenance monitoring, and evaluation. Traditional compressors are facing enormous challenges of transformation and upgrading.

Cyber Physical System (CPS) is the key technology of the new generation of intelligent manufacturing [2]. It reflects all the characteristics of the object from micro to macro and demonstrates the evolution process of the product life cycle (PLC), Then realizes the performance optimization of the whole life cycle through building the digital mirror model of physical products and integrating with external sensors.

In the stage of compressor design, the way of combining three-dimensional digital model with two-dimensional drawings is still adopted in the design scheme of compressor in China [3]. Design dimensions, process documents and test data are in the state of data island now. How to integrate multi-dimensional design data and realize intelligent design through digital twin model is the key factor to drive compressor design intellectualization.

In the process of compressor production, the data generated in the production site (operation data of instruments and equipment, production logistics data, production progress data, production personnel data, etc.) could be collected and processed in real time by digital twin technology, and the process data can be mapped and matched with the digital twins of compressor products and production line [4]. The refined management of the compressor product manufacturing process can be realized online. At the same time, it combines data mining and machine learning algorithm (cloud-based intelligent information platform, dynamic Bayesian and neural network) to realize real-time dynamic optimization and adjustment of production line, manufacturing unit, production schedule, logistics and quality.

Compressor operation and maintenance is the core of air conditioning equipment operation and maintenance, especially in new energy automotive compressor. The normal operation of compressor not only affects the car environment, but also determines the working temperature of the battery and affects the output

* Corresponding author: caojin@highly.cc

efficiency of the battery. It is the focus of affecting the operation efficiency of new energy automobiles [5]. In the process of compressor operation and maintenance, the physical space position, external environment, quality status, use status, technology and function status of compressor products are tracked and monitored in real time. The health status, life span, function and performance of compressor products are predicted and analyzed according to the actual state, real-time data, using and maintenance record data of compressor products. Meanwhile, the quality of products will be early warned. When product failures and quality problems occur, it could realize rapid positioning of product physical location, fault and quality problems record, cause analysis, parts replacement, product maintenance, product upgrade or even scrap, and decommissioning, etc.

Therefore, starting from the three aspects of compressor design, production, operation and maintenance, this paper establishes an industrial Internet platform for remote collaboration, a digital twin model for compressor product design and simulation, a digital twin model for compressor production process, and a digital twin model for compressor fault diagnosis and remote operation and maintenance to realize the digital twin system of compressor co-manufacturing in different places based on industrial Internet platform, lead the industry to transform and upgrade to intelligent manufacturing, and enhance the sustainable development ability and international competitiveness of compressor industry.

2 Literature review

Digital twin has received wide and high attention recently. Gartner, the most authoritative IT research and consulting firm, has listed digital twin as one of the top ten strategic technology development trends for two consecutive years (2016 and 2017). Lockheed Martin, the largest weapons manufacturer, has listed digital twin as the top six technologies in the future defense and aerospace industry in November 2017. On December 8, 2017, digital twin was listed as one of the top ten technological advances in intelligent manufacturing by the Intelligent Manufacturing Alliance of CAST Member Societies at the World Intelligent Manufacturing Congress [6][7].

In addition, many famous international enterprises have begun to explore the application of digital twin technology in product design, manufacturing and service. In product design, for innovative design of complex products, Dassault [8] has established a 3D experience platform based on digital twin, which continuously improves the product design model in the information world and feeds back to the physical entity product improvement by using users' interactive feedback information. In terms of production and manufacturing, Siemens [9] constructs the production system model of integrated manufacturing process based on digital twin concept, and forms the virtual enterprise model and the enterprise image based on automation technology to support the enterprise in integrating and digitalizing the entire value chain, and carry out application verification in the production process of Siemens industrial equipment Nanobox PC. In fault prediction and health management, NASA [10] combines physical system with its equivalent virtual system, studies the method of fault prediction and elimination of complex systems based on digital twin, and applies it to the health management of flight systems such as aircraft and launch vehicle. The Center for Structural Sciences of the Air Force Research Laboratory [11] combines the virtual model of ultra-high-fidelity aircraft with the structural deviation and temperature calculation models that affect flight and carries out the life prediction of aircraft structure based on digital twin. In terms of product service, PTC company [12] takes digital twin as the key link of 'intelligent interconnected products', and devotes itself to establishing a real-time connection between the virtual world and the real world to extend every action of intelligent products to the next product design cycle and realize predictive maintenance of products. Meanwhile, customers are provided with efficient after-sales service and support. In addition, workshop is the execution basis of manufacturing activities. In order to practice the concept of intelligent manufacturing and intelligent production, Fei Tao et al. [13][14] explored the concept of digital twin workshop, designed the composition and operation mechanism of DTS, expounded four characteristics and five key technologies of DTS, and provided basic theoretical support for the application of digital twin in production and manufacturing in 2017. Negri E et al. [15] analyzed the background of digital twin, expounded the connotation of product digital twin, and proposed the architecture and implementation approach of product digital twin. At the same time, he pointed out that the emergence and development of digital twin technology provided not only a new idea, method and implementation approach for the realization of Information Physics System, but also new ideas and tools for human to explore and predict the unknown world with virtual model. Schleich B [16] and others discussed the application of digital twin model in product configuration management. Zhang et al. [17] proposed a rapid customization design method for personalized production line based on digital twin, which synchronized the virtual and real of digital model and physical equipment to form a whole line digital twin system. He proposed a decoupling algorithm for coupling optimization problem in production process as an engine to drive digital twin. The performance of the whole line was evaluated and regulated by order simulation. Wang et al. [18] put forward the active scheduling mode based on the initiative of workpieces and machines, and discussed the interactive system, response mechanism, operation mode and mutual scheduling behavior of active scheduling. Leng J et al. [19] proposed an intelligent modeling method based on real-time perception and decision-making of service state of bottom manufacturing resources in CPS in order to meet the requirements of self-organizing allocation and adaptive collaborative control of manufacturing resources in intelligent factories. Ting Qu [20] et al.

extended the concept of online control in UPU manufacturing by using the idea of digital twin, and put forward the idea, mechanism and quantitative optimization method of multi-system linkage optimization control. The method was applied into three stages of production process: precise planning, real-time tracking and dynamic control. The corresponding technology and system have been applied in leading enterprises such as Carbury Chemical Group. Starting from the workshop level, the basis of manufacturing activities, in order to gradually solve the interaction and integration between the physical world and the information world in the workshop level, and ultimately achieve the goal of intelligent production and precise service, the concept of digital workshop/factory was put forward many years ago. Scholars and practitioners [21-25] at home and abroad have carried out a lot of research and construction work around digital workshop/factory, including workshop manufacturing association, digital/virtual workshop modeling, workshop data/information integration, workshop intelligent operation and precise management, etc. Many valuable theories and technologies have been put forward to analyze the problem and goal at workshop level from different angles, and put forward the corresponding solutions.

3 Research gaps

However, in the context of the new generation of information technology, there are still the following limitations and shortcomings in realizing the demand of intelligent manufacturing and production:

(1) the Internet of things in Workshop Manufacturing. Current research mainly focuses on the intelligent perception and access (including interfaces, protocols, and models), state and operation data acquisition/transmission/processing, operation status monitoring and health management of a single device. A large number of theoretical, technical and methodological studies have been carried out. At the same time, relevant devices have been developed. Therefore, it is said that the network access and intelligent operation of a single device have been realized to a certain extent. However, insufficient attention has been paid to the interconnection of heterogeneous elements in workshop (such as equipment and equipment, people and equipment, equipment and environment, people-equipment-environment, etc.). Especially, there is a lack of research on the system-level comprehensive interconnection of heterogeneous elements such as human, machine, material and environment, that is to say, there is a lack of the support of the theory of full interconnection and integration of heterogeneous elements in workshop and general devices.

(2) Digital/Virtual Workshop Modeling/ Simulation/ Analysis. Existing research mainly focuses on the construction and simulation analysis of geometric model or system model of workshop elements, lacking of models to depict and reflect the real behavior, rules and constraints of physical workshop, which makes it impossible to realize the real and complete mirror from physical workshop to virtual workshop. In addition, in the simulation analysis of workshop operation, existing research mainly relies on existing data-driven models. The factors such as workshop behavior, rules and constraints, as well as the integration of real-time dynamic data of workshop physical elements, are not considered enough, that is, the lack of support from the multi-dimensional fusion model of geometry-behavior-rules-constraints.

(3) Workshop Data/Information Integration. Integration and sharing of information/data related to workshop is mainly based on an information system or platform. Its integration scope is limited to the existing data and part of real-time data collected from information systems deployed in workshop (such as MES, enterprise resource planning, computer-aided process planning, etc.). The upstream business/process (such as market demand, product design) and downstream business/process (such as product maintenance, maintenance and overhaul data) of workshop are not considered and integrated adequately. In addition, due to the lack of support of intelligent interconnection and fusion theory and devices for heterogeneous elements in workshop, the collection and integration of real-time physical data in workshop is insufficient. Also, the two-way dynamic link and fusion of physical and information data are insufficient, that is, the lack of consideration and support for all-factor/all-business/all-process physical-information fusion data in workshop.

(4) Workshop operation optimization and precise management. Traditional ideas are mainly based on the process of 'Problem Analysis-Model Construction-Solution Algorithms Design-Optimized Analysis Control'. After the Internet of Things and big data are applied into workshop, scholars at home and abroad have proposed a new idea of 'Data Association Mining-Dynamic Evolution-Simulation/Forecasting-Intelligent Control' based on big data of workshop. Both of them depend on the operation state model and data of workshop. In the process of actual operation optimization and precise management of digitized workshop, due to the lack of 'geometry-behavior-rule-constraint' fusion model of workshop and consideration and drive of workshop physics-information fusion data, the information layer and physical layer of workshop are not synchronized, workshop information layer and physical layer are in poor consistency, low degree of intelligence, inadequate management precision and not synchronized. Moreover, there is no support of workshop application and service.

4 The distributed CPS architecture of compressor manufacturing

The distributed CPS architecture of compressor manufacturing, as show in figure 1, mainly includes industrial internet infrastructure layer, physical space entity model layer, virtual space multi-dimensional model layer, multi-dimensional model association and mapping layer, big data analysis and decision-making layer, and digital twin application layer.

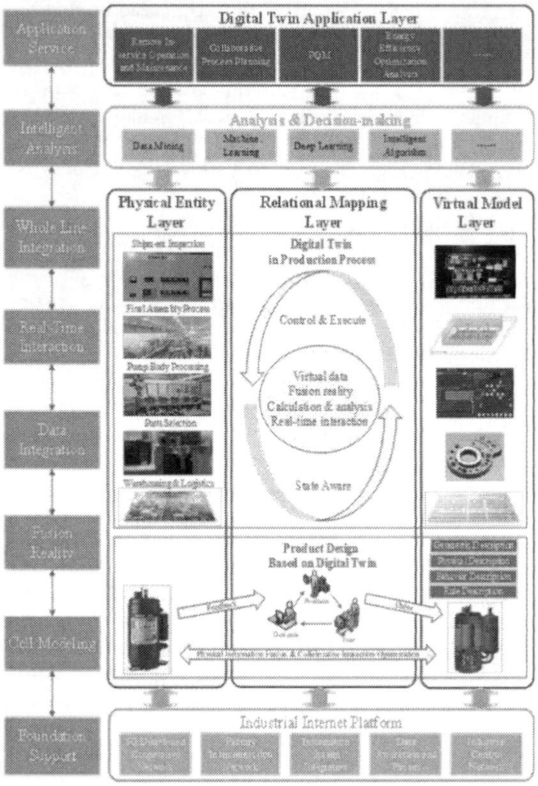

Fig.1. The distributed CPS architecture of compressor manufacturing.

The physical space entity layer includes total factor physical entities in products, equipment, workstations and production lines, such as raw materials, structure parts, assemblies, numerical control machines, robots, detection devices, and AGV. Whether it is the interactive integration of total factor physical resources, the simulation calculation of multi-dimensional virtual model, or the data analysis and processing, all of them are based on total factor physical entities. Meanwhile, the physical entities drive the operation of all parts, so that the digital twin could be realized, which is the 'skeleton' of the digital twin application.

The industrial internet infrastructure layer provides products and twin objects with unidirectional, bidirectional and one-to-many data synchronization and exchange, multi-space-time scale calculation, and closed-loop control capabilities for intelligent perception and interconnection of physical elements, realization of connection and interaction, real-time control and execution. It runs through the whole life cycle of products, which is the cornerstone of the application of digital twin system. The layout of industrial internet infrastructure in compressor assembly shop is shown in figure 2.

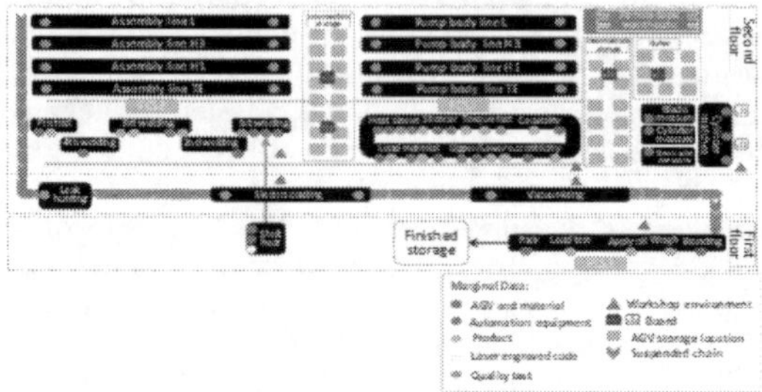

Fig.2. The layout of industrial internet infrastructure in compressor assembly shop.

The virtual space multi-dimensional model layer is used to describe the geometry, physics, behavior and rules of the above physical entities comprehensively. The geometric dimension describes geometric parameters such as size, shape and assembly relationship. The physical dimension analyses physical properties such as stress, fatigue and deformation, and the behavior dimension responds to external driving. The rule dimension is used to model the rule of the physical entity, so that the model has many functions, that is evaluation, optimization, prediction and testing. Multi-dimensional virtual model is the core component of various functions, such as product design, manufacturing, fault prediction, health management, etc. It is the 'heart' of digital twin application to transform application function from theory into reality with data-driven. The virtual space multi-dimensional model of compressor assembly is shown in figure 3.

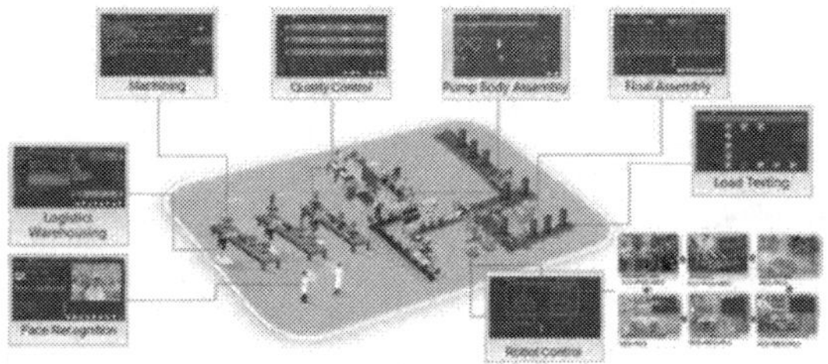

Fig.3. The virtual space multi-dimensional model of compressor assembly.

Multi-dimensional model association relation and mapping layer of physical space and virtual space is to sort out the data association relation of compressor manufacturing process, which processes industrial data of structured, semi-structured and unstructured compressor through unified data space-time dimensions such as timestamp, product number and equipment number and establish the twin data mainline of the whole life cycle with the key production factors of the production line, such as products, equipment, logistics and so on.it is used to realize the mapping of the whole life cycle of physical entity by virtual model and the symbiosis of virtual and real by combining with process flow of compressor production line and logistics route. There are various linear and non-linear, time-varying and time-invariant factors in the production process of compressors. Therefore, it is necessary to design the dynamic updating mechanism of physical entities' maintenance, upgrade and performance degradation, and establish the virtual evolutionary symbiosis neural network, to realize the mapping and motion simulation of physical entities, so that the dynamic updating capability of physical entities at each stage could always be achieved. The core element of digital twin is twin data, which originates from physical entity, virtual model and service system. At the same time, it is integrated into each part after fusion to promote the operation of each part. It is the 'blood' of digital twin application. Dynamic real-time interactive connection links physical entities, virtual models and service systems into an organic whole, so that information and data could be exchanged and transferred among different parts. It is the 'blood vessel' of digital twin application.

Based on existing historical data and specification data, data analysis decision-making layer uses deep learning technology such as deep neural network to establish intelligent prediction, analysis and decision-making model. Compressors' physical entity and its virtual model run synchronously. The generated twin data include not only real-time data-aware such as vibration, temperature, speed and current, but also simulation data such as fatigue wear, torsional vibration and thermal stress of shafting produced by virtual model, and fusion data obtained by multi-dimensional data processing and fusion. The Multi-dimensional model association relation and mapping layer of physical space and virtual space for compressor assembly is shown in figure 4.

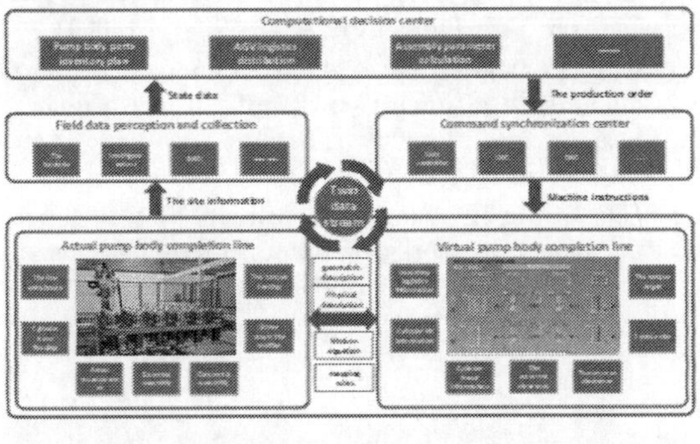

Fig.4. The Digital twin application for compressor quality detection.

In summary, establish the basic platform of industrial internet firstly. On this basis, set up the digital twin models of product and production line units from the physical space entity layer, virtual space multi-dimensional model layer, multi-dimensional model association relation and mapping layer of physical space and virtual space. Then, integrate the whole line digital twin model in the 'material-equipment-station-production line' order, and carry out thematic application development and system integration accordingly.

5 Conclusion

This paper proposes an intelligent manufacturing reference model from the three dimensions of "manufacturing with data and management". The contributions of this paper are as follows:

1.Considering the multi-source heterogeneous elements of the man-machine, machine, material, environment and other workshops, the physical layer of the full-element physical space is established, and the multi-dimensional model layer of the virtual space is constructed from the multiple dimensions of geometry, physics, behavior and rules. Then, the compressor manufacturing process data is combed to realize the mapping of the virtual model to the physical life cycle of the physical entity, and the dynamic real-time interactive connection transfers information and data.

2.The industrial Internet infrastructure layer runs through all stages of the product life cycle, and simultaneously realizes data processing and execution control of the compressor digital twinning system.

3.Using the deep learning technology like deep neural network to establish intelligent prediction, analysis and decision-making model for the existing historical data and specification data of the compressor.

4.For the orderly integration of the whole line digital twinning model according to the "material-equipment-station production line", the subject application of monitoring, early warning, prediction and optimization is realized in the compressor digital application layer.

References

1. Conry R D. Compressor: U.S. Patent **5,857**, 348(1999)
2. Boschert S, Rosen R. *Digital twin—the simulation aspect* (Mechatronic Futures, Cham, 2016)
3. Giannoulis D, Massberg M, Reiss J D. Digital dynamic range compressor design—A tutorial and analysis, Journal of the Audio Engineering Society **60(6)**, 399-408 (2012)
4. Castellanos I D, Stine J E. Compressor trees for decimal partial product reduction[C]//Proceedings of the 18th ACM Great Lakes symposium on VLSI. ACM, 107-110 (2008)
5. Prytz R, Nowaczyk S, Rögnvaldsson T, et al. Predicting the need for vehicle compressor repairs using maintenance records and logged vehicle data, Engineering applications of artificial intelligence, **41**, 139-150 (2015)
6. El Saddik A. Digital twins: The convergence of multimedia technologies, IEEE MultiMedia, **25(2)**, 87-92 (2018)
7. West T D, Blackburn M. Is Digital Thread/Digital Twin Affordable? A Systemic Assessment of the Cost of DoD's Latest Manhattan Project, Procedia computer science, **114**, 47-56 (2017)
8. Fourgeau E, Gomez E, Adli H, et al. *System engineering workbench for multi-views systems methodology with 3DEXPERIENCE Platform. the aircraft radar use case* (Complex Systems Design & Management Asia, Cham, 2016)
9. Vogel-Heuser B, Konersmann M, Aicher T, et al. Supporting evolution of automated material flow systems as part of CPPS by using coupled meta

models, 2018 IEEE Industrial Cyber-Physical Systems (ICPS), 316-323 (2018)

10. Gray D, Bowes D, Davey N, et al. The misuse of the NASA metrics data program data sets for automated software defect prediction[C]//15th Annual Conference on Evaluation & Assessment in Software Engineering. IET, 96-103 (2011)
11. Glaessgen E, Stargel D. The digital twin paradigm for future NASA and US Air Force vehicles[C]//53rd AIAA/ASME/ASCE/AHS/ASC Structures, Structural Dynamics and Materials Conference 20th AIAA/ASME/AHS Adaptive Structures Conference 14th AIAA, 1818(2012)
12. Schleich B, Anwer N, Mathieu L, et al. Shaping the digital twin for design and production engineering, CIRP Annals, **66(1)**, 141-144 (2017)
13. Tao F, Zhang M. Digital twin shop-floor: a new shop-floor paradigm towards smart manufacturing, Ieee Access, **5**, 20418-20427 (2017)
14. Tao F, Cheng J, Qi Q, et al. Digital twin-driven product design, manufacturing and service with big data, The International Journal of Advanced Manufacturing Technology, **94(9-12)**, 3563-3576 (2018)
15. Negri E, Fumagalli L, Macchi M. A review of the roles of digital twin in cps-based production systems, Procedia Manufacturing, **11**, 939-948 (2017)
16. Schleich B, Anwer N, Mathieu L, et al. Shaping the digital twin for design and production engineering, CIRP Annals, **66(1)**, 141-144 (2017)
17. Zhang H, Liu Q, Chen X, et al. A digital twin-based approach for designing and multi-objective optimization of hollow glass production line, IEEE Access, **5**, 26901-26911 (2017)
18. Wang X, Shi Z, Zhang F, et al. Mutual trust based scheduling for (semi) autonomous multi-agent systems[C]//2015 American Control Conference (ACC). IEEE, 459-464 (2015)
19. Leng J, Zhang H, Yan D, et al. Digital twin-driven manufacturing cyber-physical system for parallel controlling of smart workshop, Journal of Ambient Intelligence and Humanized Computing, **10(3)**, 1155-1166 (2019)
20. Zhang K, Qu T, Zhou D, et al. IoT-enabled dynamic lean control mechanism for typical production systems, Journal of Ambient Intelligence and Humanized Computing, **10(3)**, 1009-1023 (2019)
21. Simpson T W, Williams C B, Hripko M. Preparing industry for additive manufacturing and its applications: Summary & recommendations from a National Science Foundation workshop, Additive Manufacturing, **13**, 166-178 (2017)
22. Salter J D, Campbell C, Journeay M, et al. The digital workshop: Exploring the use of interactive and immersive visualisation tools in participatory planning, Journal of environmental management, **90(6)**, 2090-2101 (2009)
23. Orchard S, Hermjakob H, Julian Jr R K, et al. Common interchange standards for proteomics data: Public availability of tools and schema. Report on the Proteomic Standards Initiative Workshop, 2nd Annual HUPO Congress, Montreal, Canada, 8–11th October 2003, Proteomics, **4(2)**, 490-491 (2004)
24. Li B, Hou B, Yu W, et al. Applications of artificial intelligence in intelligent manufacturing: a review, Frontiers of Information Technology & Electronic Engineering, **18(1)**, 86-96 (2017)
25. Gao Q, Shi R, Wang G. Construction of intelligent manufacturing workshop based on lean management, Procedia CIRP, **56**, 599-603 (2016)

Development of advanced intumescent flame-retardant binder for fire rated timber door

Jessica Jong Kwang Yin[1], *Yew* Ming Chian [1,*], *Yew* Ming Kun[2]

*[1]Department of Mechanical and Material Engineering, Lee Kong Chian Faculty of Engineering and Science, Universiti Tunku Abdul Rahman, Cheras 43000 Kajang, Malaysia
[2]Department of Civil Engineering, Lee Kong Chian Faculty of Engineering and Science, Universiti Tunku Abdul Rahman, Cheras 43000 Kajang, Malaysia

Abstract. Intumescent flame-retardant binder (IFRB) offers a great advancement for the most efficient utilization of a wide variety of passive fire safety system at the recent development. This article highlights the fire-resistance and thermal properties of the IFRB using Bunsen burner and thermogravimetric analysis. The five IFRB formulations were mixed with vermiculite and perlite for the fabrication of fire-resistant timber door prototypes. Additionally, the fire rated door prototypes were compared under 2 hours fire test. The prototype (P2), with a low density of 637 kg/m3 showed the superlative fire-resistance rating performance, resulting in temperature reduction by up to 58.9 °C, as compared with that of prototype (P1). Significantly, an innovative fire rated timber door prototype with the addition of formulating intumescent binder has verified to be effective in stopping fires and maintaining its integrity by surviving a fire resistance period of 2 hours.

1 Introduction

The passive fire protection (PFP) system plays an important role in fulfilling the fire safety building regulations to effectively stop the propagation of a fire. PFP system includes fire or smoke dampers, fire doors and firewalls. Therefore, with regards to this research, a new and innovative fire-resistant timber door is incorporated with the intumescent flame-retardant binder (IFRB) acts as an effective PFP system for building occupants to escape safely during the outbreak of fire by trapping the fire and smoke. The intumescent binder reacts under the influence of fire and swells to many times of its original thickness, and thus produces an insulation carbonaceous char or foam that protects the substrate from the effects of the fire [1].

In this study, intumescent binders and lightweight flame-retardant materials such as vermiculite and perlite are used to construct the lightweight fire-resistant timber door prototypes (density is about 600 ±50 kg/m^3) for 2-hour fire rating. Intumescent paint is the recent trend in fire retarding product in construction building materials because of its many beneficial properties such as providing low-odor, lightweight, and environmentally friendly. Exhaustive investigations have revealed that the intumescent flame retardant coating has achieved good flammability, physical and chemical performances though many linger widely on steel structure applications [2-4]. Up to present, the performance of fire-resistant board has yet to be tested and investigated with the addition of intumescent binder for the development of fire-resistant timber door. Furthermore, this research project has also highlighted a huge potential of incorporating the intumescent materials, which consists of three main flame-retardant additives, namely ammonium polyphosphate (APP) – acid source, pentaerythritol (PER) – carbon source and melamine (MEL) – blowing agents, into the fire door. These are mixed in a weight ratio of 2:1:1 and are to be bonded together with flame retardant fillers and vinyl acetate (VA) copolymer as well as vermiculite and perlite which react together to form a protective thermal barrier at high temperature in fire door [5]. It is important to note that flame-retardant additives are useful chemical compounds for fire retardant to provide varying degrees of flammability protection.

This project aims to design and fabricate the fire-resistant timber door prototypes using the formulating IFRBs to fulfill the 2-hours fire rating.

2 Materials and sample preparation

The first part of this project is to synthesize the water-based intumescent binder using a high speed disperse mixer for 30 minutes at room temperature until it is completely homogenous. The formulations of intumescent coatings were tabulated in Table 1. APP, PER, and MEL were used as flame retardant additives. The aluminium hydroxide, magnesium hydroxide, calcium silicate, chicken eggshell (CES) powder, calcium carbonate, and titanium dioxide were used as

* Corresponding author: yewmingchian@gmail.com; yewmc@utar.edu.my

flame-retardant fillers and vinyl acetate (VA) copolymer emulsion acts as water-based polymer binder.
The fire resistance and thermal properties of the IFRBs were characterized and assessed through Bunsen burner test and thermogravimetric analysis.

Table1. Compositions of the IFRB

Samples / Ingredients	Parts by weight for formulations (wt. %)				
	J1	J2	J3	J4	J5
Flame-retardant additives					
APP	20	20	20	20	20
PER	10	10	10	10	10
MEL	10	10	10	10	10
Vinyl acetate (VA) copolymer emulsion	50	50	50	50	50
Flame-retardant filler (pigment)					
TiO_2	4.0	4.0	4.0	4.0	4.0
Flame-retardant fillers					
$Al(OH)_3$	3.0	3.0	-	-	-
$Mg(OH)_2$	-	-	3.0	3.0	3.0
$CaSiO_3$	3.0	-	3.0	-	-
CES	-	3.0	-	3.0	-
$CaCO_3$	-	-	-	-	3.0

The second part of this project is to design and fabricate the fire-resistant timber door prototype with a dimension of 300 mm (length) × 300 mm (width) × 50 mm (thickness) including 2 layers of plywood (thickness is 4 mm per layer). The fire protective performance of the fire timber door prototype was determined through a small-scale fire endurance test. **Table 2** shows the details of the experimental prototypes

Table 2. Specifications of prototypes

Fire door prototype Dimensions: 300 mm (L) x 300 mm (W) x 50 mm (T)	**P1**	**P2**	**P3**	**P4**	**P5**
Weight (g)	2279	2293	2300	2315	2318
Density (kg/m^3)	633	637	639	643	644

Moreover, the fire endurance test and the temperature rise test were conducted to compare and determine the best fire timber door prototype in terms of heat transmission and integrity failure or a significant leakage.

2.1. Bunsen burner test

The intumescent formulations were coated on a 100 mm × 100 mm × 3 mm steel plate with a thickness of 2 ±0.2 mm. The thermocouple plate is attached at the backside of the steel plate coated with intumescent flame-retardant coating and then connected to a digital handheld thermometer for temperature measurement. **Figure 1** shows the experimental setup of the Bunsen burner test for the coated samples. The sample was exposed to Bunsen burner flame spray gun blow torch for one hour at a temperature flame of about 1000 °C. The Bunsen burner gas consumption is about 160 g/h and the distance between the sample and Bunsen burner nozzle is about 7 cm.

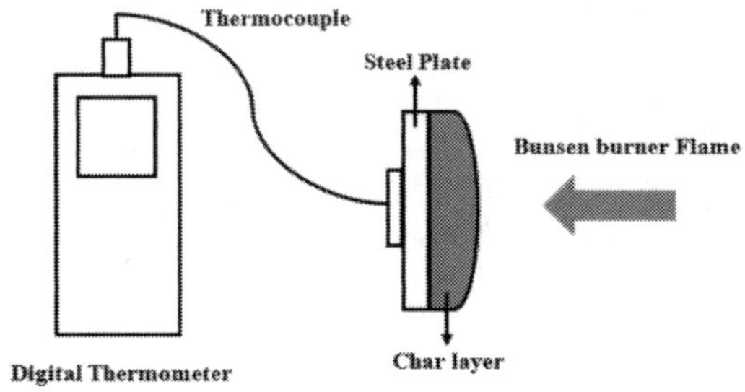

Fig. 1. Schematic of the experimental setup for coated samples.

2.2 Thermogravimetric analysis

Thermogravimetric analysis (TGA) was carried out using a Perkin Elmer STA8000 model to determine the thermal degradation by calculating the residual weight of the intumescent coatings. About 5-8 mg of thin films were placed in a crucible and heated from 30 to 1000 °C at a heating rate of 20 °C /min under nitrogen gas.

2.3. Testing for fire resistant timber door prototype

2.3.1 Small scale fire test

The small-scale fire test was conducted on prototype 1–5, the intumescent binders for each prototype which is based on the formulation J1-J5 that provided in Table 1 as shown in section 2.1. After that, the fire protection performance of the prototypes was tested using the Bunsen burner at about 1000 °C for 2 hours. The temperature profile of the backside of the fire-resistant timber door prototype is measured and recorded using a digital handheld thermometer. **Figure 2** shows a small-scale fire test experimental setup for the fire-resistant timber door prototype.

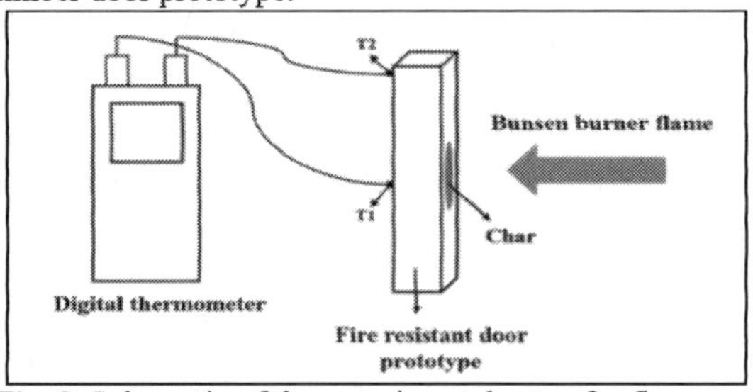

Fig. 2. Schematic of the experimental setup for fire resistant timber door prototype

For the fire endurance test, the tested prototype can be qualified as a standard fire rated timber door if it can maintain its integrity when exposed to the 2-hour fire (at about 1000 °C) without an integrity failure or displaying a significant leakage [6] .Whereas, the temperature rise is also recorded over a 30-minute time interval until 120-minute, this is to examine the heat rise on the backside of the fire-resistant timber door prototype. The thermocouples are fixed at two points of the fire-resistant timber door prototype (centre (T1) and edge (T2)) to measure the transmission of heat from one point to another point of the prototype. This is to ensure that the prototype has taken adequate fire protection by reducing the heat transmission rate and flame propagation so that it can prolong the evacuation time for the building occupants if there is a fire.

3 Results and discussion

3.1 Bunsen burner test

The purpose of this specified test is to characterize the physical and chemical reactions of the char formation of the IFRB. The temperature profiles of the steel plates coated with five IFRB formulations (J1 – J5) and the bare steel plate are compared after the 60 minutes of fire testing. In this experimental work, the temperature of 400 °C is chosen as the critical temperature for the coated sample with IFRC [7]. The fire protection results of the coated steel plates are then plotted as a function of time and presented in Figure 3.

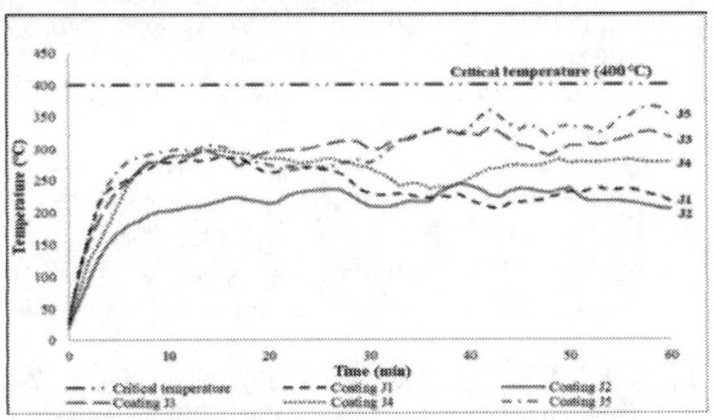

Fig. 3. Evolution of temperature on the coated steel plates

The first 10 minutes have showed that a similar pattern of the temperatures of the coated samples is below 300 °C. After that, the temperatures have started to fluctuate at 20 minutes due to the physical and chemical reactions of the coating formulations during the high temperature test. The temperature continued to rise for 30 minutes; it is observed that the coating J5 has started to rapidly increase until reaching 360 °C at 42 minutes. This may possibly be due to this coated sample has started to lose its adherence strength or interfacial bonding from the steel plate during the fire test. Whereas coating J1 has dropped rapidly after reaching 205 °C at 43 minutes, and this might due to the reducing of gas pressure from the Bunsen burner flame spray gun blow torch. It is also observed that the temperature of all the coatings has started to fluctuate again until reaching at 50 minutes. After 50 minutes of the test, the temperature has reached an equilibrium value and almost remained unchanged until the last stage of the test. The equilibrium temperature of coating J2 that is 217 °C is significantly lower than the other coatings and thus it has showed to be the best result in the fire protective performance. The char combined the positive effects of the flame retardant additives, intumescent paint and fillers (3 wt. % of Al(OH)3 and CES), promoting good fire protective barrier, and producing an excellent and the thickest char layer (i.e. about 6.0 mm) to block the fire while safeguarding the formation of a uniform foam structure which is in good agreement with other researchers, as highlighted by Wang and Yang, 2010 [8]. On the other observation, the coating J5 has the highest equilibrium temperature of 325 °C, reaching almost to the critical temperature (400 °C) after 40 minutes of the fire exposure. The equilibrium temperature of coating J5 might be due to its char layer showed porosity and lightness, resulting in the loss of adherence of the char to the steel plate or by a loss of cohesion of the char layer [9].

3.2 Themogravimetric analysis

From the result obtained in **Figure 4** under nitrogen gas, the thermal decomposition is mainly observed at the temperature in between 200-450 °C. However, the thermal decomposition can be occurred in different stages.

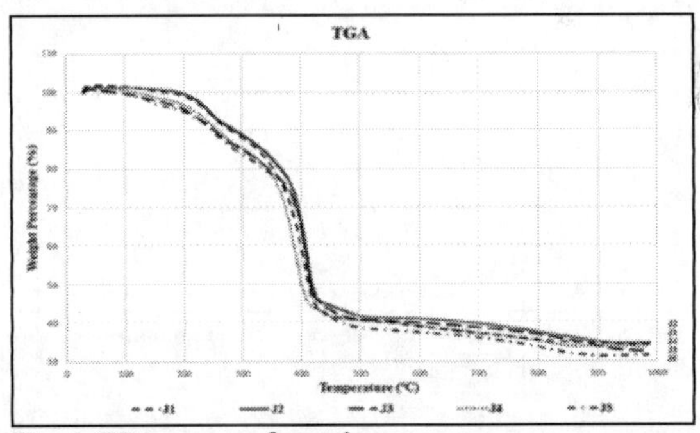

Fig. 4. TGA curves of samples

In this study, the general thermogram profiles are quite similar. However, the curves are found to be slightly shifted towards higher temperature ranges with increments in the heating rate. From 30-100 °C, around 2 % of the initial weight is loss. This might due to the possibly owing to the escape of moisture. Between 100-200 °C, J1 and J2 had slightly weight loss approximately 1 to 2 %, where J3, J4 and J5 had around 5 % of the initial weight is observed. Between 200-300 °C, a weight loss around 10 % of initial weight is observed. Between 300-450 °C, around 45 % of the initial weight loss is observed. Between 450-1000 °C, the weight loss around 10% of the initial weight is observed. This can be attributed from the degradative reactions occurring to the relatively stable components in the mixture. The residue weight of J1, J2, J3, J4 and J5 at 950 °C was 34.28 %,

34.55 %, 32.77 %, 33.82 % and 31.34 %, respectively. The TG curves demonstrated that the residue weight of coating sample J2 had the highest residue weight which indicated that the addition of aluminium hydroxide and CES fillers content in intumescent binder could enhance anti-oxidation of the coatings [9-11].

3.3 Testing for fire resistant door prototypes

This test is to characterize the fire protective performance of the fire-resistant timber door prototypes (P1-P5). The results of each sample plotted as a function of time are presented in **Figure 5**. The results show that the temperature of each fire-resistant timber door prototypes has increased gradually. From the results obtained, all the temperatures of prototypes increase significantly at the first 15 minutes revealing a similar pattern. This phenomenon is because of the good thermal insulation of vermiculite and perlite. These porous materials can dissipate the heat when exposed to fire. After 60 minutes, all prototypes were gradually increased until 120 minutes. This may be due to the physical and chemical reactions of the fire-resistant boards during the fire test. On top of that, P2 has the lowest temperatures of 140.7 °C and 32.0 °C at the T1 and T2, respectively. It has also indicated that this prototype resulted in the best fire protective performance in retarding the fire.

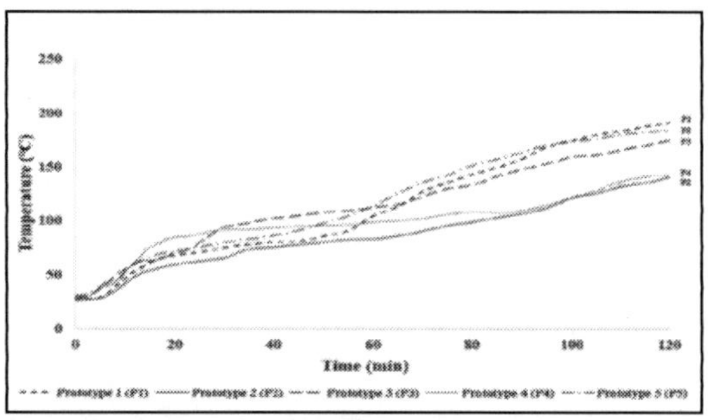

Fig. 5. Evolution of temperature of fire-resistant timber door prototypes.

4 Conclusions

From the series of experimental tests, the following conclusions can be derived based on the results presented in this paper:

(1) The coating J2 with the combination of 3 wt. % aluminium hydroxide and 3 wt. % renewable CES bio filler have showed to be the best fire protective performance with an equilibrium temperature of 217 °C.

(2) The TG curves demonstrated that the residue weight of coating sample J2 had the highest residue weight of 34.55 % at 950 °C, which indicated that the aluminium hydroxide and CES fillers content could enhance anti-oxidation of the coating.

(3) The P2 prototype with the incorporation of J2 intumescent binder has also indicated the best fire protective result (temperature reduction by up to 58.9 °C), as compared to P5 prototype.

The selection of an appropriate combination of flame-retardant materials can directly affect the fire protective performance and thermal properties of the intumescent flame-retardant binder. As a result, the lightweight fire-resistant timber door prototype (P2) has indeed demonstrated to be more efficient in reducing the heat transmission by maintaining its integrity without showing a significant leakage against the 2-hour fire test.

References

1. E.D. Weil, J. Fire Sci. **29** (2011)
2. M.C. Yew, N.H. Ramli Sulong, Mater. Des. **34** (2012)
3. M.C. Yew, N.H.R. Sulong, M.K. Yew, M.A. Amalina, M.R. Johan, Prog. Org. Coatings. **76** (2013)
4. M.C. Yew, M.K. Yew, L.H. Saw, T.C. Ng, R. Durairaj, J.H, Beh, Prog. Org. Coatings. **124** (2018)
5. Y. Xia, F. Jin, Z. Mao, Y. Guan, A. Zheng, Poly. Degrad. Stab. **107** (2014)
6. Q. Xu, L. Chen, K.A. Harries, F. Zhang, Q. Liu, J. Feng, Constr. Build. Mater. **96** (2015)
7. M.C. Yew, N.H.R. Sulong, M.K. Yew, M.A. Amalina, M.R. Johan, Prog. Org. Coatings. **81** (2015)
8. G. Wang, J, Yang. Surf. Coatings Technol. **204** (2010)
9. M.C. Yew, N.H.R. Sulong, M.K. Yew, M.A. Amalina, M.R. Johan, Sci. World J. **2014** (2014)
10. J.H. Beh, M.C. Yew, M.K. Yew, L.H. Saw, Coatings. **9** (2019)
11. M.K. Yew, M.C. Yew, L.H. Saw, J.H. Beh, Y.J. King, R. Durairaj, Mater. Sci. Forum. **947** (2019)

Numerical Investigation on Magnetically Actuated Tunable Micro Liquid Lens

Shi-Hua Wu[1], *Gui-Ping* Zhu[1*], and *Hu-Lin* Huang[1]

[1]College of Astronautics, Nanjing University of Aeronautics and Astronautics, Nanjing 210016, China

Abstract. In this paper, magnetically actuated tunable liquid lens is fundamentally studied to have further potential application scope in research, industrial, aeronautical and astronautical areas. The magnetic field, which is generated by a magnetic potential applied to the permanent magnet domain, distributes non-uniformly in the computational domain and generates magnetic field force to obtain the deformation of the ferrofluid droplet. Consequently, the light-transmitting droplet deforms due to direct contact with the ferrofluid droplet by a connecting channel. The combined effects of gravitational, frictional, interfacial and pressure-driven force make light-transmissive droplet to be a programmable liquid lens. The light-transmissive droplet exhibits changes in the curvature of the contour, which results in different focal lengths for realizing the variable-focus liquid lens. This work focus on the driven mechanism by the magnetic force for a better understanding. The performance is numerically studied for a magnetically actuated varifocal liquid lens by finite element method. The results are presented individually for the distribution of magnetic and flow field, and light refraction.

1 Introduction

In the recent decade, with the demand for various digital products such as smart phones, digital cameras, surveillance cameras, car cameras and microscopes, zoomable liquid lenses have received increasing attention in the field of imaging. Conventional imaging devices are usually composed of several convex lenses. For example, the microscope adjusts the focus by adjusting mechanical threads between the convex lenses. Such operation is very complicated, and the device is bulky, cumbersome and expensive. Conventional glass or resin convex lenses require high-precision machine tools to manufacture with high cost. The liquid lens is made of light-transmissive liquid with no mechanical processing and low cost of manufacturing. The liquid lens, which maintains its surface profile by gravity and interfacial tension, does not break under severe vibration. Moreover, the surface can be self-repairing and very smooth. Liquid lens requires only one lens to achieve zoom without polarization in the imaging.

With the development of digital microfluidics and micromechanical electronics, there have been many ways developed to control the zoom lens. The most common way is the fluid pressure actuation [1-5], which adjusts the focal length of the light-transmitting liquid by the movement of the other fluid. The other approaches for tuning the focal length of the liquid lens includes electrowetting [6-12], dielectrophoresis [13-15] and thermal control [16]. Some liquid lenses are produced by filling a thin film material with a light-transmitting liquid to prevent evaporation of the liquid [17-21]. All the controlling methods adjust the focal length of the liquid lens by changing the curvature of the liquid contour. Fluid pressure actuation methods typically require complex mechanical actuation devices that are more sensitive to vibration than electrowetting and dielectrophoresis methods. It is not suitable for applying fluid pressure actuation in high speed imaging due to a certain delay in zooming as an indirect drive method. The electrowetting and dielectrophoresis methods can operate directly on the light-transmissive liquid. By apply a high voltage on the fluid, the strong zooming ability exhibits with a relatively simple structure with no other mechanical actuating parts. However, the problem of liquid evaporation is required to be properly addressed.

The magnetic actuation [18,19,21-24] method attracts increasing attentions due to easy fabrication and integration. Fluid pressure is induced to adjust the focal length of the light-transmitting liquid. Magnetic actuation can be further divided into two categories that driven by ferrofluid and non-magnetic fluid. For example, the magnetic field generated by the electromagnetic coil was used to drive the movement of the ferrofluid to realize deformation of the liquid lens for zooming [22]. Malouin et al. [23] used ferrofluid droplets to act as liquid pistons with sinusoidal signals applied to the electromagnets to drive the liquid to zoom. Ferrofluids was actuated by electromagnets to drive

* Corresponding author: zhuguiping@nuaa.edu.cn

mechanically-wet liquid lens[24]. Oh et al. [21] used an electromagnet to drive the permanent magnet motion on the liquid and indirectly drive the liquid lens to achieve zooming. The reliability of magnetically actuated liquid lens is relatively high. However, there will be a delay of a millisecond for the deformation of the liquid lens responding to the change of the magnetic field.

So far as we know, most of work on magnetially driven liquid micro lens are realized by experimental studies. For a better understanding of the physics and process, this work focused on the numerical calcualtion of magnetically driven lens at different field strength. In addition, the delay of deformaiton of the lens is considered in our discussion. The results are systematically presented in terms of the distribution of flow field, magnetic field and the optical properties.

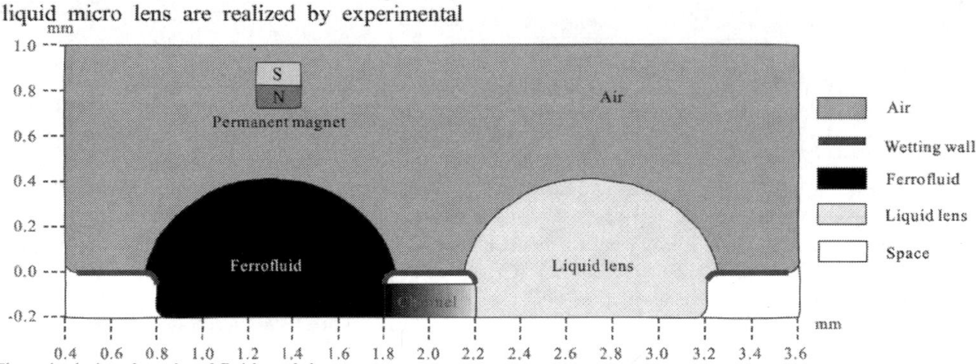

Fig. 1. The calculation domain of fluid module.

By using the finite element method, two-dimensional (2D) calculation is carried out to simulate a magnetically actuated microfluidic lens. The computational domain can be divided into fluid module, magnetic field module and optical module physically. The calculation domain of the fluid module is shown in Fig 1. Ferrofluid and liquid lens are represented by two semicircles approximately 1 mm in diameter. The fluid can flow in each other within two droplets connected by a square channel. The gray part is an air field with a density of 1.2 kg/m^3 and a dynamic viscosity of 18.1×10^{-6} Pa•s as the second phase (continue phase, CP) of the multiphase flow. As the first phase (Discrete phase, DP) of the multiphase flow, the ferrofluid and liquid lens are treated as the same medium with a uniform density of 1×10^3 kg/m^3 and a dynamic viscosity of 5×10^{-3} Pa•s. The two-phase flow of air and liquid is calculated by the level set module (LS) for tracking interface. The contact surfaces on both sides of the droplet are set to a wetted wall with a contact angle of 90 degrees. The tension between the droplet and air is set at 0.01 N/m. The space occupied by the permanent magnet and the blank space do not participate in the calculation. All the boundaries of the fluid domain are set as the outlet for eliminating the boundary effects.

The magnetic field currentless module (MFNC) is utilized for calcuting the magnetic field which only appears at the ferrofluid half region of the domain (0.4~2.0 mm) shown in Fig 1. Relative permeability of the permanent magnet domain is set at 5. With the top edge set to zero, a magnetic potential of 4-12 A is applied to the base of the permanent magnet to generate a vertical upward magnetic field in the permanent magnet area. Regardless of saturation and residual magnetization, the relative magnetic permeability of the air and ferrofluid domain is set as 1 and 1.3, respectively. The computational domain employ the magnetic shielding boundaries for reduction of the boundary influence on the distribution of the magnetic field.

2. Model and methods

The optical performance is calculated by the geometric optics module (GOP), with the calculation domain shown in Fig 2. The semicircle with a diameter of about 1 mm is a liquid lens and the rest is air. The refractive index of the lens and air is respectively set to 1.5 and 1 without considering the light reflection. The refraction boundary of the droplet is highlighted in which the arc profile is redrawn into the optical module as a parametric curve by the coordinates determined when the level set function value is 0.5. The light source release boundary on the top of the domain can uniformly emit visible light with a wavelength of 660 nm in the vertical downward direction (y-axis negative direction). The rest are the wall freezing boundaries that the light will stop automatically when passing through.

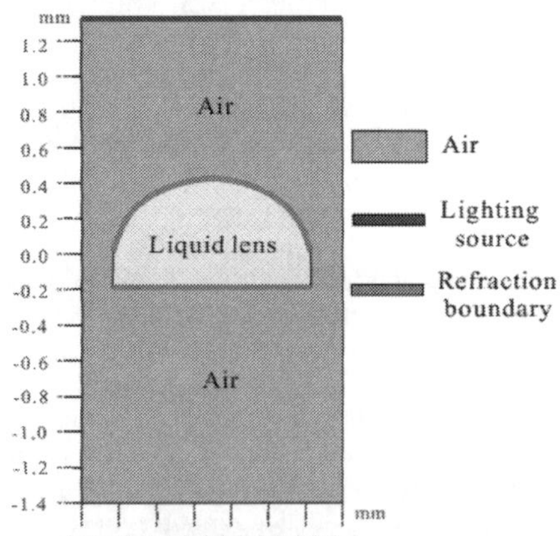

Fig. 2. The calculation domain of optical module.

In the simulation, the optical part acts as an individual calculation module. The flow field and magnetic field simulation requires iterative calculation with MFNC, SPF and LS modules by the split solver. The detailed information of the calculation process is shown in Fig 3. Two cases are calculated with the toal mesh unit number at 14028 and 28698, respectively. The calculation is considered to be independent with mesh as the relative errors are less than 5% for various parameters such as the areal integral of magnetic force, maximum velocity, maximum pressure and so on.

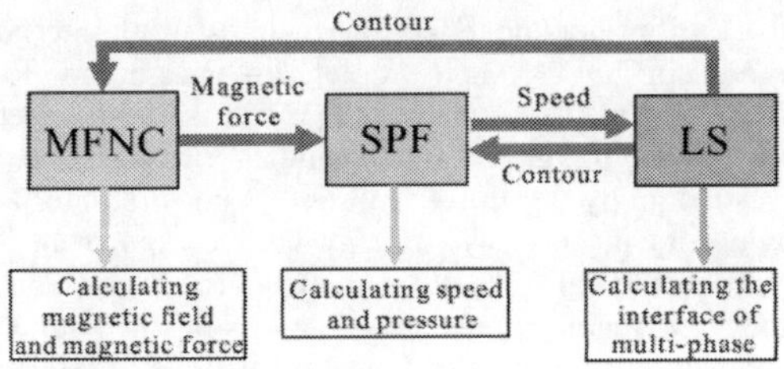

Fig. 3. The calculation process.

The magnetic field module is mainly used to calculate the magnetic field distribution without conduction current, which can be solved by the Maxwell's equations:

$$\nabla\cdot\mathbf{B}=0 \tag{1}$$

$$\nabla\times\mathbf{H}=0 \tag{2}$$

$$-\nabla V_m=\mathbf{H} \tag{3}$$

$$\mathbf{B}=\mu_0\mu_r\mathbf{H} \tag{4}$$

where $\mathbf{H}$, $\mathbf{B}$, V_m, μ_0 and μ_r are the magnetic field strength, magnetic flux density, magnetic potential, vacuum permeability and relative permeability. The electrodynamics gives the magnetic force to any medium:

$$\mathbf{F}=\oiiint_V \mathbf{J}\times\mathbf{B}dV \tag{5}$$

where $\mathbf{J}$ is the current density and V is the volume of the medium. For nonconducting ferrofluid, the magnetization current density $\mathbf{J}_m$ can be expressed as:

$$\mathbf{J}_m=\nabla\times\mathbf{M} \tag{6}$$

where M is magnetization, the magnetic force is derived as:

$$\mathbf{F}=\oiiint_V (\nabla\times\mathbf{M})\times\mathbf{B}dV \tag{7}$$

$$\mathbf{M}=\chi\mathbf{H} \tag{8}$$

$$\chi=\mu_r-1 \tag{9}$$

$$\mu=\mu_0\mu_r \tag{10}$$

where χ is the magnetic susceptibility. Regardless of the changes with the magnetic field, the magnetic permeability can be assigned to the fluid domain by the level set function:

$$\mu=\mu_0((\mu_r-1)\Phi+1) \tag{11}$$

where Φ (Eq.(25)) is the level set function. The calculation with Φ equals 0 and 1 represents the permeability of vacuum and ferrofluid, respectively. The excessive permeability distribution is smoothed when Φ varies within 0 and 1. As a result, the magnetic force acts only in the ferrofluid droplet domain in the form as:

$$\mathbf{F}_1=\frac{1}{2}\mu_0\chi\oiiint_V \nabla\mathbf{H}^2dV \tag{12}$$

The magnetic force in the ferrofluid droplet is a function of the magnetic susceptibility, the volume of the droplet and the magnitude of the magnetic intensity. In addition, the other component of the magnetic force is generated due to the magnetic permeability gradient:

$$\mathbf{F}_2=-\frac{1}{2}\mu_0\mathbf{H}^2\oiiint_V \nabla\chi dV \tag{13}$$

The applied external force in the calculation is expressed by the volume force:

$$\mathbf{f}=\frac{d\mathbf{F}}{d\mathbf{V}}=\mathbf{f}_1+\mathbf{f}_2=\frac{1}{2}\mu_0\chi_m\nabla\mathbf{H}^2-\frac{1}{2}\mu_0\mathbf{H}^2\nabla\chi_m \tag{14}$$

where f_1 and f_2 represent the volumetric forces of F_1 and F_2. Since both ferrofluid and oil are incompressible fluids, which can be solve by N-S equations and continuous equations:

$$\rho\nabla\cdot\mathbf{u}=0 \tag{15}$$

$$\frac{\partial(\rho\mathbf{u})}{\partial t}+\mathbf{u}\cdot\nabla(\rho\mathbf{u})=-\nabla P+\mu\nabla\cdot\left(\nabla\mathbf{u}+\nabla\mathbf{u}^T\right)+\mathbf{f}' \tag{16}$$

where $\mathbf{u}$ is the velocity of the fluid, P is the internal pressure, μ is the dynamic viscosity, ρ is the fluid density, and $\mathbf{f}'$ is volume force which include magnetic force, gravity and tension. The resistance of a spherical droplet in moving can be expressed as [25]:

$$f_d=6\pi r_d v_{cp}\eta_{cp}(1+2\eta_{cp}/\eta_{dp})/(1+\eta_{cp}/\eta_{dp}) \tag{17}$$

where r_d, v_{cp}, η_{cp}, and η_{dp} express the droplet radius, continuous phase velocity, continuous phase viscosity and separation phase viscosity, respectively. The unit area tension at the interface of the droplet can be expressed as [26]:

$$\mathbf{f}_\sigma=\sigma\delta\boldsymbol{\kappa}\mathbf{n} \tag{18}$$

$$\boldsymbol{\kappa}=-\nabla\cdot\mathbf{n} \tag{19}$$

$$\mathbf{n} = \frac{\nabla\mathbf{\Phi}}{|\nabla\mathbf{\Phi}|} \quad (20)$$

$$\delta = |\Phi(1-\Phi)| \cdot |\nabla\Phi| \quad (21)$$

where σ is tension. δ is the Dirac function, which is only zero at the droplet interface and the boundary of fluid. $\mathbf{\kappa}$ is the interface curvature, $\mathbf{n}$ is the unit normal vector of the interface. The wetting wall solving equation:

$$\mathbf{n}_{wet} \cdot \mathbf{u} = 0 \quad (22)$$

$$\mathbf{f}_{wet} = \sigma\delta(\mathbf{n}_{wet} \cdot \mathbf{n}_{\text{int}} - \cos(\theta_w))\mathbf{n}_{\text{int}} - \frac{\mu}{\beta}\mathbf{u} \quad (23)$$

$$\mathbf{n} \cdot (\varepsilon\nabla\Phi - \Phi(1-\Phi)\frac{\nabla\Phi}{|\nabla\Phi|}) \quad (24)$$

whrer $\mathbf{n}_{wet}$ is the wall normal unit vector, $\mathbf{f}_{wet}$ is the wall force, $\mathbf{n}_{\text{int}}$ is the unit vector of fluid motion direction, β is the slip length which value is the mesh element size. The interface capture is accomplished by the level set method with governing equation as:

$$\frac{\partial\mathbf{\Phi}}{\partial t} + \nabla \cdot (\mathbf{u}\mathbf{\Phi}) = \gamma\nabla \cdot (\varepsilon\nabla\mathbf{\Phi} - \mathbf{\Phi}(1-\mathbf{\Phi})\frac{\nabla\mathbf{\Phi}}{|\nabla\mathbf{\Phi}|}) \quad (25)$$

$$\gamma = 0.5 + V_m \cdot 10^{-2} \quad (26)$$

where γ is the initialization parameter, indicating the maximum speed of the interface movement. ε is the interface thickness control parameter, which is set to 40 μm. In this model, $\mathbf{\Phi}$=0, 1 and 0.5 means ferrofluid, mineral oil and the boundary, respectively. The calculation of the optical part is mainly carried out by the Fresnel formula:

$$\mathbf{n}_t = \eta\mathbf{n}_i + \gamma\mathbf{n}_s \quad (27)$$

$$\eta = \frac{n_1}{n_2} \quad (28)$$

$$\gamma = -\eta\cos(\theta_i) + \cos(\theta_t) \quad (29)$$

where $\mathbf{n}_t$, $\mathbf{n}_i$, $\mathbf{n}_s$, η, γ, n_1, n_2, θ_i and θ_t represent the refracted ray unit vector, the incident ray unit vector, the unit vector perpendicular to refractive surface, the refractive index, the auxiliary variable, the air refractive index, the lens refractive index, the incident angle and the refraction angle, respectively.

3. Results analysis and discussion

The magnetic flux density distribution is shown in Fig 4. The magnetic field of the permanent magnet domain can reach up to 0.7 T. White lines with arrows indicates the distribution of the simulated magnetic field. For detailed information, the distribution of magnetic field is shown in Fig 4 (b) with focus on the flux density ranging between 0 and 20 mT. The magnetic flux density close to the ferrofluid region is only a few mT.

The flow field distribution is shown in Fig 5. From Fig 5 (a), the internal velocity of the droplet does not reach equilibrium due to inertia at t=4 ms. With interface moving up, the ferrofluid droplet exhibits obvious four velocity vortex mainly due to the effect of the magnetic force. The liquid in the middle channel moves correspondingly to the ferrofluid. With the interface moving down, only two velocity vortices appear at the droplet on the right side of the domain. The high velocity values are mainly concentrated on both sides of the droplet due to the effect of wetting wall. At t=100 ms shown in Fig 5 (b), the steady state of the droplet is achieved under the stabilized flow field with no obvious movement of the droplet interface and basical disappearance of the velocity vortex. The larger velocity value mainly concentrates near the interface of the droplet due to the effect of surface tension. The corresponding pressure distribution is shown in Fig 5 (c). The larger pressure value mainly concentrates near the wetted wall. The bottom of the liquid droplet also has a large pressure value due to gravity.

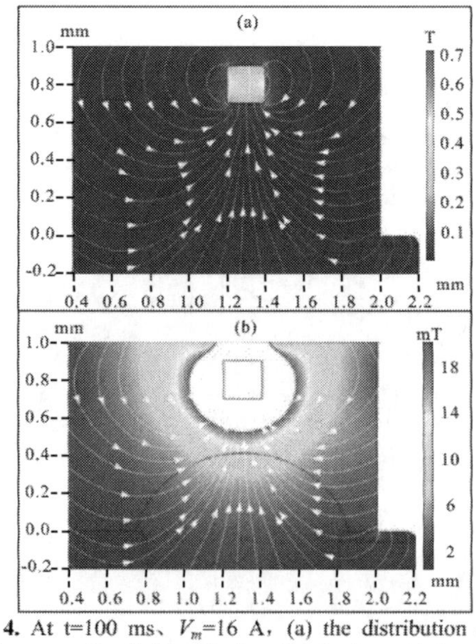

Fig. 4. At t=100 ms、V_m=16 A, (a) the distribution of the whole magnetic field; (b) the distribution of the partial magnetic field (B<20 mT)

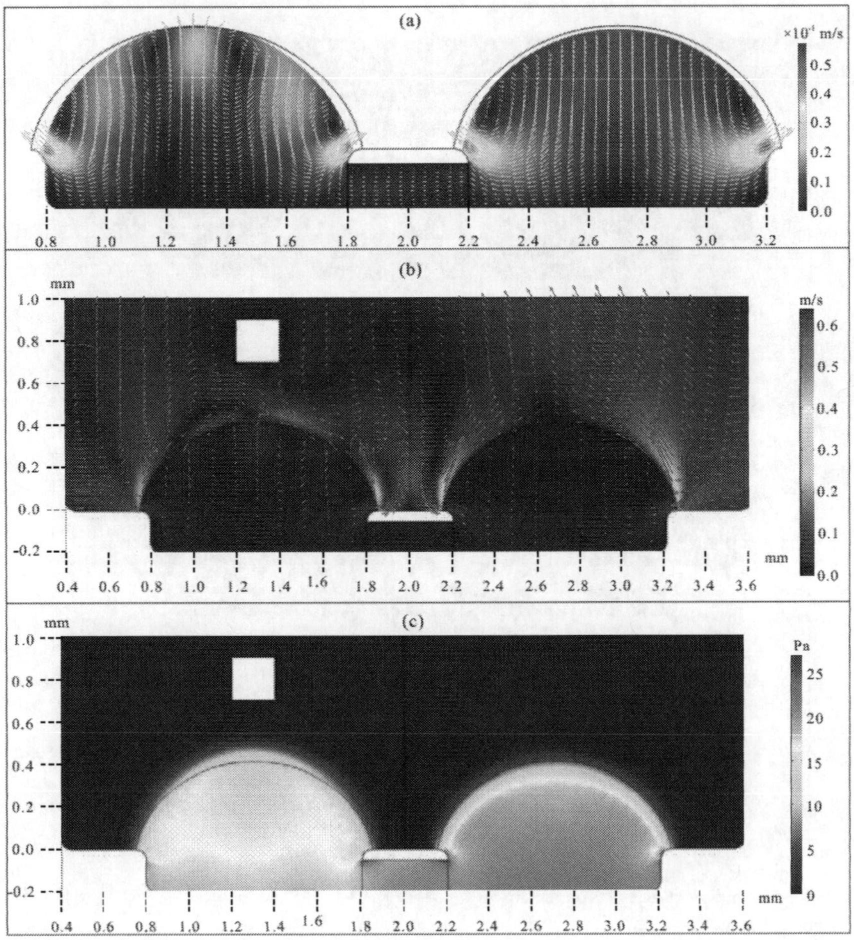

Fig. 5. At V_m=16 A, the distribution of flow field: (a) At t=4 ms and **Φ** >0.6, droplet velocity distribution; (b) At t=100 ms, global velocity distribution; (c) At t=100 ms, global pressure distribution.

The volume distribution of the droplet is shown in Fig 6, where the red domain indicates the droplet phase (DP), and the blue domain indicates air phase (CP). The color value between 0-1 indicates the interface between the droplet and the air. At V_m=4 A, there is only a slight stretching phenomenon on the ferrofluid side droplet with contour close to the one only under gravity, as shown in Fig 6 (a). At V_m=16 A, the ferrofluid droplet is stretched upward while the lens droplet is stretched downward as shown in Fig 6 (a). Fig 6 (c) illustrates the interface of the droplets at **Φ** =0.5 under different magnetic potentials. The results reveal that the height of the ferrofluid droplet gradually increases while the lens droplet gradually decreases with the increasing of the magnetic potential. As shown in Fig 7, the area of the droplet in the two-dimensional model is considered to be conserved as the variation is very small which is around 1.1×10^{-9} mm^2.

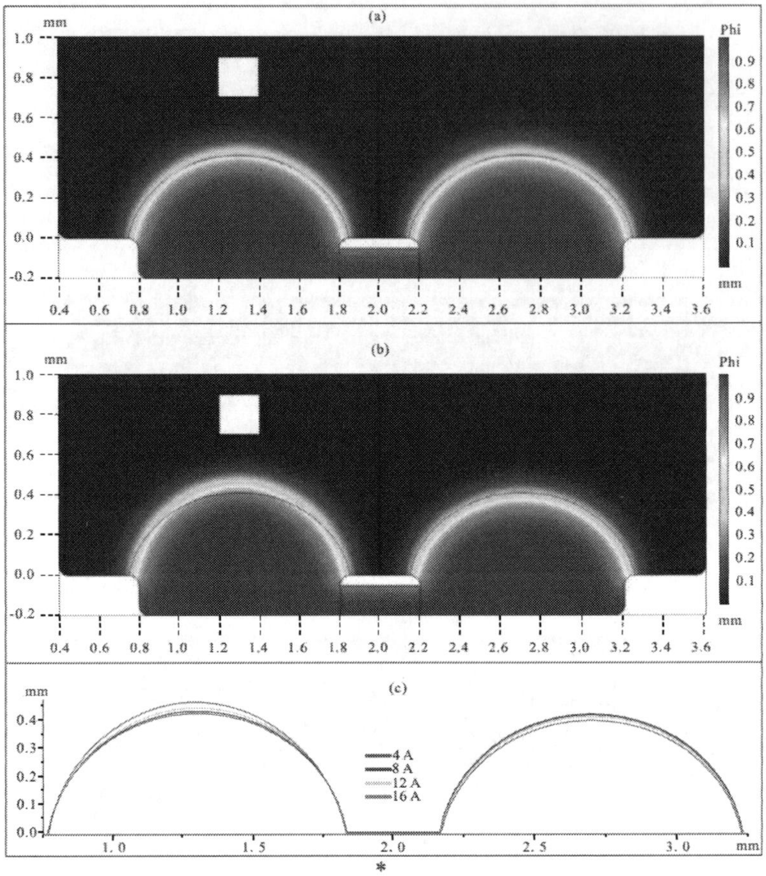

Fig. 6. At t=100 ms, the distribution of droplet volume fraction: (a) At V_m=4 A, Droplet contour; (b) At V_m=16 A, Droplet contour; (c) At V_m=4 A and Φ=0.5, Droplet interface.

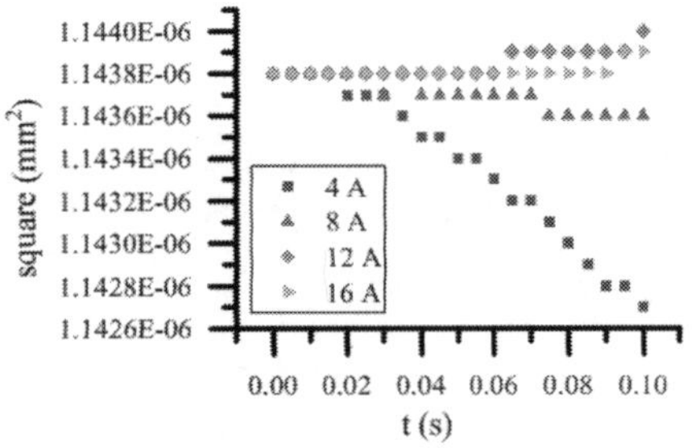

Fig. 7. The area of the droplet at different times and magnetic potential.

The results of optical module calculation are shown in Fig 8. Light is released from the incident boundary and reaches the bottom boundary of the computational domain after only 15 ns. When there are only 8 incident rays, the inner and outer focal points of the droplets are inconsistent as shown in Fig 8 (a). This may be due to the calculation of the wetting wall, resulting in irregularities in the arc of the outer contour of the droplet. When the number of incident rays reaches 80, the obvious focus of the rays can be observed, and only few of the rays outside the droplet deviates from the focus, as shown in Fig 8 (b).

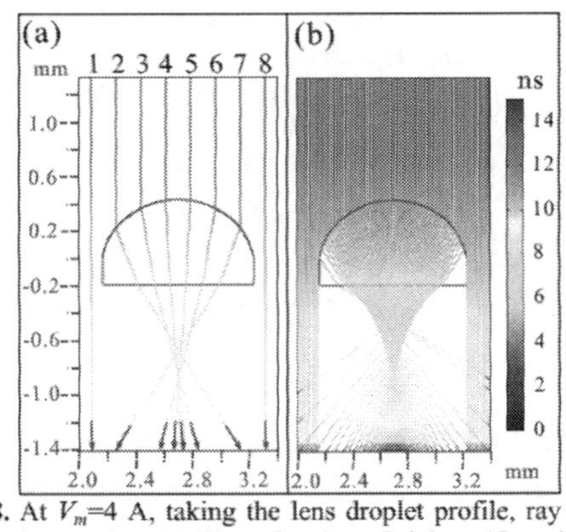

Fig. 8. At V_m=4 A, taking the lens droplet profile, ray optical calculation results: (a) The refraction of eight uniform incident light rays; (b) The refraction of 80 uniform incident rays

4. Conclusion

The magnetically actuated liquid lens, which has strong operability, can be flexibly zoomed by a magnetic field. In this work, the underlying physics of magnetically driven of ferrofluid droplet is systematically investigated and analyzed. The results are presented for flow field, magnetic field as well as pressure distribution for detailed information. The corresponding volume fraction of the ferrofluid droplet and lens droplet is discussed on the shape deformation and liquid flow through the intersection channel. The optical parameters of the lens droplet will be our futher focus for precise controlling and effective tuning under different magnetic field strength.

This work was partially supported by the Natural Science Foundation of Jiangsu Province (BK20160797), and the National Natural Science Foundation of China (11702134).

References

1. K.-H. Jeong, G. L. Liu, N. Chronis, L. P. Lee, *Opt. Express* **12**, 2494-2500 (2004)
2. P. M. Moran *et al. Appl. Phys. Lett.* **88**, 157 (2006)
3. Ren, H. & Wu, S.-T. *Opt. Express* **15**, 5931-5936 (2007).
4. N. Sugiura, S. Morita, *Appl. Opt.* **32**, 4181-4186 (1993).
5. J. Chen, W. Wang, J. Fang, K. Varahramyan, *J. Micromech. Microeng.* **14**, 675-680 (2004).
6. S. K. Cho, H. Moon, C. J. Kim. *J. Microelectromech. S.* **12**, 70-80 (2003).
7. S. K.Fan, H. Yangb, W. Hsu, *Lab Chip* **11**, 343-347 (2011).
8. J. Gong, C.-J. C. Kim, *Lab Chip* **8**, 898-906 (2008).
9. U. C. Yi, C.-J. Kim, *J. Micromech. Microeng.* **16**, 2053-2059 (2006).
10. C.-X. Liu, J. Park, J.-W. Choi, *J. Micromech. Microeng.* **18**, 035023 (2008).
11. T. Krupenkin, S. Yang, P. Mach, *Appl. Phys. Lett.* **82**, 316-318 (2003).
12. B. Berge, J. Peseux, *Eur. Phys. J. E* **3**, 159-163 (2000).
13. A. F. Chrimes, K. J. Berean, A. Mitchell, G. Rosengarten, K. Kalantar-zadeh, *Acs Appl. Mater. Inter.* **8**, acsami.5b10625 (2016).
14. C. G. Tsai *et al. IEEE Photonics Tech. L.* **21**, 1396-1398 (2009).
15. S. Xu, Y. J. Lin, S. T. Wu, *Opt. Express* **17**, 10499-10505 (2009).
16. A. Y. Malyuk, N. A. Ivanova, *Appl. Phys. Lett.* **112**, 103701 (2018).
17. N. Chronis, G. L. Liu, K.-H. Jeong, L. P. Lee, *Opt. Express* **11**, 2370-2378 (2003).
18. H. Yu, G. Zhou, F. S. Chau, S. K. Sinha, *Sensors Actuat. A Phys.* **167**, 602-607 (2011).
19. H. Ren, D. Fox, P. A. Anderson, B. Wu, S. T. Wu, *Opt. Express* **14**, 8031-8036 (2006).
20. N. Binh-Khiem, K. Matsumoto, I. Shimoyama, *Appl. Phys. Lett.* **93**, 159 (2008).
21. S. H. Oh, K. Rhee, S. K. Chung, *Sensors Actuat. A: Phys.* **240**, 153-159 (2016).
22. W. J. Xiao, S. Hardt, *J. Micromech. Microeng.* **20**, 8 (2010).
23. B. A. Malouin, M. J. Vogel, J. D. Olles, L. L. Cheng, A. H. Hirsa, *Lab Chip* **11**, 393-397 (2011).
24. H. C. Cheng, S. Xu, Y. F. Liu, S. Levi, S. T. Wu, *Opt. Commun.* **284**, 2118-2121 (2011).
25. F M, C. I. White, *Viscous Fluid Flow.* (McGraw-Hill, 2006).
26. J. U. Brackbill, D. B. Kothe, C. Zemach, *J. Comput. Phys.* **100**, 335-354 (1992).

Implementation and investigation of a force control for a forming demonstrator with coupled servo drives

André Sewohl[1,*], *Armin* Schleinitz[1], *Holger* Schlegel[1] and *Matthias* Putz[2]

[1]Chemnitz University of Technology, Department of Control and Feedback Control Technology, 09126 Chemnitz, Germany
[2]Fraunhofer Institute for Machine Tools and Forming Technology, Production Department, 09126 Chemnitz, Germany

Abstract. In the area of production engineering, there are ongoing efforts to improve manufacturing strategies and processes in terms of stability, quality and efficiency. One possibility for ensuring stable process conditions and reducing rejected parts is closed loop control of quality determining parameters. For many applications, the machining forces often represent the limiting factor for the design of the processes and the choice of parameters. As a controlled value, they are predestined for ensuring process stability and safety. For the tasks and problems of forming technology, control of process forces is not yet mature and offers significant potential for improvement. The drafting of sophisticated control concepts, which enable a mutual use or a combination of force and position control, is the focus of future developments. The design and implementation of a force control for a forming demonstrator with coupled servo drives is described in this publication. The combination of force and position control is realized by direct switching. In addition, further combination options and structures are explained. The focus here is first on the parameterization of the force control and the assessment of the control performance using appropriate comparison criteria in the time domain.

1 Introduction

In modern production systems, there is a trend to replace mechanical motion solutions with electrical ones. There are many strategies for controlling machine-specific quantities, such as the position or speed of electromechanical axes. The concept of cascade structure, also called servo control, has become established in this field [1]. The use of controlled electromechanical drive systems can meet the increasing demands on the machines in terms of dynamic behavior, as well as higher productivity and accuracy.

Nevertheless, in the area of production engineering, there are ongoing efforts to improve manufacturing strategies and processes in terms of stability, quality and efficiency. One possibility for ensuring stable process conditions and reducing rejected parts is closed loop control of quality determining parameters [2]. The development of suitable control concepts at the process level, in which significant process variables are taken into account as controlled values, offers considerable room for improvement at this point. Process control in terms of manufacturing technology corresponds to influence process variables for controlling the process [3, 4, 5]. This requires the integration of sensors for the detection of the current process status.

Besides the determination of state variables, the use of sensors also makes it possible to set up additional control loops which react to changing boundary conditions in the process and allow specific manipulation. This is the basis for the development of intelligent production machines, which react to unknown disturbances and automatically adapt to process fluctuations. As a result, even complex manufacturing processes can be safely operated at their technological limits and productivity as well as process reliability can be further increased.

At the higher process level, suitable variables must be selected to influence the process. There are many process variables which have an influence to the quality of a part. However, usually it is very difficult to control these values. The metrological acquisition of corresponding parameters constitutes a further challenge. But it is necessary in order to obtain precise knowledge about the current process status. For a large part of the processes in the field of production technology, the machining forces are of particular relevance. As a controlled variable, they are predestined for ensuring process stability and safety. Machining forces are often the limiting factor for the design of the processes and the choice of parameters. Excessive loads can cause damage and defects to the workpiece, tool or machine. In the worst case, they even lead to their destruction. In addition, process forces provide important information about the process state and allow conclusions about deviations in the production process, the machine, the tool, the workpiece or material.

Particularly in processes from the field of machining technology [3] and forming technology [2], the targeted influencing of process forces is of outstanding

* Corresponding author: andre.sewohl@mb.tu-chemnitz.de

importance. In the area of metal-cutting manufacturing processes, a large number of concepts and algorithms for controlling the process forces were investigated and developed on both the research and industry side [6, 7]. Also in the area of forming technology extensive investigations took place. Nevertheless, there is still a clear need for action in the future, as shown in [2].

In the next chapter, the concept of hybrid force-position control will be presented as one possible approach. This is followed by a description of the selected test rig for forming technologies, with regard to its mechanical structure and the existing control architecture. Subsequently, a force control is designed for this demonstrator as a first step on the way to the hybrid force position control. Finally, various adjustment rules are experimentally investigated and compared on the basis of corresponding quality criteria.

2 Control Concept

While the control of process forces has considerable potential for improvement, in many applications and processes the position control is at the same time essential for compliance with defined position specifications. Thus, the drafting of sophisticated control concepts, which enable joint use or the combination of force and position control, is the focus of future developments. For the tasks and problems of production engineering, the control of process forces is not yet mature and offers significant room for improvement. The concept of a hybrid force position control represents a promising approach at this point. The basic structure and functionality is illustrated in Figure 1.

The core component of the concept is the simultaneous usage and operation of a force and position controller. Various control algorithms and controller structures can be used here. It makes sense to implement the position control for electromechanical drive systems in the form of the established and extensively studied cascade structure. A variety of different embodiments comes in question for the force controller. These range from simple PID-controllers to complex methods from the field of so-called higher control technology. For example, adaptive or predictive controllers, as well as artificial neural networks or fuzzy controllers in the domain of cognition. Also a linkage and combination of the different methods is conceivable. The individual areas and variants were presented in more detail in [8].

For use in production machines, suitable concepts that allow simultaneous control of force and position still need to be developed. The main challenges are the concurrent use of the controllers, the selection of design principles and structures as well as the fusion of the two manipulated variables. One possible approach is to introduce a weighting function that prioritizes the controllers and their manipulated variables. To clarify these questions, extensive investigations are necessary. Against this background, simulation models are helpful, which can be successively extended to examine the complex relationships step by step. On the basis of the models, possibilities for the combination of force and position control should be considered. Starting point for this is a forming demonstrator with coupled drives. A servo control in cascade structure was initially designed for the test rig so that it can be operated in a position-controlled manner. The next chapter will give a more detailed description.

The demonstrator will be used for the field trial and practical implementation of the concepts, as well as the examination of the feasibility and limitations. As a first step, the enhancement of the control structure with a force controller takes place. At the test rig, the design is

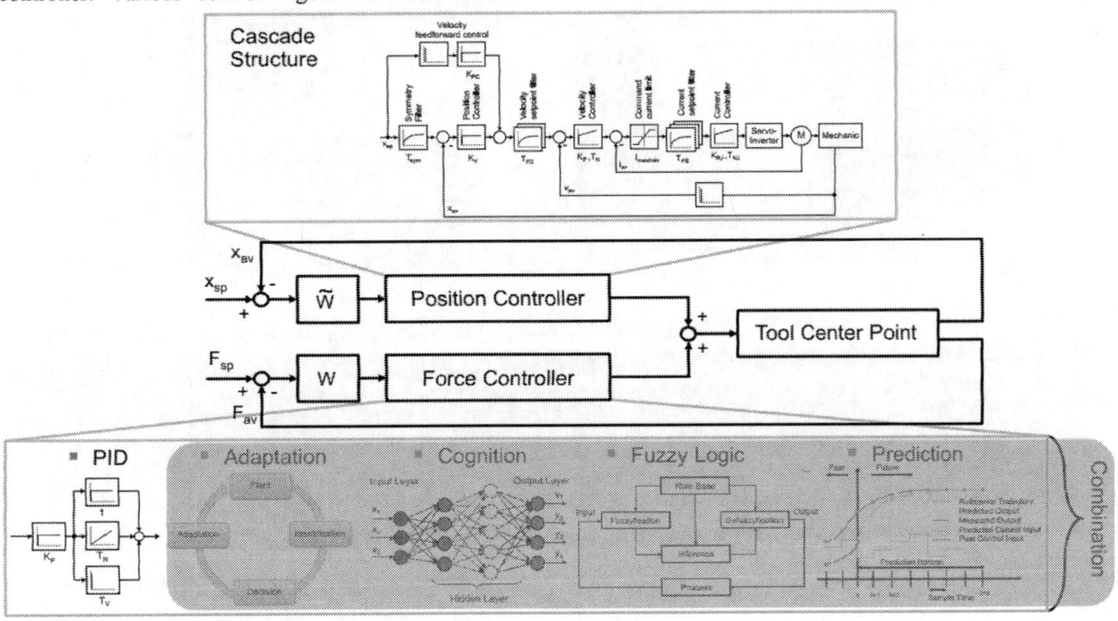

Fig. 1. Schematic concept for hybrid force-position control.

initially carried out with a PID-structure for the force controller. There are various options for integrating the controller in the existing architecture, which are explained in more detail in Chapter 4. In addition, the combination with the position control is realized by a switchover for experimental proving.

3 Servo forming unit

3.1 Mechanical components and structure

The core component of the investigations is a self-driven press tool, which offers a very high degree of flexibility and process stability. The basis construction of the demonstrator is similar to a servo screw press with two drives. A general description of the demonstrator with its set-up of basis tool, single modules, actuators and control technology is given in [9]. The schematic structure of the mechatronic system is shown in Figure 2. The mechanical components are illustrated as CAD-model and electrical components as well as the control structure are described in more detail in the following sections.

The basic tool (grey) consists of two plates, which are connected to the direct driven actuators (blue) on opposite sides. The top plate is coupled via tilting joints with the two drives. A piezoelectric measuring ring (green) is integrated in each drive train, in order to detect the force. Parallel to the actuator-basis tool-connection two linear measuring systems (also green) are utilized to monitor the relative movement of the upper and lower plate of the basis tool. This value can be used in the control concept in order to avoid unwanted tilt. Additionally a column guide with four elements (pink) is mounted in the workspace. This is a sliding guide with a diameter of 25 mm and with a backlash between 18 µm to 25 µm. Both servo-electric motors are driven by individual drive units and offer a maximum in power and momentum at every position, which is required to avoid tilt. Since tilt is not avoided by the guiding elements of their own, maximum forming force depends on the load eccentricity. The influence of an eccentric load case on the basis module can be evaluated with the demonstrator. The single load modules (brown) can be placed eccentrically as well as centrally in the basis module. Thus an unbalanced load can cause tilt in the guidance which needs to be compensated by the feedback control system. Furthermore it is possible to remove the guiding elements to allow a larger tilt. The single load modules are different in their construction and accordingly different in their force-displacement-behavior. The force-stroke curve from these modules is so far adaptable that several real processes can be replaced perfectly. Major advantages are an exact reproducibility, no need in further material and the force-stroke curve can be adjusted individually for many kinds of tools. For example, the spring modules are comparable to real processes like deep drawing or bending and replace real deep drawing or bending tools.

3.2 Electrical components and structure

The core components of the demonstrator are the two linear actuators of the type GSX 40-0601-MFA-IG4-258-SR-RB manufactured by Exlar. In order to generate the linear movement, planetary roller screws are integrated directly in the drive. They possess a larger contact surface, as well as a longer lifetime and higher load than a ball screw system. The schematic construction is shown in Figure 3.

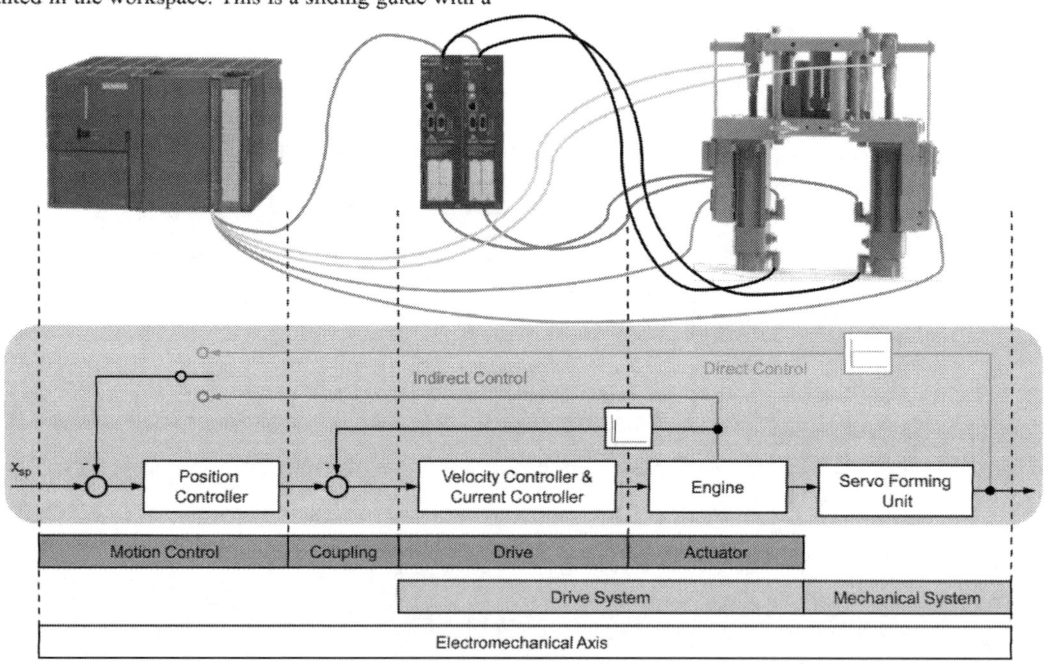

Fig. 2. Mechatronic system of the demonstrator.

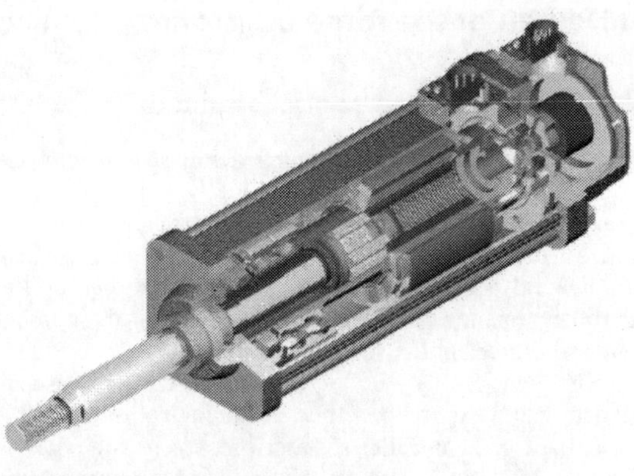

Fig. 3. Linear actuator with planetary roller screw [10].

The actuators are each connected to an ISR 2310-FS servo-controller from Metronix. These drive controllers are characterized by the smallest possible cycle times in the two-digit µs range up to the speed and position controller level. This allows a very fast system response, which is especially required for force control. Furthermore, that ensures a parallel position in the tool plane. In the field of control technology, this can only be realized with low dead times by fast clocking, in order to ensure a robust velocity controller with dynamic setting. A Motion Control Simotion C240 from Siemens is superimposed on the inverters. The movement specification and position control takes place here. In addition, the control of the mechanically coupled drives is implemented via an appropriate interface as axis arrangement with a master-slave structure. In this way, synchronization functions can be realized and the slave drive receives its setpoints from the master axis and follows them via its internal control. Although a direct control of parallelism is not possible in these functions, this configuration is a common implementation for an operation with synchronous motion. A superimposed parallelism control includes delaying communication times. That means their advantages are compensated in this setup by small delay times.

The use of a superordinated motion control ensures a maximum in flexibility. The Simotion C240 enables the integration of additional force sensors and position measuring systems. An adaptation and extension of the control structures can also be applied. Because of extensive I/O-modules, the sub-functions interpolation, position control and speed control can be divided as required between the superordinated motion control and the drive controllers from Metronix. In addition, the appropriate Simotion Scout software from Siemens is available for the operation, configuration and parameterization of test series. The program also offers possibilities for recording and evaluating the measuring signals.

The incremental position measuring systems LS177 from Heidenhain are connected to the Simotion C240 over digital I/O modules. They possess a resolution of 1 µm. This enables the implementation of a direct position control. Furthermore, piezoelectric measuring rings 9051A from Kistler are preloaded in the power flow of the push rods. Their measurement signals are evaluated via a permanently installed industrial 2-channel charge amplifier. This one is also connected to the C240, so that the measuring signals can be used for a force control.

3.3 Commissioning of the test rig

In a first step, a position control in cascade structure was first implemented for the commissioning of the demonstrator and the execution of practical experiments with the load units. As a result, real machining processes can be simulated. At this point, the variant was selected, that interpolation and position control takes place on the motion control. The velocity and current control is implemented in each of the drives. This has the advantage of a flexible parameterization and data acquisition. The commissioning of a cascade control always takes place starting with the inner control loop outwards. In this context, the first step was to configure the parameters of the feed axes based on the motor data. This was followed by the parameterization of the current control loop. Based on this, the speed control loop could be designed. The automatic controller tuning and optimization system of the commissioning tool from the drive controllers was used for this purpose. The individual drives were not mechanically linked at this point. The determined parameters of the control loops are summarized in Table 1.

Table 1. Settings of the closed-loop control system.

Velocity Controller	Amplification constant	K_P	25	[Nms/rad]
	integration constant	T_n	2.5	[ms]
	actual value filter time constant	$T_{f,av}$	0.21	[ms]
	Synchronous speed filter time constant	$T_{f,syn}$	0.6	[ms]
Current Controller	Amplification constant	K_i	7.1	[A/Nm]
	integration constant	T_{ni}	1.46	[ms]

The position control was designed for aperiodic transmission behavior with a damping of D = 1 in order to avoid overshooting. For this purpose, [11] proposes the calculation of the gain factor K_v according to the following formula:

$$K_v = 1 / (4 * T_{sum}) \quad (1)$$

Here, T_{sum} corresponds to the total time constant, into which the time constants and dead times of the subordinated current and speed control loops are incorporated. It amounts to:

$$T_{sum} = T_{dead} + (T_r - T_{dead}) / 3 + T_{sub} \quad (2)$$

The corresponding time constants were determined experimentally and are given in Table 2. In addition, the position control has been supplemented with a velocity feed forward control in order to improve the dynamic behavior and to reduce the contouring error. For this, the weighting factor K_{pc} and the symmetry filter time constant T_{sym} must be specified. In this regard, no sizing rule has yet been established. There are only suggestions within certain limits, so that offers an empirical dimensioning. The parameters selected for this specific application are listed together with the calculated characteristic values in Table 2.

The resulting controller structure is shown schematically in Figure 2. In this configuration, forming processes respectively their force-displacement behaviour can be simulated with the demonstrator. An extension of the controller structure is necessary for the force-controlled operation. The approach followed is described in the next section.

Table 2. Settings of the position closed-loop control system.

Amplification constant	K_v	45	[1/s]
Dead time	T_{dead}	2	[ms]
Substituted time constant	T_{sub}	2.7	[ms]
Control rise time	T_r	4.4	[ms]
Weighting factor	K_{pc}	60	[%]
Synchronous filter time constant	T_{syn}	3.3	[ms]
Total time constant	T_{sum}	5.5	[ms]

4 Design of the force controller

4.1 Extension of the control structure

The force control on the forming demonstrator with the servo-electric drives must be basically implemented in the form of a cascade structure. Accordingly, it makes sense to adapt or expand the available structure. There are several possibilities to integrate the force controller into the cascade structure. The complexity or the number of subordinate control loops can vary.

One possibility is to use the existing cascade structure and expand it with a fourth control loop. In this case, the force controller is superimposed on the position control loop and outputs the path as a manipulated variable. However, here should be noted that the control loops are getting slower from inside to outside due to sampling times, computation times and communication times. This is a major drawback, especially for force control, which must react quickly to changes. Another variant is to use the force controller instead of the position controller. This structure is superimposed on velocity control loop. In addition, it is also possible to dispense with the position and speed controller and to superimpose the force controller directly to the current control loop.

As already explained, the combination of force and position control on the demonstrator should first be realized by a switching operation. This results in problems in terms of signal processing, which must be taken into account for the controller design. The changeover causes jumps in the manipulated variable profile, which can lead to unstable behavior. To ensure a continuous process during switching, an exchange or transfer of manipulated variables should take place. For this reason, it is advisable to use the force controller superimposed on the speed control loop. Here, the force and position controller can use the same manipulated

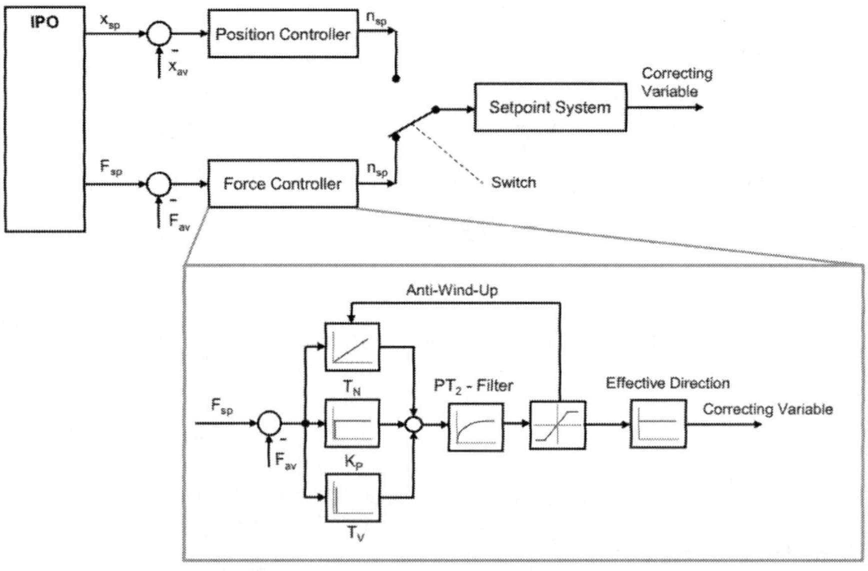

Fig. 4. Changeover and structure of force controller.

variable. As a result, the velocity or speed continues to be transferred as the setpoint for the subordinate control loops in the inverter. The basic structure is shown in Figure 4.

For switching from position-controlled movement to the force-controlled state and vice versa, corresponding criteria must be defined. These can be defined in specific activation commands. The activation can be done directly or automatically thereby. Direct activation is only possible when the machine is at a standstill with activated position controller. Furthermore, the force control can be automatically activated with a corresponding condition. For automatic activation, a separate command is available. A check of the conditions and the switching takes place in the servo clock. Possible conditions are force, position, time and input values. These parameters can be specified and linked in several stages with the command. By cancelation of the command before the conditions are met, the conditions can also be switched in operation. Parameters are stored at the moment of switching. These values are available through the system variables. The force setpoint or profile specification is then also made via corresponding commands in the C240. There, the force control can be realized via an adjustable PID-controller with filter and anti-wind-up. The corresponding structure is illustrated in Figure 4.

The I-component of the controller is initialized in such a way that the new manipulated variable from the force controller corresponds to the estimated manipulated variable from the position controller, in order to obtain a constant manipulated variable when switching from position control to force control. The estimated manipulated variable is the last manipulated variable of the position controller plus the filtered actual acceleration multiplied by the cycle time. In order to obtain a continuous curve of the manipulated variable when switching from force control to position control, the setpoint generation and the position controller are reset. The currently actual value, measured by the encoder, is the new value for the position setpoint, modified by the estimated contouring error. The estimated contouring error is calculated from the actual manipulated variable, multiplied by the substituted time constant of the position control loop. Then, the position controller outputs the same manipulated variable in the first cycle after switching as the force controller before switching over.

4.2 System identification

For the design and parameterization of the controller, it is necessary to perform a system identification. In force control, the process which causes the load is part of the controlled system and must be included in the design. Thus, the controlled system is composed of the current and speed control loop, the mechanical components and the process. The spring pack was selected and used in order to simulate a load situation respectively a forming process. The introduction of the load is done centric, so that the forces are evenly distributed on both drives.

The identification and excitation of the controlled system takes place by means of a jump function. Setpoint value is the velocity, that is the manipulated variable of the force controller. For practical execution, the position controller was deactivated and the drives were decoupled. The velocity jump was switched to both drives simultaneously by a setpoint input. So that static friction effects are not included in the identification, both drives were initially moved at a speed of 1 mm/s. Then the jump of 10 mm/s took place. In accordance with this procedure, the step response of drive 1 is shown in Figure 5.

It can be clearly seen that this is a controlled system with integrating behavior and one delaying time constant. Since the movement takes place downwards and there are pressure forces, these are associated with a negative sign. Accordingly to Figure 5, characteristic parameters of the controlled system can be determined. The gain factor K_{SI} is here 170 N/mm and the delaying time constant T_u is 10 ms. For drive 2 the same values could be determined. Thus, the theoretically known spring constant of 333 N/mm, which is split between the two drives during a load case, could also be validated. Based on these characteristics, the controller parameters can be calculated according to various adjustment instructions, which are described in the next section.

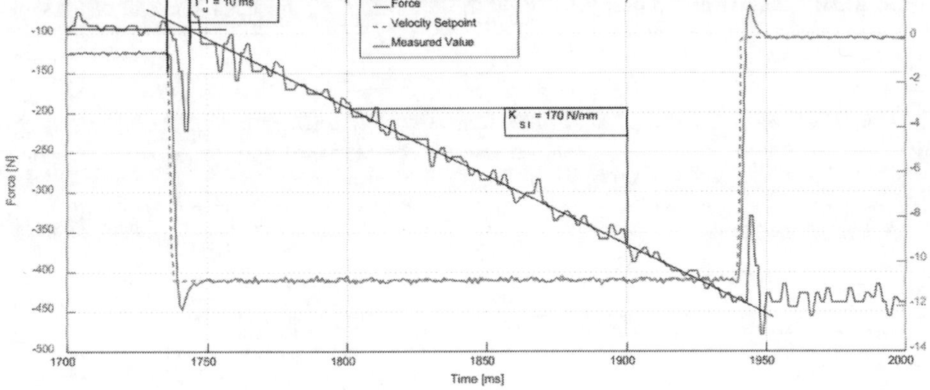

Fig. 5. Identification of the controlled system.

 https://doi.org/10.1051/matecconf/202030603001

4.3 Adjustment instructions

There is a variety of methods and procedures available for adjusting the controller. However, corresponding specifications and boundary conditions must be taken into account. An integrating behavior with one delaying time constant was determined in the identification of the controlled system. Compliant with the calculation according to the symmetrical optimum, PI-controllers or PID-controllers are suitable for this purpose [12]. The adjustment instruction according to the symmetrical optimum is advantageous in the case of disturbances and is often used, in particular, for force control loops. If the parameters of the controlled system are known, the method can be used without much mathematical effort. On the basis of the identified system parameters, the controller parameters could be calculated. These are summarized together with the formulas in Table 3.

Table 3. Calculated controller parameter.

Controller type	Setting rule	Controller parameter
PI	$T_N = a^2 * T_u$	$K_I = K_P/T_N =$ 4.819 [mm/(N*s²)]
	$K_P = 1 / (a * K_{SI} * T_u)$	$K_P =$ 1.205 [mm/(N*s)]

First, a PI-controller was selected for the experiments. The factor a is a calculation coefficient, in which the damping of the system is included. It is calculated to:

$$a = 2 * D + 1 \quad (3)$$

The damping D was determined as 2. This shall allow a fast balancing with moderate overshoot.

5 Experimental and evaluation

The force control can be realized on the mechanically coupled drives with various embodiments. On the one hand, it is possible to control both drives separately to a force setpoint. However, positional differences occur here with high probability, which leads to a tilt of the plunger. This variant is therefore only suitable for applications in which tilting is permitted or where the differences are very small respectively tolerable. The demonstrator is suitable for such scenarios.

In the second variant, the master axis is in the force-controlled state, while the slave axis is controlled to the actual position value of the master axis. As a result, the parallelism of the plunger can be ensured. This is the much more relevant use case. Therefore, the investigation is first focused on this variant. The comparison and evaluation of the parameterization for the controller is based on quality criteria (control rise time, overshoot, settling time) in the time domain. For this purpose, a force jump is switched to the closed loop and the step response is detected. Thus, there is first a direct activation respectively a changeover of the force control. For this reason, the spring is preloaded. Subsequently, the jump with a height of 300 N is switched on manually.

In this way, the determined parameters for the force controller, summarized in Table 3, were examined on the demonstrator. It turned out that this parameter setting is not suitable for force control. There were strong vibrations and the control loop was unstable. For this reason, an empirical adjustment of the force control parameters was carried out. For this purpose, a P-controller and two PI-controllers, which were supplemented by an actual value filter with a time constant T_f = 10 ms, were analyzed. Their characteristic values are summarized in Table 4.

Table 4. Controller parameter.

Controller type	Controller parameter
P	K_P = 7.28e-002 [mm/(N*s)]
PI (1)	K_P = 5.6e-002 [mm/(N*s)]
	K_I = 3.4e-002 [mm/(N*s²)]
PI (2)	K_P = 7.28e-002 [mm/(N*s)]
	K_I = 6.8e-002 [mm/(N*s²)]

Figure 6 compares the curves of the three controller

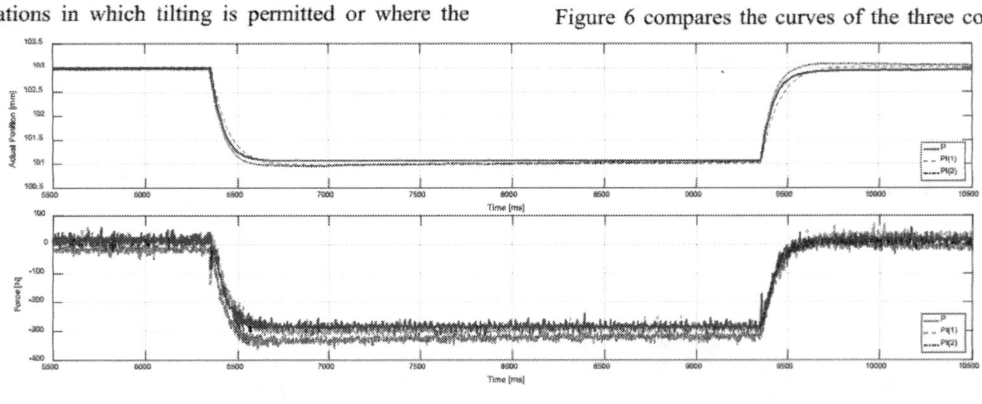

Fig. 6. Comparison of controller performance.

settings in the time domain for drive 1. Both, the actual position and the force are displayed. A force jump of 300 N was applied to the system again, in order to assess the controller behavior. The force jump was applied to the preloaded spring assembly. This results in slight deviation in the starting value of the force jump. A difficulty for the investigation and evaluation of the course of forces is the noise. This results from the measuring range and the resolution of the force sensors.

Both PI-controller show a slight overshoot in the actual position as well as in the force value. In addition, the setting PI(2) is faster due to its higher I-component. However, the I-component in the PI-controller is relatively low, which is why only small differences occur in the overshoot and in the dynamics compared to the P-controller. Therefore, they can also be labeled as a PI-controller with dominant P-component. The exact values for assessing the control performance can be found in Table 5.

Table 5. Comparison of quality criteria.

Controller typ	Control rise time	Settling time	overshoot
P	220 [ms]	220 [ms]	0 [N]
PI (1)	220 [ms]	1320 [ms]	20 [N]
PI (2)	166 [ms]	1650 [ms]	25 [N]

On the basis of these results, it is recommended to implement the P-controller for further investigations at the demonstrator with the described control structure.

6 Summary and future work

Constantly increasing demands on the productivity and the quality of manufacturing processes always require new research efforts. The control of process forces has considerable potential for improvement in the area of forming technology. At the same time, position control is essential for compliance with defined position specifications in many applications. Thus, the drafting of sophisticated control concepts, which enable joint use or the combination of force and position control, is the focus of future developments.

The approach presented in this paper aims to develop a hybrid control concept for forming operations that is suitable to control process forces and the position simultaneously. Therefor a force control was designed at a servo forming unit with two coupled drives in a first step. It could be shown that a force control, which is superimposed on a speed control loop with subordinate current control loop, can be integrated into the control architecture. In addition various controller types and adjustment rules were experimentally investigated and compared on the basis of corresponding quality criteria. It turned out that an empirical design achieves the best results. Furthermore, a P-controller prevailed against the investigated PI-controllers in terms of performance.

Despite the promising results, there is a clear need for further investigation in the future. Examination fields are especially methods and possibilities for the combination of force and position control as well as the controller design and parameterization. The positioning and integration of the force controller in the cascade structure is of interest, too. In future work, the other variants and their effects will be examined. In this context, novel control algorithms and methods such as prediction, adaptation or cognition should be considered, too. Furthermore, the effects of problematic load situations, such as an eccentric force application, must be investigated. Moreover, an integration of sensors into the tool is necessary for different use cases, in order to detect the current process variable. In this context, suitable concepts for the integration of the measuring signals must be investigated and developed.

Acknowledgement

European Union
European Regional Development Fund
European Social Fund
Europe funds Saxony.
EFRE
ESF

Funded by the European Union (European Social Fund) and the Free State of Saxony.

References

1. D. Schröder. Elektrische Antriebe – Regelung von Antriebssystemen 2. Auflage. Springer Verlag. Berlin (2001)
2. J. M. Allwood, et al., Closed-loop control of product properties in metal forming. *CIRP Annals – Manufacturing Technology*, **65**: 573–596 (2016)
3. A.G. Ulsoy, Y. Koren. Control of Machining Process. *Journal of* Dynamic *Systems Measurement and Control* **115**: 301-308 (1993)
4. G. Chryssolouris, S. Patel. In Process Control for Quality Assurance. In: D. Tijunelis, K. McKee. *Manufacturing High Technology Handbook*, Marcel Dekker, New York, 609-643 (1987)
5. R.G. Landers, A.G. Ulsoy, R.J. Furness. Process monitoring and control of machining operations. In: O. Nwokah, Y. Hurmuzlu. *The mechanical systems design handbook*, CRC Press, Boca Raton, 85-101 (2002)
6. Y. Liu, T. Cheng, L. Zuo. Adaptive control constraint of machining processes. *The International Journal of Advanced Manufacturing Technology*, **17**(10): 720-726 (2001)
7. R.G. Landers, A.G. Ulsoy, Y.-H. Ma. A comparison of model based machining force control approaches. *International Journal of* Machine *Tools and Manufacture* **44**(7–8): 733-748 (2004)
8. A. Sewohl, M. Rehm, H. Schlegel, M. Putz. Konzepte zur hybriden Kraft-Lageregelung an Produktionsmaschinen. Kolloquium *vom Werkstoff*

zum ressourcenschonenden Produkt, Chemnitz (2018)

9. D. Landgrebe, et al. Tooling Flexibility = Flexibility in Forming Technology?!, *Proceedings of 5th International Conference on Accuracy in Forming Technology*, Chemnitz, November: 10-11, (2015)

10. Exlar Cooperation. GS Series Linear Actuators with Integrated Motor manual, *https://exlar.com/de/produkt/gsx-serie/.*Exlar Cooperation (2013)

11. O. Zirn. Machine Tool Analysis - Modelling, Simulation and Control of Machine Tool Manipulators. Habilitation, ETH Zürich (2008)

12. H. Lutz, W. Wendt. Taschenbuch der Regelungstechnik, Wissenschaftlicher Verlag Harri. Frankfurt am Main (2012).

MATEC Web of Conferences **306**, 03002 (2020) https://doi.org/10.1051/matecconf/202030603002
ICMME 2019

Research on load simulator control strategy based on BP neural network and PID method

Yong Zhou, *Yubo* Zhang*, and *Tianhao* Yang

School of Aeronautics, Northwestern Polytechnical University, Xi'an, China

Abstract. In the research of load simulator control method, PID control is the most widely used control strategy, but PID controller's three parameters is difficult to set. This paper proposes a BP neural network feedforward PID controller system which uses BP neural network for setting these parameters, and in order to make the network learning speed up the convergence speed and not fall into local minimum, the adaptive vector method is adopted to improve the algorithm. The simulation and experimental results show that this method is good at avoiding the primeval shock and the sine tracking performance of the system has also been improved.

1 Introduction

The flight control system is an important system of modern aircraft, and its performance plays a decisive role in the performance and safety of the aircraft. In actual engineering, the load simulator is often used to test its performance. With the rapid development of aerospace technology, the performance requirements of flight control systems are getting higher and higher in the industry. How to improve the performance of load simulators become a difficult and important point in this field.

At present, researches have been done much on this issue at home and abroad. In general, it can be divided into two aspects: improving hardware or optimizing control strategies. For example, Jacazio G and Balossini G propose a spring correction method [1] in which the loading part and the measured part are connected by elastic elements to buffer the influence of the positional disturbance of the measured part. Baghestan K propose a robust control method [2], that is selecting an appropriate correction network to reduce the system sensitivity.

Due to the high cost of improving hardware, this paper proposes a feedforward PID controller system which uses BP neural network for setting parameters and adopt an adaptive vector method to improve the algorithm.

The rest of this paper is organized as follows: The feedforward PID control method is introduced in Section 2, Section 3 provides a brief introduction of back-propagation neural network (BPNN) and proposed a BPNN-based feedforward PID control method. In Section 4 and Section 5, simulation and experiment show that this method has good performance.

2 Feedforward PID controller

2.1. PID and extra force

The PID controller consists of three correction links: proportional link P, integral link I, and differential link D. The roles of each correction link are as follows:(1)P: The output of the regulator is proportional to the input deviation, as long as the deviation occurs, it can play a proportional adjustment effect in time.(2)I: It can eliminate the steady state error of the system and improve the control accuracy.(3)D: It can predict the trend of error changes in advance and overcome the adverse effects caused by the error changes in advance.

The system under test is connected to the loading system through a rigid structure. When the system under test moves actively, the loading system will move together and generate interference forces, which are called excess forces [3].

When only the traditional PID control is used, the output of the system will have a certain degree of lag relative to the input, and the excess force generated during this period cannot be eliminated in time, which has a great influence on the system performance.

2.2. Feedforward controller

The feedforward compensation link is a controller designed to overcome excess forces. The role of the feedforward controller is to compensate the system before the active motion of the steering gear interferes, thus overcoming the effects of excess force. The control schematic diagram is shown in Fig. 1. In the figure, G_b is the transfer function of the feedback link, G, G_1, G_2 are the

* Corresponding author: 1030869628@qq.com

transfer function of the load simulator, F_r is the target command signal, F is the output feedback force signal, and d is the excess force.

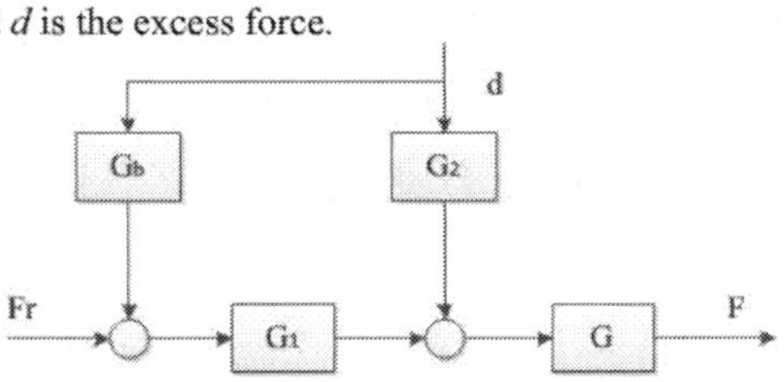

Fig. 1. Feedforward compensation controller schematic

The control effect of the feedforward compensator is related to the input signal. The specific analysis is as follows:

The input of the feedforward controller selects the actual displacement of the servo to be tested: because the correction of the phase and gain of the system output needs to be based on the actual motion of the steering gear, and only the actual displacement of the steering gear is collected by the grating scale. The operation is convenient and the compensation effect is good. Therefore, the actual displacement of the steering gear is used as the input of the feedforward compensator.

3 PID controller based on BPNN

3.1 Back-propagation neural network (BPNN)

The BP neural network (BPNN) is also known as the backpropagation neural network, which was proposed by the research team headed by Rumelhart and McCelland in 1986. BP neural network is a multi-layer feedforward network trained according to the error back propagation algorithm. It is one of the most widely used neural network models in practical applications. Fig. 2 illustrates the BP architecture in details. It consists of an input layer, one or more hidden layer, and an output layer. Layers are connected sequentially starting from the input layer through the hidden layers to the output layer. Where the connections between layers contain weights and each layer includes one or more neurons [4].

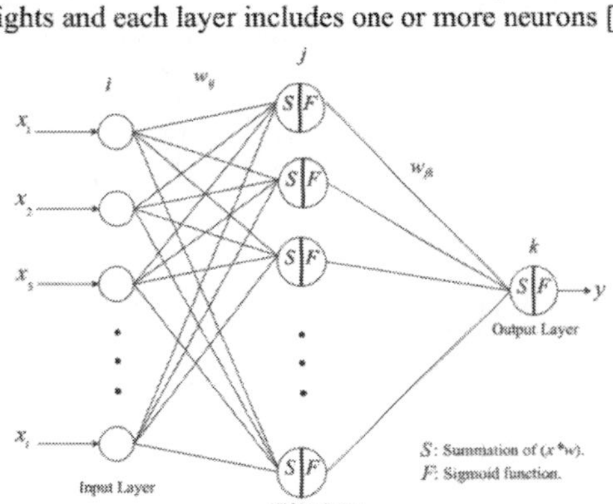

Fig. 2. Back-propagation architecture.

3.2 PID control algorithm based on BPNN

The formula of the traditional incremental digital PID control algorithm is as follows.

$$\begin{aligned}\Delta u(k) &= k_p[e(k)-e(k-1)]+k_I e(k)\\ &+k_D[e(k)-2e(k-1)+e(k-2)]\\ &= k_p\Delta e(k)+k_I e(k)\\ &+k_D[\Delta e(k)-\Delta e(k-1)]\end{aligned} \tag{1}$$

Where:

k_P means proportional coefficient.

k_I means integral coefficient.

k_D means differential coefficient.

And taking a three-layer BP neural network structure as an example which is shown in Fig. 3.

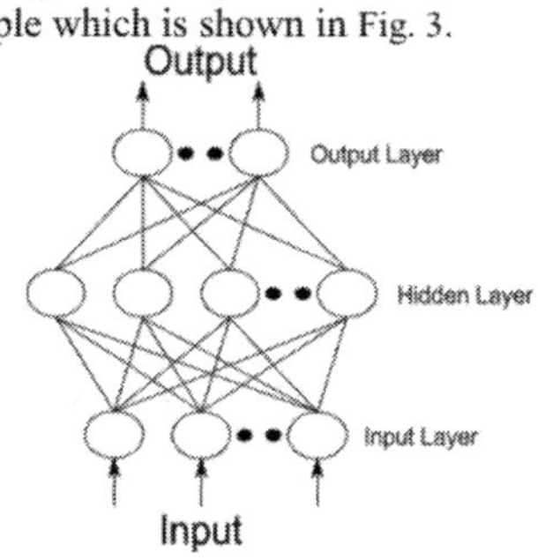

Fig. 3. Three-layer BP neural network.

3.2.1 Forward algorithm

The title is set in bold 16-point Arial, justified. The first letter of the title should be capitalised with the rest in lower case.

The input to the input layer is:

$$O_j = x(j) \quad (j = 1, 2, \ldots. M) \tag{2}$$

In the formula, M is determined by the complexity of the controlled system. Here M is assumed to be 4.

$$x(2) = e(k) \tag{3}$$

$$x(3) = e(k) - 2e(k-1) + e(k-2) \tag{4}$$

$$x(4) = u(k-1) \tag{5}$$

The input and output expressions of the hidden layer neurons of the network are as follows:

$$net_i^{(2)}(k) = \sum_{j=0}^{4} w_{ij}^{(2)} O_j^{(1)} \tag{6}$$

$$O_i^{(2)}(k) = f\left(net_i^{(2)}(k)\right) \quad (i = 1, 2, 3, 4, 5) \tag{7}$$

Where $w_{ij}^{(2)}$ is the connection weight between the hidden layer and the input layer; the superscripts (1), (2), and (3) represent the input layer, the hidden layer and the output layer.

Suppose the activation function f is a Sigmoid function, and the expression is as follows:

$$f(x) = \frac{1}{1+e^{-x}} \tag{8}$$

The input and output expressions of the output layer of the neuron network are as follows:

$$net_l^{(3)}(k)=\sum_{i=0}^{5}w_{li}^{(3)}O_i^{(2)} \tag{9}$$

$$O_l^{(3)}(k)=g\left(net_l^{(3)}(k)\right) \tag{10}$$

The output values of the output layer correspond to the scale factor, integral coefficient, and differential coefficient, respectively. The formula is as follows:

$$O_1^{(3)}(k)=K_P \tag{11}$$

$$O_2^{(3)}(k)=K_I \tag{12}$$

$$O_3^{(3)}(k)=K_D \tag{13}$$

It is also assumed that the activation function of the output layer neuron is a sigmoid function.

$$g(x)=\frac{1}{1+e^{-x}} \tag{14}$$

3.2.2 Backward algorithm

Suppose the performance indicator function is:

$$E(k)=\frac{1}{2}(r(k)-y(k))^2 \tag{15}$$

The gradient descent method is used to correct the weight of the neural network, that is, the search is adjusted along the direction in which E(k) has the largest negative gradient of the weight, and a global minimum inertia term that can quickly converge the search process is attached.

$$\Delta w_{li}^{(3)}(k)=-\mu\frac{\partial E(k)}{\partial w_{li}^{(3)}}+\alpha w_{li}^{(3)}(k-1) \tag{16}$$

Where μ is the learning rate and α is the inertia coefficient.

$$\frac{\partial E(k)}{\partial w_{li}^{(3)}}=\frac{\partial E(k)}{\partial y(k)}*\frac{\partial y(k)}{\partial \Delta u(k)}*\frac{\partial \Delta u(k)}{\partial O_l^{(3)}(k)} *\frac{\partial O_l^{(3)}(k)}{\partial net_l^{(3)}(k)}*\frac{\partial net_l^{(3)}(k)}{\partial w_{li}^{(3)}} \tag{17}$$

Where:

$$\frac{\partial E(k)}{\partial y(k)}=e(k) \tag{18}$$

$$\mathrm{sgn}(k)=\frac{\partial y(k)}{\partial \Delta u(k)} \tag{19}$$

$$\frac{\partial \Delta u(k)}{\partial O_1^{(3)}(k)}=\Delta e(k) \tag{20}$$

$$\frac{\partial \Delta u(k)}{\partial O_2^{(3)}(k)}=e(k) \tag{21}$$

$$\frac{\partial \Delta u(k)}{\partial O_3^{(3)}(k)}=\Delta e(k)-\Delta e(k-1) \tag{22}$$

$$\frac{\partial O_l^{(3)}(k)}{\partial net_l^{(3)}(k)}=g'\left(net_l^{(3)}(k)\right) \tag{23}$$

$$\frac{\partial net_l^{(3)}(k)}{\partial w_{li}^{(3)}}=O_i^{(2)}(k) \tag{24}$$

After substituting Equation (40~46) into Equation 39, the connection weights of the output layer and the hidden layer can be obtained as follows:

$$\Delta w_{li}^{(3)}(k)=-\mu\varphi_l^{(3)}O_i^{(2)}(k) \tag{25}$$

Where:

$$\varphi_l^{(3)}=e(k)\mathrm{sgn}(k)\frac{\partial \Delta u(k)}{\partial O_l^{(3)}(k)}g'(net_l^{(3)}(k)) \tag{26}$$

Similarly, the connection weight between the hidden layer and the input layer can be derived as follows:

$$\Delta w_{li}^{(2)}(k)=\mu\varphi_i^{(2)}O_j^{(2)}(k) \tag{27}$$

Where:

$$\varphi_i^{(2)}=f'(net_i^{(2)}(k))\sum_{l=1}^{3}\varphi_l^{(3)}w_{li}^{(3)}(k) \tag{28}$$

The specific steps of the BP neural network PID controller algorithm are as follows:

(1) Determine the structure of the BP neural network, that is, determine the number of neurons in the input layer and the hidden layer, find the appropriate initial weight, and choose the learning rate and the inertia coefficient.

(2) The input value $r(t)$ and the output value $y(t)$ of the system at time k are obtained by sampling, and the error at the moment is calculated as $e(k)=r(t)$-y(t).

(3) According to the forward algorithm, the input and output of each layer of neurons are obtained. At this time, the output values of the output layer correspond to the parameters of the PID controller: proportional coefficient, integral coefficient, differential coefficient

(4) Adjust the connection weight according to the gradient descent method, and realize the adaptive adjustment of the three parameters of the PID controller through neural network self-learning.

(5) Set $k=k+1$ and return to step (2)

3.3 Controller optimization

The choice of the learning rate has a great impact on the system. If the learning rate is very large, it will cause system fluctuations and even the system will collapse; but if the learning rate is small, the training time will be extended. In the actual design, the fixed learning rate does not satisfy the whole learning process very well. In order to make the system speed up the convergence without falling into the local minimum, the adaptive learning rate method is used here, which means the rate can be automatically adjusted according to the change in the error value throughout the learning process.

The algorithm of adaptive learning rate is as follows:

$$w^{k+1}=w^k-\mu g(w^k) \tag{29}$$

Where w^k is the training weight at time k, and $g(w^k)$ is the gradient vector.

4 Controller design and simulation

The block diagram of the load simulator control system of the feedforward PID controller based on BP neural network is established in Matlab/simulink, as shown in Fig. 4. The part in the dotted line is the BP neural network PID controller.

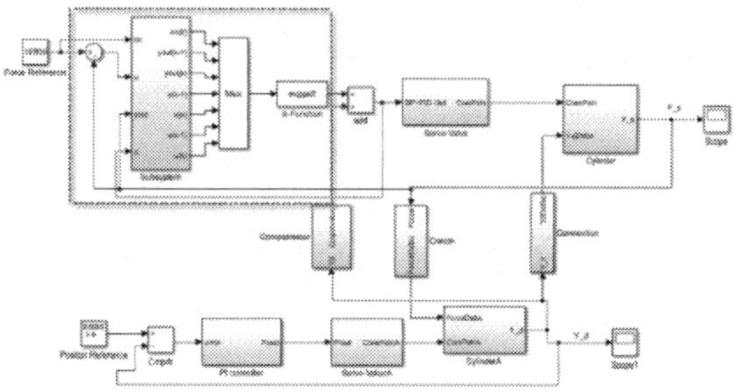

Fig. 4. Three-layer BP neural network.

The system is affected by the extra force at work, and good control strategy can suppress the excess force. Therefore, the strength of the control strategy is measured by the amount of excess force received by the system during the test.

Under no-load conditions, the system is sinusoidal with an amplitude of 2 mm and a frequency of 10 Hz, the actual loading force is shown in Fig. 5. Compared with the traditional PID controller, the BP neural network feedforward PID controller reduced the excess force from 1250N to 230N, which means the excess force was reduced by 82% based on the traditional PID controller.

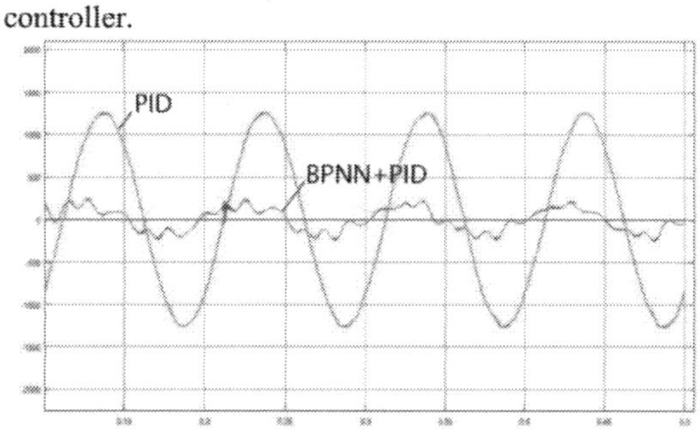

Fig. 5. 0N,2mm,10Hz loading force.

At this point, a constant load of 10kN is added to the system. The actual loading force curve is shown in **Fig. 6.** Obviously, the loading force of the BP neural network feedforward PID controller tends to be more predictive. At 0.18s, the loading force becomes stable, and the error between it and the load command is 230N, which means that the error rate is only 2.3%.

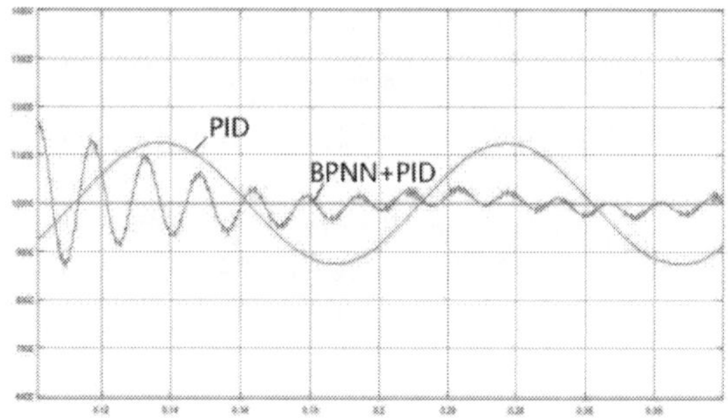

Fig. 6. 10kN,2mm,10Hz loading force.

The above simulation comparison test shows that BP neural network feedforward PID control strategy does have excellent control effects.

5 Experiments

The actual test bench is shown in Fig.7. The right side is the loading device and the left side is the measured device. During the test, the controller parameters are input to the controller through the input command. After being processed, the device under test is driven to move actively, and the hydraulic power is supplied by controlling the servo valve opening to achieve loading.

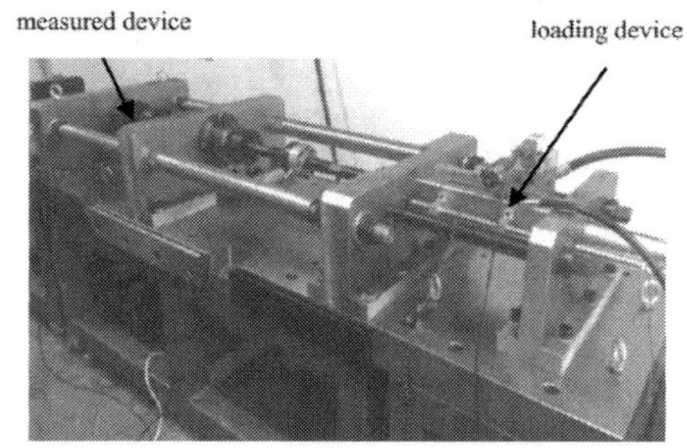

Fig. 7. Test bench

The displacement input is a 15mm step command, and the load command is 16kN. The test results are shown in Fig. 8 and Fig. 9.

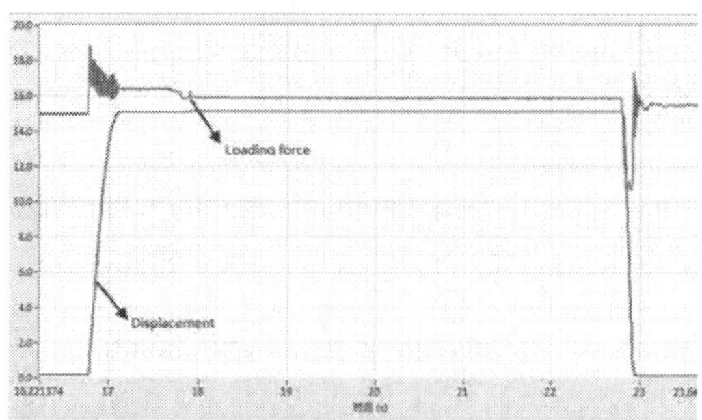

Fig. 8. Traditional PID control method

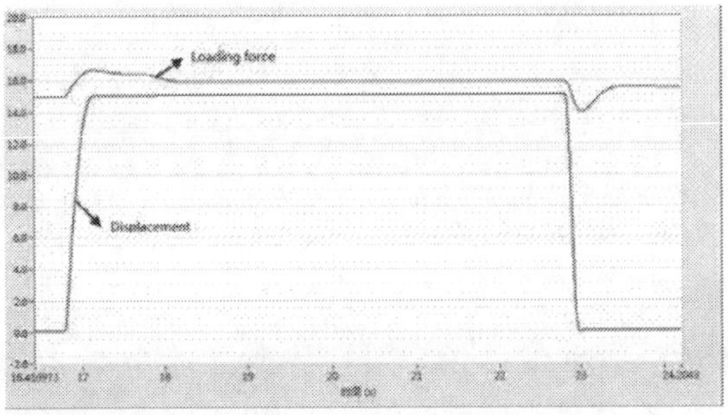

Fig. 9. BPNN feedforward PID control method

It can be seen from the above Fig. 8 and Fig. 9 that the loading system using the traditional PID control method will generate excess force of up to 5.5kN, while the loading system using the BP neural network feedforward PID control method will generate only 2kN of excess power, which is 63.6% lower than the traditional PID method. It can be seen from the above test results that under the action of constant force, the feedforward PID control method based on BP neural network can effectively suppress the extra force.

6 Conclusions

The load simulator is mainly used to test the flight control system, but it is often affected by excess force during operation. Traditional PID control is the most widely used control method in the industry, but its suppression effect on excess force is poor.

In this study, a BP neural network feedforward PID control method has been developed. This method adds feedforward control based on PID control and uses BP neural network to automatically find three parameters of PID. Simulation and experiments show that this method has better control effect than traditional PID control, and can suppress the excess force generated during the experiment. All in all, this method has a strong engineering application value.

This research is supported by "the Fundamental Research Funds for the Central Universities", No. 3102019O504003 and No. 3102016ZY002.

References

1. Jacazio G, Balossini G. Real-time loading actuator control for an advanced aerospace test rig. Proceedings of the Institution of Mechanical Engineers, Part I: Journal of Systems and Control Engineering, 221(2): 199-210 (2007)
2. Baghestan K, Rezaei S M, Talebi H A, Zareinejad M. Robust force control in a novel electro-hydraulic structure using polytopic uncertainty representation. Isa Trans, 53(6):1873-1880 (2014)
3. Han S , Jiao Z , Yao J , et al. Compound Velocity Synchronizing Control Strategy for Electro-Hydraulic Load Simulator and Its Engineering Application. Journal of Dynamic Systems, Measurement, and Control, 136(5):051002 (2014)
4. Hameed A A, Karlik B, Salman M S. Back-propagation algorithm with variable adaptive momentum. Knowledge-Based Systems, 114:79-87 (2016)

Design of the Landslide Multi-factor Monitoring System Based on the GNSS Technology

Meng Qingjia[1,*], *Wang* Chenhui[1], and *Guo* Wei[1]

[1]Center for Hydrogeology and Environmental Geology Survey, CGS, 1305 Qiyi Zhong Road Baoding, China

Abstract. Based on the multi-element monitoring of landslide, this paper adopts the GNSS technology to build the three-dimensional space monitoring system to research the landslide multi-element three-dimensional space monitoring technology. Through collecting the rainfall, soil moisture content, slope, pore water pressure, stress, single point surface deformation etc, the system has many key technologies such as multi-parameter, information acquisition, data fusion analysis and real-time early warning of landslides and integrate various elements by using the professional technology. This research provides an important reference for the landslide disaster prevention.

1 Introduction

There are many problems in geological hazards, such as the complex process of disaster pregnancies and disaster-causing patterns, the difficulty of early identification and monitoring and early warning, and insufficient technical support for risk prevention. Therefore, the national disaster prevention and mitigation strategy has a great demand for the technical equipment of geological hazard monitoring. The key technology of three-dimensional space monitoring can integrate the monitoring technology of rainfall, soil water content, tilt, pore water pressure, stress and other landslide factors, and realize the integration of professional monitoring technology system. In this technical system, the multiple information processing at each monitoring point is transformed into a time series problem. By using data fusion technology, a set of comprehensive information is synthesized, and the landslide is monitored and predicted directly from point to surface[1].The main task of landslide monitoring is to grasp the characteristic information of landslide deformation and failure by combining deformation monitoring and induced factors monitoring, analyze the dynamic law, and realize the reliability and validity of monitoring in space, time and scale[2].

Based on the key technologies of the landslide multi-dimensional and multi-parameter information acquisition about rainfall, soil moisture content, tilt, pore water pressure, stress and others, data fusion analysis and real-time disaster warning, this paper carries out the research of landslide multi-factor three-dimensional spatial monitoring technology. This research is helpful toimprove the application effect of geological hazard monitoring technology and provide technical support for geological hazard monitoring and early warning.

2 Overall design of the monitoring system

The integration of the high-precision three-dimensional spatial positioning technology and multi-factor landslide real-time monitoring technology can increase the spatial dimension information of monitoring data on the basis of the original time dimension. This integration effectively improve the effective information quantity and use value of monitoring data. In addition, through multi-node deployment, the original monitoring mode of point monitoring can be extended to the overall monitoring. Through data fusion analysis technology, a large number of basic monitoring data are obtained, such as rainfall, soil moisture content, slope, single point surface deformation, stress and other key elements of landslide, as well as three-dimensional coordinates of monitoring points, and used to analyze and obtain the overall three-dimensional spatial deformation of landslide. The above data fusion analysis technology, combined with the location of monitoring points and other means, can be simple and efficient to achieve the overall control of the landslide condition. In addition, the representation of the spatial location makes the original monitoring parameters have the three-dimensional space vector attributes. This is of great significance for data mining and analysis, which can improve the value of application scenarios such as water, stress, soil moisture transmission direction and speed research.

2.1 High Precision Three-Dimensional Space Positioning Technology

Based on the basic data of three-dimensional coordinates, velocity and precise timing provided by GNSS, a low-

* Corresponding author: menmen-00@163.com

https://doi.org/10.1051/matecconf/202030603003

power and high-reliability satellite signal receiving device suitable for geological hazard monitoring scenarios is developed to realize the functions of receiving, filtering, storing and transmitting satellite signals. Research high-precision three-dimensional space positioning technology and related algorithms, research three-dimensional space position calculation algorithm and implementation method based on GNSS, carry out static RTK, dynamic RTK, multi-satellite and multi-source data joint algorithm research and compile related software. By means of algorithm analysis, error propagation mechanism analysis, real scene testing and other means, the performance and effect of the algorithm are continuously improved, and finally the positioning accuracy of millimeter level is achieved. According to the satellite trajectory and the characteristics of satellite signal, the antenna and hardware structure of the receiving device are optimized. According to the actual needs of disaster monitoring, the working mode, positioning accuracy, calculation frequency, power consumption and volume are considered and coordinated. Through collecting monitoring data of GNSS landslide deformation under monitoring point environment, the observation range, signal-to-noise ratio and multi-path effect sequence of different environmental satellites are analyzed and corrected[3]. Key technologies such as real-time cycle slip detection, fast ambiguity fixing and multi-system joint positioning are adopted to eliminate system errors under short baseline and medium baseline conditions to achieve high-precision positioning [4].

2.2 Multi-factor Real-time Monitoring Technology

The occurrence of landslides is not only related to topographic and geological conditions, but also to the inducing factors of geological conditions. According to the main characteristics of landslide, the related factors are summarized: rainfall, stress, slope, single point surface deformation, soil moisture content, etc [5]. Landslides are time series events occurring at different stages of development. Single factor monitoring landslide, such as displacement monitoring, is often aimed at local monitoring, which can't fully reflect the dynamic change of the overall landslide. This system integrates many factors, takes the key related factors of high-level long-distance landslide and actual monitoring needs as the guidance, integrates the corresponding monitoring variables, and gradually realizes from quantitative analysis to qualitative analysis..

2.3 Data fusion technology

Landslide monitoring multi-factor system composed of Multi-heterogeneous sensors involves data determination, data management complexity processing and evaluation of multiple models in data fusion. The basic combination model is established to comprehensively analyze many influencing factors of landslide deformation, so as to improve the possibility of landslide prediction. In this model, the artificial neural network model is used as input variable, and the genetic algorithm model is combined with GM grey prediction model[6-7].

The system integrates high-precision three-dimensional space positioning technology and multi-factor real-time monitoring technology from hardware and software levels to form a unified monitoring equipment with strong practicability, stability and reliability. Through embedded software control and bus-level data exchange, the system realizes high-speed fusion and calculating of multi-factor real-time monitoring data and three-dimensional spatial positioning data, so as to obtain real-time dynamic data of deformation speed and acceleration on the basis of basic monitoring data.

3 Hardware Circuit Design of Monitoring Equipment

The hardware design of monitoring equipment adopts the idea of generation by generation optimization. In the initial stage, hardware design was carried out in the way of independent functional modules and unified baseboard. In the design of functional modules, priority is given to the realization of functions and the effect of implementation. In the design of the unified baseboard, it first meets the operational requirements of each functional module and the realization of data fusion. In terms of hardware resources, more abundant configuration is adopted to ensure that monitoring equipment has all the functions and requirements required by the project.

The basic components of the hardware: high-precision three-dimensional positioning module, multi-element real-time monitoring module, equipment infrastructure board

3.1 High-precision three-dimensional positioning module

The module mainly performs data interpretation, analysis and processing on monitoring data. The module mainly includes an antenna, a clock, a pseudo random code generator, a digital filter circuit, and a signal analysis processor.

3.2 Multi-element real-time monitoring module

This module is responsible for front-end data acquisition. The module is connected to a variety of sensors (the rainfall, soil moisture content, slope, pore water pressure, stress, single point surface deformation etc.), and the data is packaged and sent to the cloud. The module mainly includes an AD conversion circuit, an analog signal conditioning circuit, a 485 serial port circuit, and a general-purpose interface.

3.3 Equipment infrastructure board

This module mainly covers power supply, data storage, remote communication and embedded processing. The module is mainly composed of an embedded processor, a power supply module, a local storage circuit, a near field/far field communication circuit, and a clock circuit.

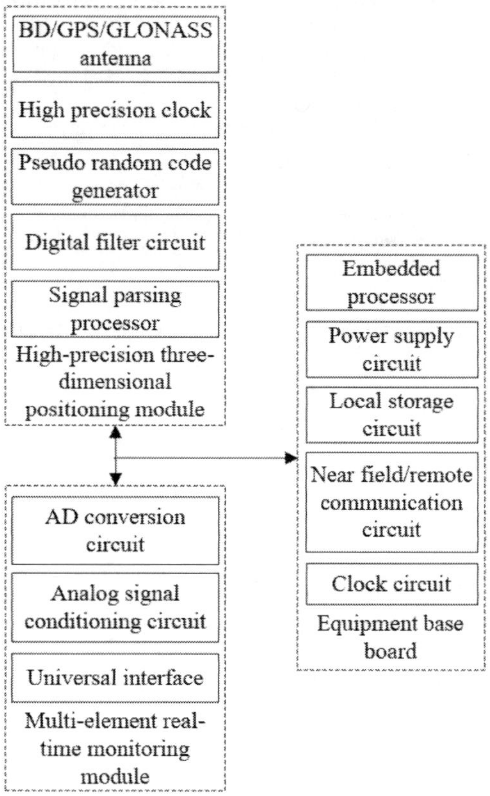

Fig. 1. Hardware system design.

4 Research on Three-Dimensional Space Location Algorithms

GNSS is gradually used in high-precision deformation monitoring and analysis, and the accuracy can reach millimeter level[8].In the three-dimensional space location algorithm, the ambiguity resolution method and the normal equation superposition method are used. Fuzzy fixing is carried out by Z transformation matrix construction method and decision function, and the fuzziness is analyzed and screened comprehensively. Pseudo-range phase combination combined cycle slip detection is used to reduce the number of cycle slip combination, and short-range and medium-range GNSS baseline is used to monitor landslide deformation. [4]

The important stage of the design of the program is the processing of monitoring data. This system first imports the monitoring data into the database for pre-processing. In the next step, the data is shunted, repaired, calculated, and standardized. The key problem of the design is to analyze and solve the satellite positioning data, model correction, floating point solution, fixed ambiguity, and baseline settlement. The data solving process is based on regional network adjustment and time series analysis. The settlement method removes the possibility of error in the data. Eventually, the final amount of deformation of the landslide object is obtained.

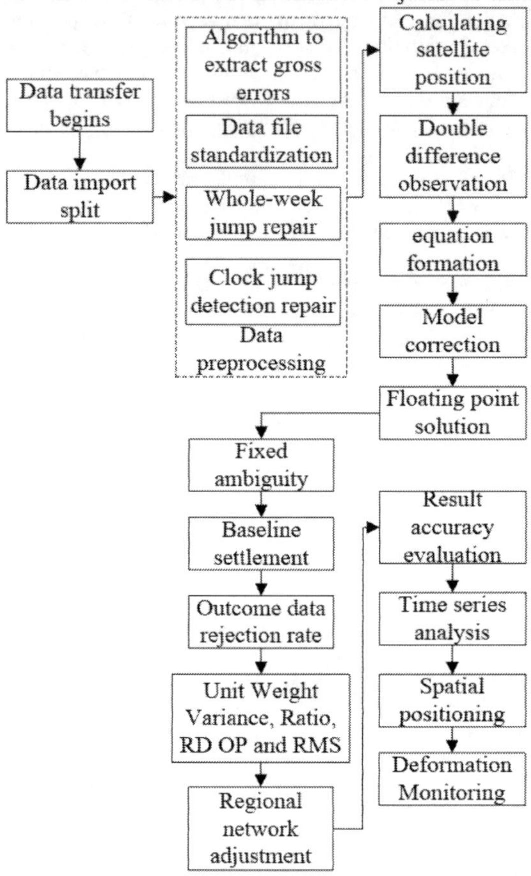

Fig. 2. GNSS monitoring system work flow chart.

5 Conclusion

The landslide multi-factor monitoring system based on GNSS technology integrates the collected monitoring data, satellite remote sensing data, geological structure model of disaster body and standardized early warning analysis model to form an efficient, reliable and stable intelligent command and dispatch management system. Through the algorithm analysis and trend discrimination of the system, the automatic data report forms of periodic, periodic and unexpected events can be realized. The system can provide effective data decision analysis and technical support for landslide monitoring.

This research is supported by the development of intelligent monitoring equipment for key elements of landslides in Karst Mountains (2018YFC150480502), and also supported by geological disaster monitoring technology integration and application projects.

References

1. Guo Ke, Peng Jibing, Xu Qiang, Realization of the extraction of comprehensive information of multi-

station monitoring data of landslide with the technique of multi-sensor target tracking, PROGRESS IN GEOPHYSICS, 20, 808-813 (2005)

2. Liu Huan, Chen Hai, Application Research of Integrated Monitoring Technology in Landslide Prediction and Forecas, Transportation Science & Technology, 3, 10-12 (2018)
3. Han Junqiang, Huang Guanwne, Li Zhe, Multipath Effect analysis and Processing Method of GNSS Landslide monitoring Under Complicated environment, Journal of EarthSciences and Environment, 40, 355-362(2018)
4. Xiao Yugang, Jiang Weiping, Chen Hua, Research and Realization of Deformation Monitoring Algorithm with Millimeter Level Precision Based on BeiDou Navigation Satellite System, Acta Geodaetica et Gartographica Sinica, 45, 16-21(2016)
5. Wan Ling Chuan, GNSS Real-Time Relative Positioning Technology and its Application in Landslide Monitoring, Chang'an University, 20-50(2018)
6. Wang chunhua, Appication of complex event processing to landslide monitoring sensor data integration research, East China Normal University, 15-20(2016)
7. Fan Junqing, Research on multi-source heterogeneous sensor information fusion for landslide monitoring, China University of Geosciences, 20-25(2015)
8. Sun Bo, Method research on comprehensive Treatment and Data Fusion of Multiple Monitoring Data, China University of Geosciences, 19-27(2012)

Modelling of Artificial Neural Network to control the cooling rate of a Laboratory Scale Run-Out Table

Biswas Prabir[1,*], *Mondal* Md Safwan[1], *Mookherjee* Saikat[1] and *Mandal* Pranibesh[1]

[1]Jadavpur University, Mechanical Engineering Department, Kolkata - 700032, India

Abstract. Run Out Tables (ROTs) have been used for long time in order to achieve different microstructure of steel in the industries. The microstructure of steel controlled by the cooling rate which in turn depends on various factors like the plate velocity, nozzle bank distance, coolant flow rate, and many others. Achieving new steel grade thus demand a proper combination setting of all such parameters. The observed data like upper nozzle distance, lower nozzle distance and mass flow rate of coolant from the laboratory scale ROTs are used to find out the cooling rate which is important parameter for achieving desired properties in steel. An Artificial Neural Network has been used here to creating an empirical relation between the observed data and thermodynamics parameter which will determine the cooling rate and validate it.

1 Introduction

Run Out Tables (ROTs) have been used for long time to achieving different microstructure of steel in the industries. The microstructure primarily depends on the cooling rate which in turn depends on various factors like the plate velocity, nozzle bank distance, coolant flow rate, and many others of Run Out Tables (ROTs). Thermo-metallurgical phase transformation taking place during the cooling process so cooling rate is the important parameter for having the desired mechanical properties of steel. In ROT high flow rates of coolants such as air, water, air water mist etc. impinged on a uniformly distributed surface area in motion is apply for Ultra-Fast Cooling (UFC).

Run Out Tables (ROTs) and cooling rate control corresponding to different properties of steel has been an important topic of discussion among researchers for a long time. Suebsomran et al. [1] study to determine the effective cooling parameters for the run-out table (ROT) of strip steel in a hot rolling process. Mukhopadhyay et al. [2] develop an on-line model for an HSM so as to obtain the appropriate cooling rates required to predict the mechanical properties of HR coil. Serajzadeh et al. [3] developed a mathematical model to prediction of temperature variations and kinetics of austenite phase change on the run-out table. Nam Han et al. [4] developed a model for deformation, temperature and phase transformation behavior of steels on run-out table in hot strip mill. Wang et al.[5] A thermal, micro structural and mechanical coupling analysis model for predicting flatness change of steel strip during the run-out table cooling process was established using ABAQUS Finite Element Software. In this model, Esaka phase transformation kinetics model was employed to calculate the phase transformation, and coupled with temperature calculation by means of the user subroutine program HETVAL. Xiao-dong et al. [6] studied Thermal, Micro structural and Mechanical Coupling Analysis Model for Flatness Change Prediction during Run-Out Table Cooling in Hot Strip Rolling. Zhou et al. [7] Numerical analysis of residual stress in hot-rolled steel strip on the run-out table. Suwanpinij et al. [15] make fast algorithms for phase transformations in dual phase steels on a hot strip mill run-out table to study the modelling and simulation of the evolution of phases on the ROT. Weisz-Patrault et al.[16] study the evolution of residual stresses during cooling on the run out table (after hot rolling) that is a very non linear and multiphysics process. A cooling pattern is imposed to the strip on the run out table in order to obtain a specific cooling path and phase transitions.

In this work a Laboratory scale ROT with a furnace and nozzle bank has been used. Different experimental data are achieved from Laboratory scale ROT [14]. The cooling rate, which is important parameter for achieving desired properties in steel are trying to find by using the observed data. Some part of the observed data are taken as input parameter and cooling rate as an output parameter. The inverse problem is significant i.e., for a particular cooling rate as an input what will be the corresponding output parameter. The inverse problem is quite difficult because the multidimensional data are achieved from the experiment. We need apply Artificial Neural Network (ANNs) to train in order to generate more data which will enable to creating an empirical relation between the input and the output. In this present work some observed data are trained in ANNs to predict the output i.e., parameter which will determine the cooling rate and validate it.

* Corresponding author: prabirbiswas.kgec@gmail.com

Table 1. Nomenclature

Symbol	Description
Q	Mass flow rate of air(m^3/Sec)
du	Upper nozzle distance(m)
dl	Lower nozzle distance(m)
T	Temperature(°C)
t	Time(Sec.)
Þ	Thermodynamic parameter(°C.Sec)

It has been a good practice to apply Artificial Neural Network in research work from long time. N.Xie et al. [8] applied Artificial Neural Network (ANN) for heat transfer analysis of shell-and-tube heat exchangers with segmental baffles or continuous helical baffles. Ermis et al.[9] proposed a feed-forward back-propagation artificial neural network (ANN) algorithm for heat transfer analysis of phase change process in a finned-tube, latent heat thermal energy storage system. Islamoglu et al.[10] study presents an application of artificial neural networks (ANNs) to predict the heat transfer rate of the wire-on-tube type heat exchanger. Pacheco-Vega et al.[11] consider the problem of accuracy in heat rate estimations from artificial neural network (ANN) models of heat exchangers used for refrigeration applications. Jambunathan et al.[12] successfully applied neural networks based on the back propagation algorithm to predict heat transfer coefficients from a given set of experimentally obtained conditions. Tan et al.[13] use of artificial neural network models to simulate the thermal performance of a compact, fin-tube heat exchanger with air and water/ethylene glycol anti-freeze mixtures as the working fluids. Hassan et al. [17] study the prediction of density, porosity and hardness n aluminum–copper-based composite materials using artificial neural network. Patel G.C et al.[18] study the applications of Back propagation genetic and recurrent neural network in modelling and analysis of squeeze casting process. Junghui Chen et al.[19] monitor dynamic process fault based on neural network and PCA.

2 System Description

The laboratory scale ROTs are consist of a furnace, couple of nozzle bank which are connected with air supply line and water supply line, an electro-hydraulic actuation system for proving the reciprocating motion to the specimen plate, fixture with roller fitted rail to hold the specimen before cooling bay and mechanical handling region in between cooling table and furnace. Couples of nozzle banks, each consisting of twenty nozzle units, are on the both sides of the cooling bay so that both the upward and downward surfaces of the specimen plate can be sprayed upon at the same time. In this work only air is used for cooling. The chamber type furnace has 18 nos. of GLOBAR SD Silicon Carbide heating elements lining on the two inner side walls of the chamber. Separately installed control panel of furnace has current and voltage indicator with safety controller to limit the maximum achievable temperature by the thermocouples which have been attached to the sample MS plate for the temperature. The operator can control the heating process with the help of an Automatic PID type programmable temperature indicating controller. Distance of the nozzle banks from the specimen plate are adjust using simple screw and nut mechanism to have the different cooling rate.

3 Modeling of Artificial Neural Network

Artificial Neural Network (ANN) is a part of Artificial Intelligence. A network or a computer system composed of artificial neurons modeled after biological nervous system, where advanced computing techniques are used to solve complex and nonlinear relationship between inpus and output variables. Through the algorithm the output signals of the neurons are expressed as the function of the input signals. Here, with our input vectors 'p_i' respective weights 'w_i' are multiplied and they are fed to the summing function where bias value 'b' is added to produce the net input 'n'.

$$\text{i.e.,}\quad n = \sum_{i=1}^{R}(p_i w_i + b), \forall i = 1(1)R$$

where, w_i = weight, b = bias value.

The net input is fed to the tansfer function to generate the output.

i.e, $a = f(n)$

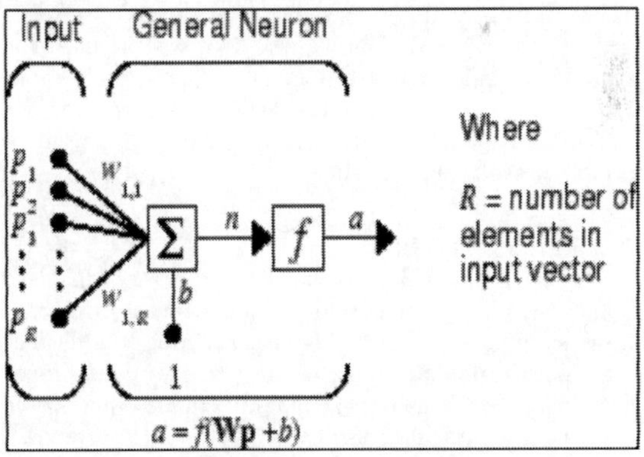

Fig. 1. A typical neuron of Artificial Neural Network

There are various types of Transfer Functions used by Artificial Neural Network where Log-Sigmoid Transfer Function ('logsig'),Tan Sigmoid Transfer Function ('tansig'),Linear Transfer Function ('purelin') are the mostly used ones. After Initialization of the bias and weights the network is become ready for training purposes which involve tuning of the values of the biases and the weights for the optimization of the network performance. There are performance functions to analyze the network performance. Mean Square Error ('mse') is used as the default performance function which is defined as below:

 https://doi.org/10.1051/matecconf/202030603004

$$\text{mse} = \frac{1}{N}\sum_{i=1}^{N}(e_i)^2 = \frac{1}{N}\sum_{i=1}^{N}(t_i - a_i)^2$$

Where , e_i = error , t_i = target , a_i = output

In this paper a multilayer feed-forward neural network consisting of an input layer, single hidden layer and an output layer is proposed to fit between a numeric dataset of inputs and a dataset of outputs. Weights and biases for the inputs are updated by using standard back propagation learning algorithm. NEURAL NETWORK FITTING TOOL has been used to find the best combination of the number of hidden neurons along with the training function and transfer function. Followings steps are incorporated for modeling of ANN:

Step 1. Selection of Data

A number of 29 samples of upper nozzle distance (du), lower nozzle distance(dl) and mass flow rate of air(Q) are served as the inputs where the same number of samples of $\int_0^t Tdt$ i.e., a thermodynamic parameter Þ served as the target values defining the network outputs. So, input is a '29x3' matrix and the target is a '29x1' matrix.

Step 2. Divison of Data

Input vectors and target vectors are divided into three data sets as follow- (i)Training: These data will be used for purpose of training of the network. (ii)Validation: These data will be used to validate Network generalization. (iii)Testing: These data will be used for completely independent test of network generalization. Different combination of this division has been tried out and 60%-20%-20% division is found to give the most accurate result respectively.

Step3.Creation of Network

In building the network structure different combinations of number of hidden neurons, different transfer function has been used. It was found that sigmoid transfer function ('tansig') in the hidden layer and linear transfer function ('purelin') in the output layer give the most suitable result when the hidden layer size is 20.

Step 4.Training Network

Several algorithms are tested, out of which Levenberg-Marquardt training algorithm of MATLAB ('trainlm') is found to be the most accurate when Mean Squared Error (MSE) is considered to be the performance function. The training stopped when validation error is increased six iterations.

Step 5. Evaluation

Mean Squared Error (MSE) is used as the performance function. There are various tools like performance plot, regression plot, histogram plot which are used to validate and analysis the network performance.

4 Results & Discussion

The number of neurons in the hidden layer isvaried to find out the optimal size of the hidden layer. MSE and correlation coefficient(R) are chosen as the performance index. Table 2 shows the values of MSE and correlation coefficient(R) for different sizes of hidden layer.

Table 2. Comparison of Different Network Structure

ANN Structure	3-15-1	3-20-1	3-25-1	3-30-1
MSE	$4.89x10^{-16}$	$9.97x10^{-23}$	$9.97x10^{-23}$	$9.97x10^{-23}$
Overall Correlation Coefficient	0.1455	0.7929	0.5829	0.3256

It can be concluded from table 2 that the optimal structure of the neural network is 3-20-1 which is shown in figure 2.

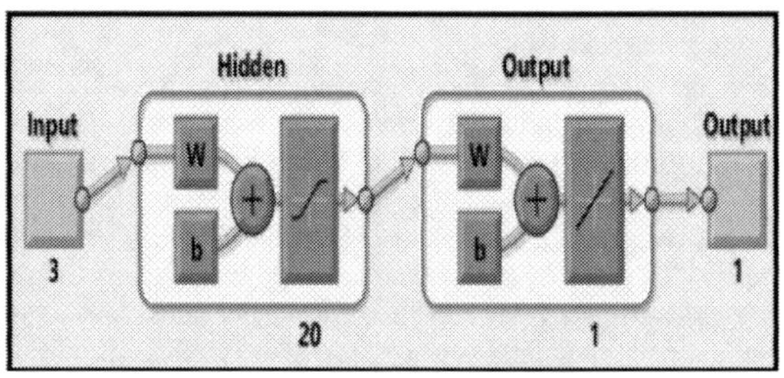

Fig. 2. Neural Network Architecture

When the network was trained in this structure using Levenberg-Marquardt training function of MATLAB, we obtain respective output for each input. Training is stopped at iteration 6 when the MSE is $9.97x10^{-23}$. Figure 3 showing a plot of training error, validation error and test error which signify that the result is reasonable since (i) the final Mean Sqare Error is very small (ii) training error and Validation error have similar nature & characteristics (iii) no significant over fitting is occurred by iteration 3 where the best validation performance is occurred.

 https://doi.org/10.1051/matecconf/202030603004

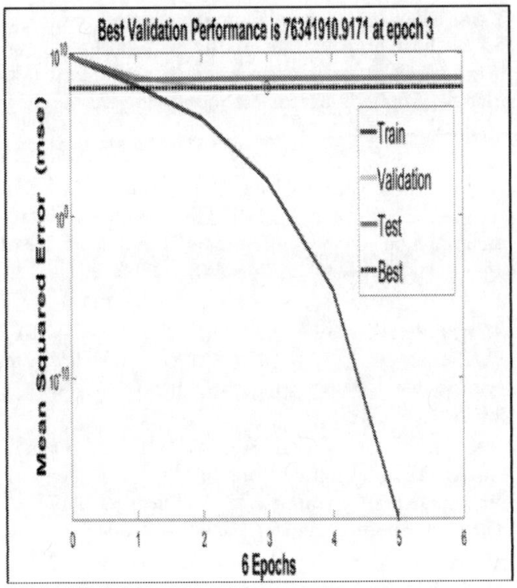

Fig. 3. Performance Plots

Figure 4 shows the regression analysis between the network outputs and the corresponding targets. Overall Coefficient of Correlation is found to be R= 0.79293.

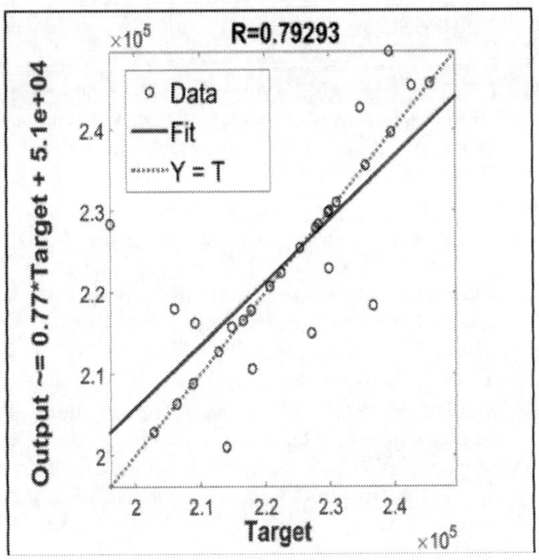

Fig. 4. Regression Analysis between outputs &Targets

Additional verification of the Nework performance is obtained from Error Histgram which shows the outliers where the fit is significantly worse than the majority of the other datas . Error Histogram is shown in Figure 5.

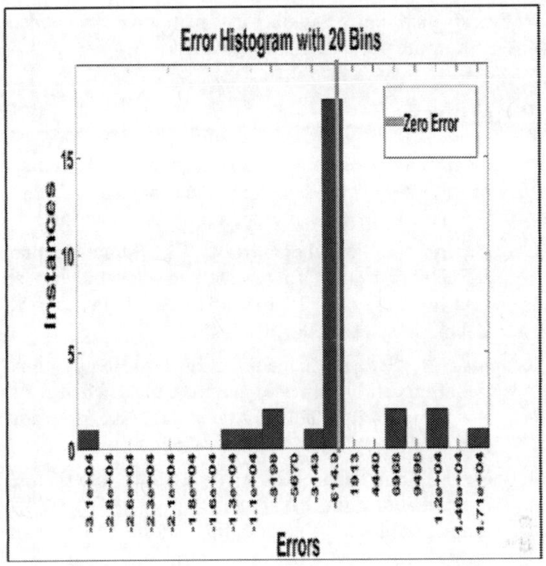

Fig. 5. Error Histogram

A relative comparison is done between the predicted outputs and real measured targets which are shown graphically in Figure 6. Where the curve is significantly overlapping which indicates the accuracy of the performance of the Artificial Neural Network.

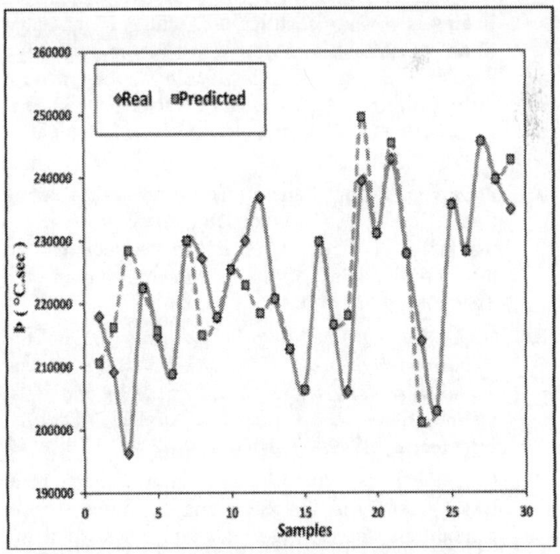

Fig. 6. Predicted outputs Vs Real Measured Targets

5 Conclusion

Heat transfer from a heated MS plate in a laboratory scale ROT under different cooling conditions has been experimentally observed. An empirical relation between the thermodynamic parameter defining the overall heat transfer rate with the affecting cooling condition parameters have been obtained by Artificial Neural

Network mapping. Satisfactory matching has been observed in the predicted and actual data.

References

1. Anan Suebsomran, Suthep Butdee, Cooling process on a run-out table by the simulation method, Cas Studies in Thermal Engineering 1 (2013) 51–56
2. Ananya Mukhopadhyay, SudiptaSikdar, Implementation of an on-line run-out table model in a hot strip mill, Journal of Materials Processing Technology 169 (2005) 164–172
3. Siamak Serajzadeh, Prediction of temperature variations and kinetics of austenite phase change on the run- out table, Materials Science and Engineering A 421 (2006) 260–267
4. Heung Nam Han, JaeKon Lee, Hong Joon Kim, Young-Sool Jin, A model for deformation, temperature and phase transformation behavior of steels on run-out table in hot strip mill, Journal of Materials Processing Technology 128 (2002) 216–225
5. Xiaodong Wang, Fei Li, Quan Yang , Anrui He, FEM analysis for residual stress prediction in hot rolled steel strip during the run-out table cooling, Applied Mathematical Modelling 37 (2013) 586–609
6. WANG Xiao-dong, LI Fei , JIANG Zheng-yi, Thermal, Micro structural and Mechanical Coupling Analysis Model for Flatness Change Prediction During Run-Out Table Cooling in Hot Strip Rolling, JOURNAL OFIRON AND STEEL RESEARCH, INTERNATIONAL. 2012, 1 9 (9) : 43-51
7. Zhongqing Zhou, Peter F. Thomson, Yee Cheong Lam, Daniel D.W. Yuen, Numerical analysis of residual stress in hot-rolled steel strip on the run-out table, Journal of Materials Processing Technology 132 (2003) 184–197
8. G.N. Xie, Q.W. Wang , M. Zeng, L.Q. Luo, Heattransferanalysis for shell-and-tube heat exchangers with experimental data by artificial neural networks approach, Applied Thermal Engineering 27 (2007) 1096–1104
9. Kemal Ermis, Aytunc Erek, Ibrahim Dincer, Heat transfer analysis of phase change process in a finned-tube thermal energy storage system using artificial neural network, International Journal of Heat and Mass Transfer 50 (2007) 3163–3175
10. Yasar Islamoglu, A new approach for the prediction of the heat transfer rate of the wire-on-tube type heat exchanger—use of an artificial neural network model, Applied Thermal Engineering 23 (2003) 243–249
11. Arturo Pacheco-Vega, Mihir Sen, K.T. Yang, dney L. McClain, Neural network analysis of fin-tube refrigerating heat exchanger with limited experimental data, International Journal of Heat and Mass Transfer 44 (2001) 763-770
12. K.JA MBUNATHAN, S. L. HARTLE, S. ASHFORTH-FROST and V. N. FONTAMA, Evaluating convective heat transfer coefficients using neural networks, 0017-9310(95)00332-0
13. C.K. Tan , J. Ward, S.J. Wilcox, R. Payne, Artificial neural network modelling of the thermal performance of a compact heat exchanger, Applied Thermal Engineering 29 (2009) 3609–3617
14. Abhyuday Aditya, Prahar Sarkar and Pranibesh Mandal, Advances in Materials, Mechanical and Industrial Engineering.
15. P.Suwanpinij , U.Prahl , W.Bleck , R.Kawalla, Fast algorithms for phase transformations in dual phase steels on a hot strip mill run-out table (ROT), A R C H I V E S O F C I V I L A N D M E C H A N I C A L E N G I N E E R I N G 1 2 (2 0 1 2) 3 0 5 – 3 1 1
16. Daniel Weisz-Patrault , Thomas Koedinger, Residual stress on the run out table accounting for multiphase transitions and transformation induced plasticity, Applied Mathematical Modelling 60 (2018) 18–33
17. Adel Mahamood Hassan, Abdalla Alrashdan, Mohammed T. Hayajneh, Ahmad Turki Mayyas, Prediction of density, porosity and hardness in aluminum–copper-based composite materials using artificial neural network, JOURNAL OF MATERIALS PROCESSING TECHNOLOGY 2 0 9 (2 0 0 9) 894–899
18. Manjunath Patel G.C, Arun Kumar Shettigarb, Prasad Krishnaa, Mahesh B. Parappagoudarc, Back propagation genetic and recurrent neural network applications in modelling and analysis of squeeze casting process, Applied Soft Computing 59 (2017) 418–437
19. Junghui Chen, Chien-Mao Liao, Dynamic process fault monitoring based on neural Network and PCA, Journal of Process Control 12 (2002) 277–289

Design of Dynamic Controllers for Continuous Paths on Parallel Platforms (Slide Modes and PD+)

Angie Valencia[1], *Mauricio* Mauledoux[1*], and *Claudia* Castañeda[1]

[1]Militar Nueva Granada University, Mechatronics Engineering Program, Bogota Colombia

Abstract. A parallel robot is defined as a mechanism of closed kinematic chains with rigid movements, high speed, precision and better inertias, in respect to those with structures composed by open links. besides these configurations describe highly nonlinear behaviours, reflected in the complexity of their kinematic and kinetic models. Finally, these systems are integrated with control techniques that obtain the desired trajectories in this paper for continuous 3D printing applications. For this reason, the development of control techniques based on dynamics by torque control law using drivers for the actuators maintaining the desired torque are described, these control techniques allow to obtain continuous behaviours in the final effector in order to make uniform impressions for biopolymer scaffolds.

1 Introduction

Additive manufacturing is the basis of bio-printing systems for tissue engineering, in which an extruder head deposits material in successive layers, with movement guided by computational algorithms. In the development of pieces, it is important to correlate the geometric properties of the deposited material with the mechanical structure that will orient the printing system, because the inertia of the mechanism must be minimized in order to generate continuous paths reflected in the final effector [1,2].

There are mechanisms with links, joined one to one using linear or rotational actuators, for example there are XYZ Cartesian robots that combine independent linear movements in three-dimensional space, or serial manipulators that base their operation on the anatomy of human structure. Its main advantage lies in its large work space, and the number of configurations the final effector can reach, giving up the precision by the accumulated error in the joints, in addition to high control efforts by the sum of the moments of inertia generated by each element, which is reflected in actuators of higher torque and cost, compared to closed kinematic chains[3].

On the other hand, there are closed kinematic chains, characterized by having rigidity in high speed applications and accelerations of up to 50 times the value of gravity [4, 5, 6]. For its design, complex kinematic and dynamic calculations are required, due to non-linear behaviours, internal singularities, reduced workspace and non-acted joints [7, 8]. Among the most commonly used configurations are delta-type mechanisms consisting of three fixed motors located on a stationary base that move a terminal platform. The tripteron which is a cartesian parallel actuator capable of reaching linear configurations [9]. Finally, there is the hexa configuration with a structure like the delta-type, but it consists of six fixed motors that allow the position and orientation of the final effector [2].

The analysis of the structures is accompanied by the kinematic and dynamic models that linked to controllers allowing to obtain the desired behaviours on the final effector. As for the dynamic calculation, the implementation of the position, velocity and acceleration derivatives of the final effector is considered to know the value of the forces to be exerted to perform a movement or vice versa [10]. Then, control algorithms such as PIDs (proportional, integral and derivative) are designed that work with linear models [11], or techniques such as Sliding Modes that base their operation on the non-linear model, considering system uncertainties and external disturbances [12, 13].

Therefore, this document presents two control techniques for a Hexa-type platform; because it performs its movements according to the position of rotational actuators, which allows continuous behaviour in the final effector with respect to those whose transmission is carried out by mechanisms such as an endless screw, since these discretize the movements in function of the screw pitch. Finally, the requirements of additive manufacturing such as continuous printing and precision will be analysed, compared to the control efforts in order to select the technique that best suits the needs of the project.

2 Robot Dynamics

In Figure 1, the configuration of a Hexa-type robot is presented that allows to obtain desired behaviours in position $(x, y, z)$and in orientation defined by Euler's angles (α, β, γ). For this, it is necessary to design control

* Corresponding author: mauricio.mauledoux@unimilitar.edu.co

https://doi.org/10.1051/mateccconf/202030603005

techniques based on the dynamic model that considers the actuators' effort to generate movement.

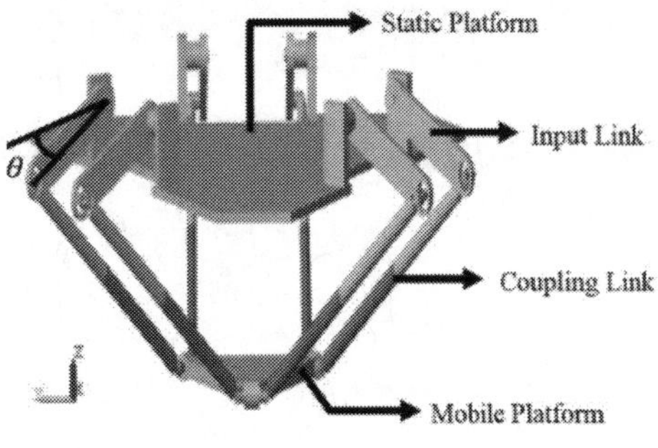

Fig. 1. Robot structure

In equation (1), it is present in a matrix form the positions, internal and external forces that determine the torque of the six generalized coordinates ($\theta_1, \theta_2, \theta_3, \theta_4, \theta_5, \theta_6$), obtained according to the platform geometry and material, which define the lengths and weights of the links.

$$M(q)\dot{q} + V(q,\dot{q}) + G(q) + F_v = \tau \quad (1)$$

Where M denotes the mass or inertia matrix, V the matrix of centrifugal and Coriolis forces, G gravitational terms and F_v represent the coefficients of viscous friction.

3 Control Systems

In this section, two types of control are shown, which are aimed at tracking the path based on the dynamics of the error.

3.1. PD+

This technique bases its control law on signals that are modified according to the weighting of the error and its derivative as shown in equation (2) and (3) and figure 2.

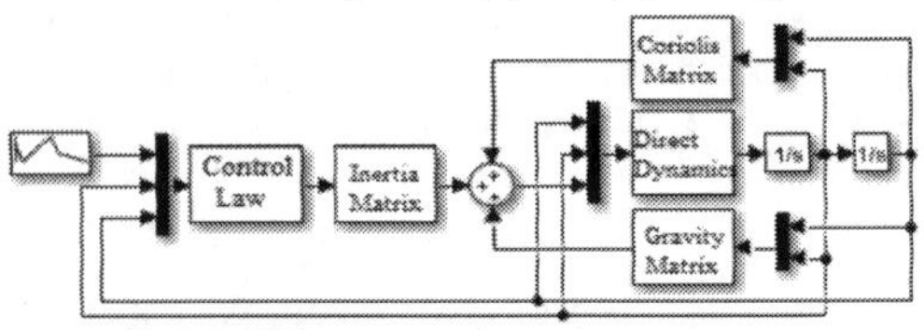

Fig 2 PD+ Control scheme

Additional centrifugal, Coriolis and gravitational forces are considered, as a parameter that modifies the force on the actuators.

$$k_v(\dot{q_d} - \dot{q}) + k_p(q_d - q) = \tau \quad (2)$$

$$M(q,\dot{q})(k_v\dot{e} + k_p e) + V(q,\dot{q}) + G(q,\dot{q}) = \tau \quad (3)$$

3.2. Sliding Modes (SM)

It consists of guiding the trajectories of the system to a sliding surface in the phase diagram, which exists if in the vicinity the tangent or velocity vectors point in the direction of said surface as described in equation (4) and (5).

$$\sigma = q + k_1 e + k_2 \int e + k_3 \iint e \quad (4)$$

$$\dot{\sigma} = k_1\dot{e} + k_2 e + k_3 \int e \quad (5)$$

Then, the control is determined as the sum of the equivalent control (τ_{eq}) (equation (6)) and the attractive control (τ_N) (equation (7)), to obtain the control law described in equation (8).

$$\tau_{eq} = G(q) + V(q,\dot{q}) + M(q)\left(k_2 e + k_1\dot{e} + k_3 \int e\right) \quad (6)$$

$$\tau_N = M(q)L \cdot sgn\left(q + k_1 e + k_2 \int e + k_3 \iint e\right) \quad (7)$$

$$\tau = G(q) + V(q,\dot{q}) + M(q)\left(k_2 e + k_1\dot{e} + k_3 \int e\right) + M(q)L \cdot sgn\left(q + k_1 e + k_2 \int e + k_3 \iint e\right) \quad (8)$$

Equating equation (8) and robot dynamics, the error dynamics in equation (9) are obtained.

$$k_2 e + k_1\dot{e} + k_3 \int e + L \cdot sgn(q + k_1 e + k_2 \int e + k_3 \iint e) = 0 \quad (9)$$

4 Analysis of results

To corroborate the design of the controllers, the platform is subjected to a helical path in its final effector, from which the behavior of Figure 3 to 6 is obtained for the PD + control, with initial conditions 0 for x ,0.01 for Y, -0.15 for Z, 0.1 for α, β and γ

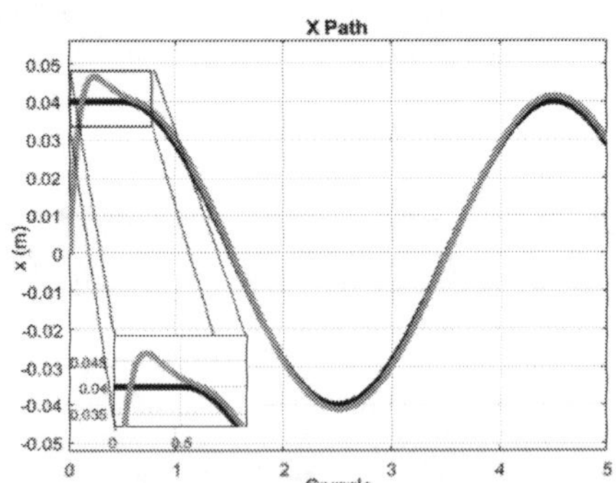

Fig 3 X-axis reference vs output PD+ behavior

 https://doi.org/10.1051/matecconf/202030603005

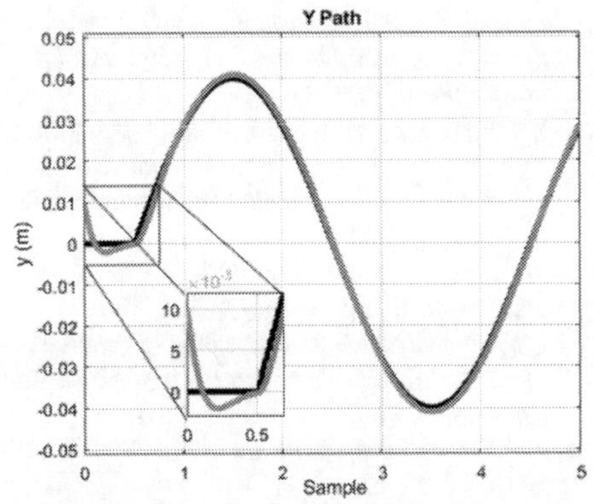

Fig 4 Y-axis reference vs output PD+ behavior

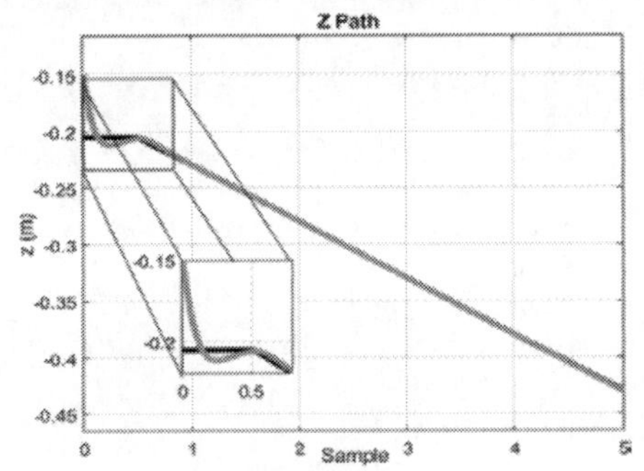

Fig 5 Z-axis reference vs output PD+ behavior

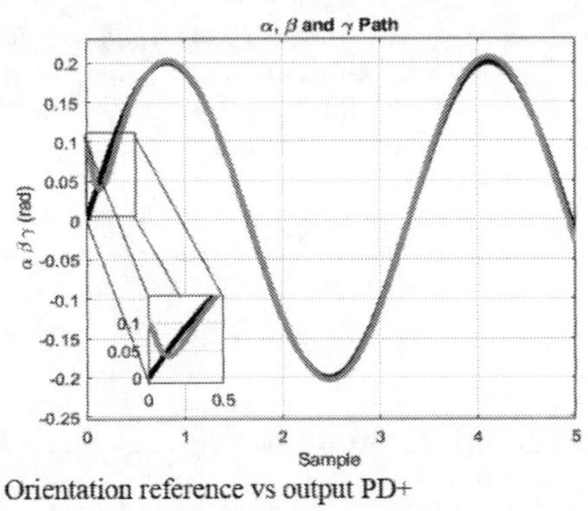

Fig 6 Orientation reference vs output PD+

Figure 7 to 10, the behaviors of the platform with the controller are obtained by sliding modes, with initial conditions 0 for x ,0.01 for Y, -0.15 for Z , 0.1 for α, β and γ

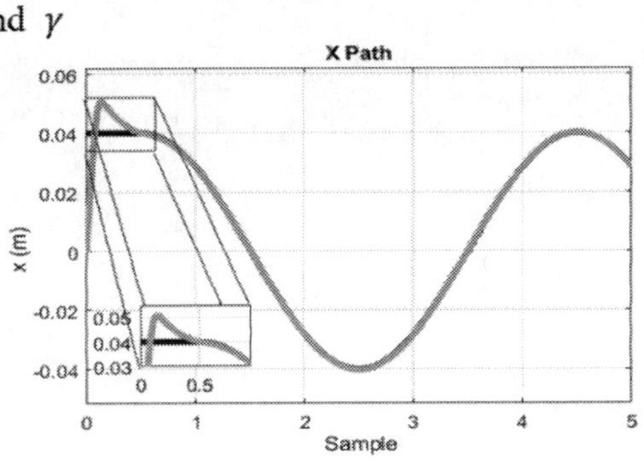

Fig 7 X-axis reference vs output Sliding Modes behavior

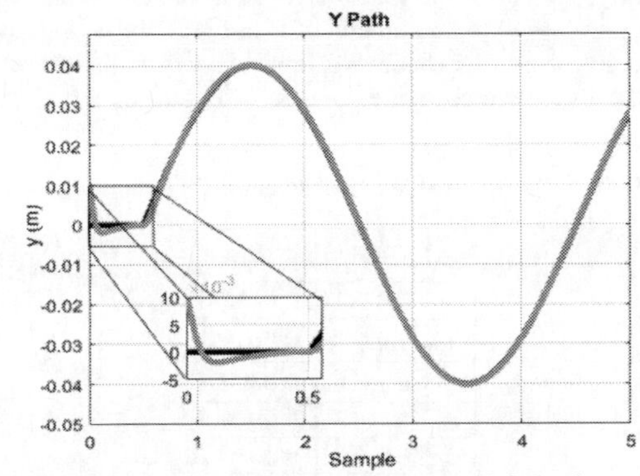

Fig 8 Y-axis reference vs output Sliding Modes behavior

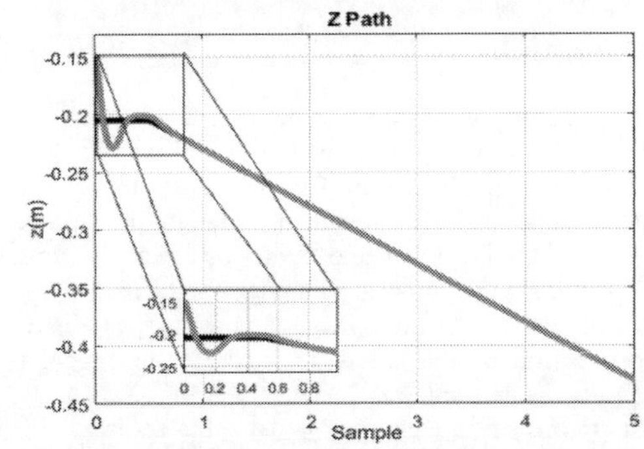

Fig 9 Z-axis reference vs output Sliding Modes behavior

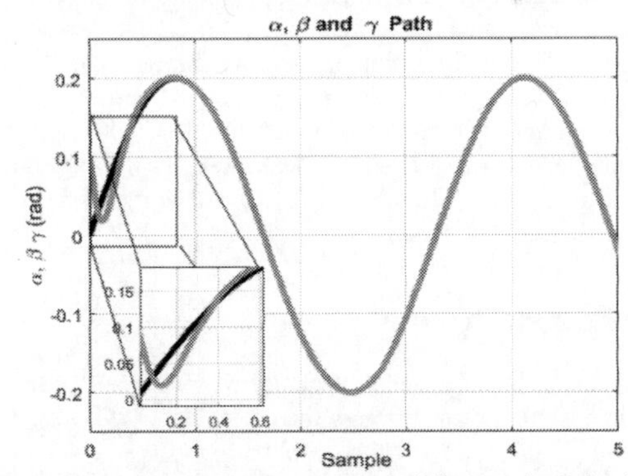

Fig 10 Orientation reference vs output Sliding Modes behavior

From the above, the robustness of the control by SM is observed, and how this modified the robot behavior to get the desired trajectory in less than 0.4 seconds compared to the PD + that makes more than 0.5 seconds. Then Then the error present in the controls is observed in table 1, concluding that the control law by SM is better at tracking paths.

Table 1. Mean square error

	PD+	SM
x	0.0017	3.8921e-08
y	0.000248	1.9359e-08
z	0.0031	2.9866e-04
α β γ	0.0162	0.0791

Finally, the effort of each controller on the articular variables are obtained on table 2. And it is observed that the control by SM requires a lower energy with respect to the PD + which makes it more efficient in terms of energy consumption.

Table 2. control effort

	PD+	SM
1	0.0271	0.0038
2	0.0210	0.0019
3	0.1378	0.0494
4	0.1033	0.0898
5	0.0980	0.0836
6	0.1373	0.0868

5 Conclusions

It is possible to conclude that the non-linear controllers based on cancellation of the mathematical non linearities on the dynamical system (in this case the parallel mechanism HUNT-RUS) achieve excellent performance in terms of the error tracking the trajectories obtaining a very similar MSE of the error signals, nevertheless is very important to have a complete mathematical representation of the model to successfully cancel the nonlinear behaviours

In terms of the control signals the sliding modes control obtain a less torque, using less energy, but whit a high-speed overshoot caused by the attractive control that act saturating the actuators. In the PD+ controller is possible to conclude that uses more energy and higher torques than the sliding modes control technique, but avoiding the overshoots mentioned.

In terms of the implementation is important to mention that the sliding modes control is more difficult to implement.

Acknowledgement

The research for paper was supported by Nueva Granada Military University, through the project ING-IMP-2658.

References

1 Yin-Yen Kuo, Chiao-Hua Cheng and Shao-Kang Hung, "3D-Printed Linear Positioner with Micrometer Accuracy", ICMME (2016)

2 Hermes Giberti, Luca Sbaglia and Marco Silvestri "*Mechatronic Design for an Extrusion-Based Additive Manufacturing Machine*", Machines, (2017)

3 Husein Zilullah Toresano, La Ode And Prajitno, Prawito., "*Design of 3-Axis Cartesian Robot Manipulator*" (2017)

4 I. Zabalza, J. Ros, J.J. Gil, J.M. Pintor, J.M. Jimenez, Tri-Scott. *A new kinematic Structure for a 6-dof decoupled parallel manipulator, Workshop on fundamental issues and future research directions for parallelmechanisms and manipulators*, pp.12-15, Québec, (2002).

5 E Mirshekari, A Ghanbarzabeh, K Heidari Shirazi, *Structure Comparison and Optimal Design of 6-RUS Parallel Manipulator Based on Kinematic and Dynamic Performances*, LAJSS, **Vol. 13**, pp 2414 – 2438, 2016.

6 Vázquez H. Jesús, Cuenca J. Francisco, *Cinemática Inversa y Análisis Jacobiano del Robot Paralelo Hexa,* Memorias Somim, **Vol. 1, No**. 4, pp 800-810, Septiembre, (2009), Sonora. México.

7 A Zubizarreta, I Cabanes, E Portillo, D Orive, M Marcos *Modelo dinamico de robots paralelos considerando sensorica redundante* universidad del Pais Vasco 2006.

8 M Díaz., H Quintero, L A Mejía, G Holguín, MH errera, C Mesa, G Daraviña ., *Aplicación de los robots paralelos", En Manipuladores paralelos: Síntesis, análisis y aplicaciones*, Universidad Tecnológica de Pereira.

9 A. Arian, B. Danaei, M. Tale-Masouleh, *Kinematic and Dynamic Analyses of Tripteron, an Over-Constrained 3-DOF Translational Parallel Manipulator, through Newton-Euler Approach*" AUT Journal of Modeling and Simulation, (2018), pp 61-70

10 J. Ruiz-García, D. Chaparro-Altamirano, R. Zavala-Yoe and R. Ramírez-Mendoza, *Direct and Inverse Dynamics Modeling of a 6-PUS Parallel Robot,* ICMEAE, Morelos, (2013), pp. 21-26.

11 J. Uniyal, M. Joshi and P. K. Juneja, "*Set point tracking capability and comparitive response of PI-PD controller with PI, PD and PID controller,*" ICACCA (Spring), Dehradun, (2016), pp. 1-6.

12 L. Jun, W. Xiaoguang, W. Yuqi and L. Qi, *"Continuous terminal sliding mode control of a 6-DOF wire-driven parallel robot*, ROBIO, Macau, (2017), pp. 1757-1762.

13 Z. Caihong and Z. Hongtao, *Dynamic Sliding Mode Control Applied in Parallel Robot System*, DCABES, Wuxi, (2011), pp. 405-408.

Design of rapid monitoring system of geological disaster based on LoRa

Wang Chenhui[1,*] and *Meng* Qingjia[1]

[1]Center for Hydrogeology and Environmental Geology Survey, China Geological Survey, Baoding, Hebei, China

Abstract. According to the characteristics of the geological environment and disasters, this paper uses microelectronics, wireless communication, thin-film solar power supply and other technologies, combined with lightweight process design, a new scheme for rapid geological disaster monitoring system based on LoRa is proposed. The system is based on embedded microprocessor of STM32F103 and SX1278 module of LoRa, which uses self-organizing network design of star-type and constructs a monitoring system with long communication distance and stable and reliable data transmission. The system can realize real-time data collection of multiple monitoring parameters of the disaster body and transmit the monitoring data to the data center or dedicated data receiving terminal through LoRa/GPRS/BeiDou satellite, which can provide data support for expert analysis and decision-making. The system has the features of low power consumption, long transmission distance, ad hoc network, stable and reliable communication, which has wide application prospect in the field of geological disaster monitoring.

1 Introduction

Geological disaster is natural disaster caused by geological dynamic activities or abnormal changes in geological environment. Geological disasters will cause huge losses to human life and property. The most common geological disasters include collapse, landslide and debris flow, etc. The monitoring methods for the above types of geological disasters mainly include monitoring methods such as rainfall monitoring, surface crack displacement monitoring, and water content monitoring. At present, there is a lack of means for rapid monitoring of geological disasters, especially the rapid monitoring systems for geological disasters that can be deployed on-site before or after disasters.

At this stage, the conventional deployment method of geological disaster monitoring is to install and deploy relevant geological disaster monitoring and early warning instruments at the hidden points of geological disasters. Geological disaster monitoring and early warning instruments are mainly composed of data acquisition system, transmission system, power supply system and protection system.

Currently, the shortcomings of this type of conventional geological disaster monitoring and early warning instrument are that the instrument has large weight, large volume, high installation and transportation cost, long time to obtain disaster body monitoring data, and can't be replaced in a short time, that has brought many inconveniences to geological disaster monitoring. Therefore, it is urgent to develop a geological disaster monitoring system that can be easily carried, quickly deployed, and quickly acquire data.

The development of the geological disaster rapid monitoring system and the application of related technologies can meet the monitoring requirements of hidden geological hazards and critical geological disasters. Its design features of light and easy to carry, easy to use, fast deployment, complete functions, and reliable performance are accordance with the requirements of rapid disaster monitoring.

Once the geological disaster occurred, monitoring personnel can carry the rapid monitoring system to enter the site according to the results of the rapid survey. Furthermore, the equipment was installed and evacuated, that the collected data was transmitted to the monitoring center in real-time through the data link of the BeiDou and GPRS communications, which can be provided to the leaders and experts for decision analysis. The rapid geological disaster monitoring system can simultaneously collect monitoring parameters such as disaster point rainfall, surface crack displacement, and video monitoring.

For the sudden geological disasters, rapid monitoring sensors can be realized rapid deployment and rapid monitoring, which need to be improved so that the sensor functional structures can be flexibly layout during the disaster phase and the corresponding monitoring parameters can be quickly achieved. The monitoring data can truly and effectively reflect the monitoring factors such as rainfall, surface deformation, and status change of the source area.

At present there are several problems in the rapid monitoring: (1) Insufficient signal coverage, many devices are deployed in areas with sparse population or complex environments. The operator network covers

* Corresponding author: 522297135@qq.com

blind spots or lacks signal strength, which making it difficult to ensure stable data transmission. (2) High power consumption, a large number of equipment require battery power. If a cellular network is used and frequent battery replacement is required, which is difficult to implement in many harsh environments. (3) Low cost effectiveness, the device transmits a small amount of data and transmission frequency is very low, at present, cellular networks are designed for high bandwidth. The use of cellular networks takes up network and code number resources, and it also generates monthly traffic fees.

The following key problems are need to be solved in this paper: (1) The lightweight design of the rapid geological disaster monitoring system, which can be achieved to improve the rapid monitoring system and process of monitoring the sensor shell, which is more in line with the characteristics of emergency monitoring, achieving a compact size and complete functions. (2) Wireless sensor design, it is need to eliminate the wired connection scheme of the existing sensors and the intelligent monitoring terminal for geological disaster fast monitoring, furthermore, it is need to realize the wireless connection between the terminal and sensor, and realize fast access monitoring data. (3) Appearance of the sensor process design, according to the characteristics of the monitoring elements, the structure of the sensor is improved to realize rapid development.

2 Methods and Approach

The system mainly includes front-end monitoring sensor, RF transmission modules of LoRa, network nodes, GPRS networks, geological disaster monitoring and warning cloud severs[1]. The network nodes include on-site acquisition nodes, relay nodes (if necessary) and aggregation nodes. The on-site acquisition nodes are responsible for the collection of geological disasters sensor data (such as rainfall, displacement, and tilt sensor). The aggregation node is mainly responsible for sending and receiving data of the acquisition node and uploads the data to the geological disaster monitoring and warning cloud server through GPRS or BeiDou.

One side, acquisition nodes collects the on-site monitoring data accurately, stably, and reliably according to the selection of sensor types input, on the other side, acquisition nodes can communicate with the aggregation node by using long distance wireless communication of LoRa networks and transmit the data to the cloud server through the aggregation node.

The geological disaster monitoring and warning cloud server not only can make analysis of the on-site data, but also can intelligently judge the status of the connected sensors, and integrate the monitoring data through the established data algorithm to analyze the specific conditions and deformation trends of the geological disaster.

The system can realize rapid deployment and fast access to data. Furthermore, the system can be powered on and its network architecture is simple. It not only provides data query and management for experts, also provides intelligent decision-making basis and effectively improves the efficiency of monitoring and early warning of geological disasters[2-4]. The diagram of rapid monitoring system is shown in Figure 1.

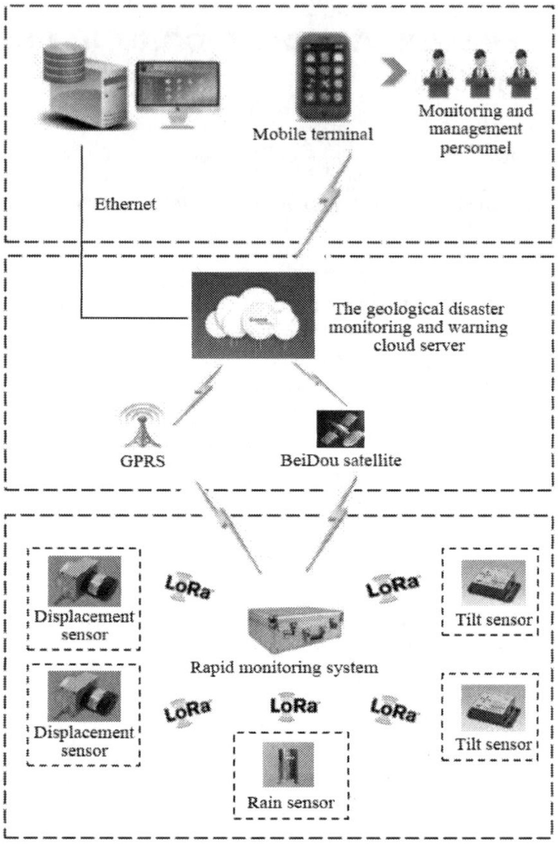

Fig. 1. The diagram of rapid monitoring system

3 Hardware Design of Rapid Monitoring System

3.1 Microprocessor

STM32F103 devices use the Cortex-M3 core, with a maximum CPU speed of 72 MHz. The portfolio covers from 16 Kbytes to 1 Mbyte of Flash with motor control peripherals, USB full-speed interface and CAN. It integrates a variety of high-performance industrial interconnect standard interfaces, which has 6-12 clock cycles and realizes fast nested interrupts. In addition, STM32F103 has MPU protection set access rules and perfect compatibility and can adapt to a variety of IOT applications.

3.2 Selection of LoRa Module

LoRa is a kind of LPWAN communication technology. It is an ultra-long-distance wireless transmission scheme based on spread spectrum technology, which is adopted and promoted by Semtech of the United States. It effectively solves the problem that the traditional wireless sensor network cannot have both the transmission distance and the low power consumption.

The main technical features of LoRa are as follows:

• Strong compatibility, all applications that meet the LoRa WAN protocol can access.

• Access is flexible, single gateway can access dozens to tens of thousands of nodes, nodes can randomly access the network and the number of nodes can be extended.

• Strong concurrency, the gateway can support at least 8 frequency points, while random 8-way data concurrency, frequency point can be extended.

• Full-duplex communication can be achieved, with no conflict between uplink and downlink, and strong practical effect.

• Network topology is simple, the star network has higher reliability and lower power consumption.

In this article we select KT-F1278 as communication module of LoRa. KT-F1278 wireless module adopts Semtech company SX1278 device, the device adopts the LoRa TM frequency-hopping spread spectrum modulation technology. Its communication distance and receiving sensitivity are far above the FSK, GFSK modulation, and its multiple transmission of signal occupy the same channel and the channel is unaffected. KT-F1278 can significantly extend the transmission distance. It can reach 15 km in sparse environment and 3 km in densely populated areas. Therefore, there is no need for relays and complicated communication infrastructure in this system.

3.3 Design of the acquisition node

According to the characteristics of the geological disaster rapid monitoring system, the acquisition node is mainly responsible for the data collection of each monitoring sensor at the geological disaster site[5-7]. The schematic diagram of the acquisition node structure is shown in Figure 2.

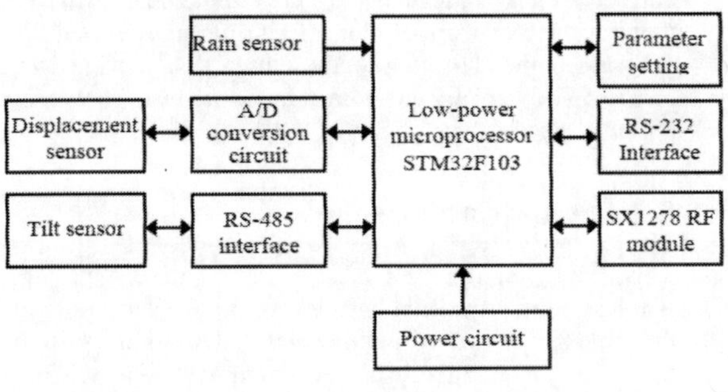

Fig. 2. The schematic diagram of the acquisition node structure

On the one hand, the acquisition node is responsible for the collection and control of data such as on-site rainfall sensor, surface displacement sensor, and tile sensor. On the other hand, it needs to respond to the control instructions issued by the convergence nodes and upload the collected dat a to the convergence nodes to achieve real-time data acquisition, control and storage of geological disasters on the spot.

The rain sensor is connected to the IO port of microprocessor for data acquisition. The surface displacement sensor accesses the ADS1256 for data acquisition. The tilt sensor uses the RS-485 interface for data acquisition[8-9].

3.4 Design of the convergence node

The convergence node is responsible for the wireless network instruction issued, data reception and upload, detection of the system and management, and other functions.

The convergence node receives the field data uploaded by the acquisition nodes in the area through the LoRa wireless network, and uploads the data to the geological disaster monitoring and warning cloud server via the GPRS network or the BeiDou satellite network; at the same time, it sends the acquisition and control commands to any collection node of the LoRa network. The schematic diagram of the convergence node structure is shown in Figure 3.

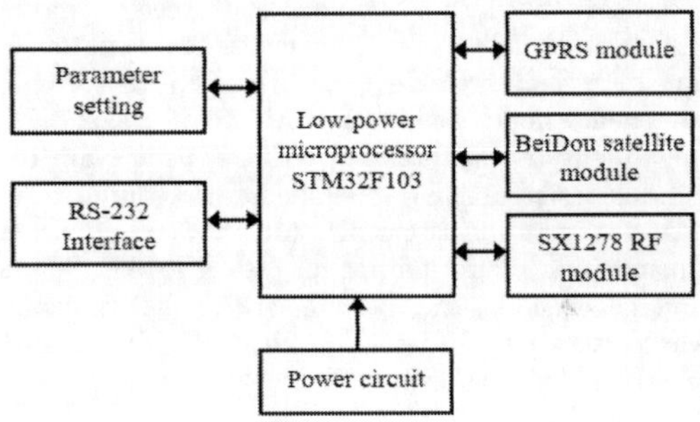

Fig. 3. The schematic diagram of the convergence node structure

4 Software Design of Rapid Monitoring System

The system software design refers to the μcos real-time operating system, which is mainly based on the task scheduling mechanism and can guarantee the concurrent execution of multiple tasks.

4.1 Communication protocol design of the application layer

The pros and cons of the protocol design directly determine the degree of intelligence of the system. In order to reflect the degree of intelligence of the system, a protocol design was performed on the application layer of the system to realize the transparent transmission of data packet packing and parsing[9-10]. The data transmission mode adopts the data frame mode, and the transmission sequence is a binary byte stream.

At the same time, the CRC16 checksum algorithm is used for data in data transmission so that the accuracy of the data transmission is correct. During data acquisition, the cloud sever sends protocol commands through the convergence node and the acquisition node. After the sensors respond, the data is returned to the acquisition node. The acquisition node packages the data and uploads it to the convergence node, and finally the convergence node upload the data to the cloud sever.

The system can realize three kinds of data model: timing acquisition, broadcast acquisition, and threshold wake-up acquisition. Once the monitoring network is established, a one-to-many mapping relationship can be established between the convergence node and the acquisition node. The acquisition node can set a unique device number and assign a unique device address to each acquisition node according to each SX1278 RF module. For example, when the convergence node sends a message with the device number 0x55AA, the acquisition node address is the device number and it needs to be immediately make a response.

4.2 Software design of the acquisition node and the convergence node

The acquisition node not only can collect the data of on-site sensors, but also can upload and receive data and commands through the LoRa network[11-13]. After the acquisition node receives the command sent by the convergence node, it first performs CRCl6 check on the data to ensure the accuracy of the data, and then performs corresponding operations according to the frame command in the packet. After the execution, the corresponding data information is uploaded to the convergence node, and the software of the acquisition node is executed. The workflow chart of software of the acquisition node is shown in Figure 4.

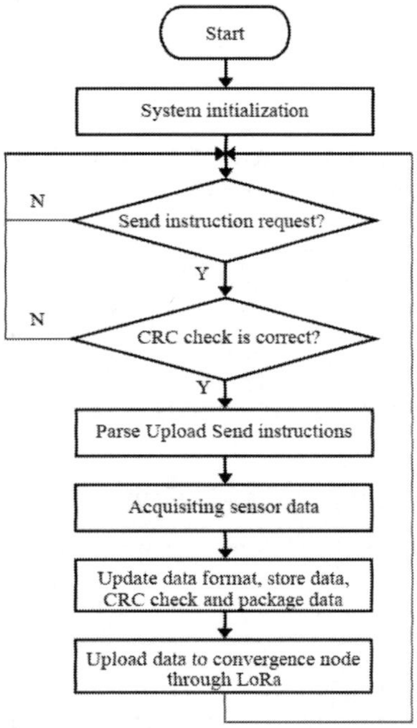

Fig. 4. The workflow chart of software of the acquisition node

The convergence node needs to communicate with the LoRa wireless network and also communicate with the GPRS public network. The flow chart of software design is shown in Figure 5. When the convergence node receives the data sent by the cloud server or the data uploaded by the acquisition node, it also performs a CRC check on the data to ensure the accuracy of the data[14-15].

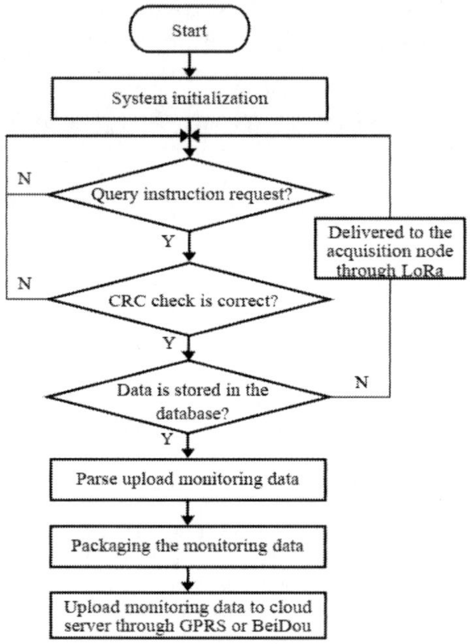

Fig. 5. The workflow chart of software of the convergence node

At the same time, the command in the packet determines whether the packet is sent by the cloud server or uploaded by the acquisition node. If the message is delivered, the device number is used to judge whether it is still operating on the acquisition node. If the message is uploaded and the CRC check of the received data is correct, it is uploaded to the geological disaster monitoring and warning cloud server through the GPRS network or the BeiDou satellite network. The workflow chart of software of the convergence node is shown in Figure 5.

5 Conclusion

This paper applies technologies of microelectronic, wireless communication and thin-film power technology to geological disaster monitoring system. This system uses low-power microprocessor, remote LoRa wireless network and BeiDou/GPRS public network to achieve real-time dynamic acquisition of geological disaster monitoring data. Specific data of geological disaster can be queried real-time in the remote data monitoring center or dedicated data receiving terminal, which provide data support for expert analysis and decision-making. Through the comparison and analysis, the new design of geological hazard monitoring system compared with the previous monitoring system has the following characteristics, such as shown in table 1.

Table 1. The Rapid monitoring system and Conventional system feature comparison table

System characteristics	Rapid monitoring system	Conventional sysytem
Instrument weight	Light(≈4Kg)	Heavy(≈30Kg)
Instrument volume	small	large
Instrument installation time	1 hour	4 hour
The instrument response time	30s	1min
Instrument transmission way	2G/3G/4G/LoRa/BeiDou	2G/3G

The overall design of the system, hardware and software circuit design are introduced in detail. The system has the technical features of low power consumption, ad-hoc network, stable and reliable communication, and long transmission distance. The system completely meets the technical requirements for data collection of geological disaster and has a wide application prospect in the field of geological disaster monitoring.

ACKNOWLEDGMENT

This research was supported by the China Geological Survey project (DD20160283, DD20190639) and the National Key Research and Development Program of China (2018YFC150480502). We appreciate the help and support of Xiuding Cao, Yue Wu, Honglei Wang, Wei Guo of the Center for Hydrogeology and Environmental Geology Survey, China Geological Survey.

References

1. Hui-fu Zhang, Wei Kang, " Design of the Data Acquisition System Based on STM32 ", Procedia Computer Science, vol.17, 2013, pp.222-228.
2. Dan Chen, Zhixin Liu, Lizhe Wang, Minggang Dou, Jingying Chen, Hui Li, " Natural Disaster Monitoring withWireless Sensor Networks: A Case Study of Data-intensive Applications upon Low-Cost Scalable Systems ", Mobile Networks and Applications, vol. 18, 2013, pp. 651–663.
3. Yonghua Song, Jin Lin, Ming Tang, Shufeng Dong, "An Internet of Energy Things Based on Wireless LPWAN ", Engineering, vol 3, 2017, 460-466.
4. Mahmoud MS, Mohamad AAH, " A study of efficient power consumption wireless communication techniques/modules for Internet of Things (IoT) applications ", Adv Internet Things, vol 6 2016, 19–29.
5. Joong-Soo Lim, " Design of High Speed Data Acquisition and Fusion System with STM32 Processor ", Journal of the Korea Convergence Society, vol. 7, 2016 , pp. 9–15.
6. Jing Luo Cai, Guan Li, Xiang Li Bu, Xian Guo Liu, Hua Lu Xing, " The Design of Landslide Hazard Remote Monitoring System Based on STM32 MCU and GPRS ", Applied Mechanics and Materials, vol. 599-601, 2014, 1115-1119.
7. Dimitrios Loukatos, Ioannis Manolopoulos, Evangelia-Sofia Arvaniti, Kostas G. Arvanitis, Nick A. Sigrimis, "Experimental Testbed for Monitoring the Energy Requirements of LPWAN Equipped Sensor Nodes", IFAC-PapersOnLine, vol 51, 2018, 309-313.
8. Wang Chenhui, Wu Yue, Yang Kai. " Design of multi - channel data acquisition system based on STM32 ". Application of Electronic Technique, 2016, 42 (1) : 51–53, 57.
9. He Man-chao, " Real-time Remote Monitoring and Forecasting System for Geological Disasters of Landslides and ITS Engineering Application. Chinese Journal of Rock Mechanics and Engineering ", 2009, 28(6): 1081–1090.
10. Yi-Bing Lin, Yun-Wei Lin, Chung-Yun Hsiao, Shie-Yuan Wang, " Location-based IoT applications on campus: The IoTtalk approach ", Pervasive and Mobile Computing, vol 40, 2017, 660-673.
11. Renato Macciotta, Michael Hendry, C. Derek Martin, " Developing an early warning system for a very slow landslide based on displacement monitoring ", Natural Hazards, vol. 81, 2016, pp. 887–907.
12. Rashmi Sharan Sinha, Yiqiao Wei, Seung-Hoon Hwang, "A survey on LPWA technology: LoRa and NB-IoT ", ICT Express, vol 3, 2017, 14-21
13. Lishan Baoa, Lei Weib, Chengling Jianga, Weiwei Miaoa, Bo Guoa, Wei Lia, Xiangdong Cheng, Rui Liu,Jun Zou, " Coverage Analysis on NB-IoT and LoRa in Power Wireless Private Network" Procedia Computer Science, vol 131, 2018, 1032-1038.
14. W. Yang, M. Wang, J. Zhang, J. Zou, M. Hua, T. Xia and X. You, "Narrowband Wireless Access for Low-Power Massive Internet of Things:A Bandwidth Perspective," in IEEE Wireless Communications, vol. 24, 2017 , 138-145.
15. Matthias Jakob, I T. Owen, T. Simpson, " A regional real-time debris-flow warning system for the District of North Vancouver, Canada ", Landslides, vol. 9, 2012, 165–178.

Simulation and disturbance estimation of speed-controlled mechatronic drive systems

Chris Schöberlein[1,*], *Manuel* Norberger[1], *Holger* Schlegel[1], and *Matthias* Putz[1,2]

[1]Institute of Machine Tools and Production Processes, Chemnitz University of Technology, Reichenhainer Straße 70, 09126 Chemnitz, Germany
[2]Fraunhofer-Institute for Machine Tools and Forming Technologies, Reichenhainer Straße 88, 09126 Chemnitz, Germany

Abstract. In the field of modern production systems, the process-parallel measurement of time-varying process forces and torques for high-level process monitoring is becoming increasingly important. Commonly utilized methods are based on additional sensors placed close to the working area. Unfortunately, this often leads to a higher complexity and additional costs due to the necessity of external hardware. An alternative approach is to evaluate the already available machine-internal signals of the subordinate drive systems. The process forces act on the load side of the machine axes in the form of disturbance forces and influence the drive-internal signals such as motor current or motor speed. To extract these disturbances, additional superimposed forces (e.g. friction and acceleration forces) and the influence of the feedback control as well as the mechanical system must be considered. Therefore, in the present paper, the application of various observer-based approaches for the estimation of load side disturbances on speed-controlled mechatronic drive systems will be examined. The investigations are performed on a simulation model of a speed-controlled machine axis. After an introduction, three different disturbance observers will be presented. Subsequently, the paper describes the structure of the simulation model including its parameterization based on test rig measurements. Afterwards, the performance of the disturbance observers will be investigated and compared taking selected influencing factors like changing controller and inertia parameters into account. The paper closes with a summary and an outlook.

1 Introduction

In the field of production engineering, the prevailing process forces are a sensitive parameter for the indirect evaluation of the process state and its stability [1]. Usually, their measurement is realized by installing near-process sensors, for example in the form of piezo-based force measuring sensors or by means of capacitive or inductive displacement sensors [2, 3]. These sensors have advantages in terms of accuracy, dynamics and sensitivity due to the proximity to the working area. However, the process conditions (chips, temperature, cooling lubricant) may have negative effects on the lifespan of the sensors. In addition, an increased complexity of the overall system due to additionally required evaluation units and high requirements with regard to sensor integration have to be considered.

On the other hand, there are various approaches which are based on the already available control-internal signals and do not require any manipulation of the existing machine structure. However, one major obstacle is the reduced accuracy of this concepts which is highly dependent on the mechanical system and its state of motion. In addition, changes in friction torque and moment of inertia are affecting the quality of the estimation. Eventually, the quality of the measured signals is affected by the internal system clock of the machine control or the drive system, respectively.

In recent years, a variety of methods for the drive-based estimation of process forces has been developed. Common concepts were based on the evaluation of the motor currents [4-7] or the use of different kinds of state observers [8, 9] and were investigated for various motor types and mechanical systems. An alternative is provided in the form of model-based approaches based on disturbance observers, which consist of an order-reduced model of the mechanical part of the drive system. In this paper three disturbance observers are examined including a performance evaluation under variation of different influencing factors.

The paper has the following structure: in chapter 2 several disturbance observers will be presented. In addition to the conventional disturbance observer (DOB), the load side disturbance observer (LDOB) and the multi-encoder based disturbance observer (MEDOB) will be introduced. In chapter 3, the modeling of the drive system based on a speed-controlled rotary drive test rig in MATLAB® Simulink® will be carried out. In particular, the model consists of the mechanical subsystem including effective friction torques and different model orders as well as the control loops for speed and current. Chapter 4 compares the observers by

* Corresponding author: chris.schoeberlein@mb.tu-chemnitz.de

simulating different disturbance effects and influencing factors. In former research, especially the effects of the drive controller settings or deviations in the parameters of the mechanical system were not examined in detail. Therefore, several settings of the speed controller as well as the robustness of the observers under variation of the load side inertia and the elastic coupling between input and output will be investigated.

2 Types of Disturbance Observers

In the following section, three common concepts of disturbance observers are examined in detail. With regard to the mechanical system, a two-mass system with one elastically coupled load inertia is assumed (figure 1, top). This system consists of a motor side moment of inertia J_m and a load side moment of inertia J_l. The elastic coupling is defined by the torsional rigidity c and the mechanical damping constant d. Friction torques act on the motor side ($T_{f,m}$) and the output side ($T_{f,l}$), respectively. Relevant disturbance torques T_{dist} usually act on the load side only.

2.1 Disturbance observer (DOB)

The elementary structure for estimating load torques on mechatronic drive systems is the conventional disturbance observer as described in [10, 11] (figure 1, left). The estimation of the load side disturbance torque is based on the simplified assumption of a rigid coupling between motor and load and thus a single-mass system. The equation of motion of such a system is shown in equation (1).

$$J\frac{d\omega_m}{dt}+T_l= I_m K_t \quad (1)$$

J denotes the total moment of inertia, ω_m the motor angular speed, T_l the load torque including disturbance and total friction torque, I_m the motor current and K_t the torque coefficient. Usually it is assumed that fluctuations of the moment of inertia and the torque coefficient are negligibly small and therefore the nominal values of both parameters are utilized. How changes in the load side moment of inertia influence the estimation accuracy is examined in chapter 4. Due to the assumption of a rigid coupling between the input and output side, the disturbance torque is calculated as shown in equation (2). Note that $\hat{T}_{f,m}$ and $\hat{T}_{f,l}$ denote the models for the friction torques and $J_{m,n}$ and $J_{l,n}$ determine the moments of inertia on motor and load side, respectively.

$$\hat{T}_{dist}\approx I_m K_{t,n}-\hat{T}_{f,m}-\hat{T}_{f,l}-(J_{m,n}+J_{l,n})\frac{d\omega_m}{dt} \quad (2)$$

In addition, it is possible to implement a low-pass filter with the cut-off-frequency g_{dist} to filter signal noise which arises by the discrete-time differentiation of the speed signal. Due to the reduced model order, the DOB is suitable for disturbance estimation on linear direct drives. However, electromechanical axes are usually characterized by an elastic coupling between input and output sides. This results in a reduced bandwidth and estimation accuracy of the DOB.

2.2 Load disturbance observer (LDOB)

To negotiate the disadvantages of the DOB, an observer structure with focus on the load side is suggested in [12, 13]. In contrast to the DOB, the structure shown in the center of figure 1 is based on the assumption of an elastically coupled load inertia and therefore utilizes the nominal spring constant c_n and the angular difference between the input and output angles φ_l-φ_m. In addition, the observer includes a model of the load side friction torque $\hat{T}_{f,l}$ as well as a load side moment of inertia $J_{l,n}$. This results in the following estimation equation for the disturbance torque:

$$\hat{T}_{dist}\approx c_n(\varphi_m-\varphi_l)-\hat{T}_{f,l}-J_{l,n}\frac{d\omega_l}{dt} \quad (3)$$

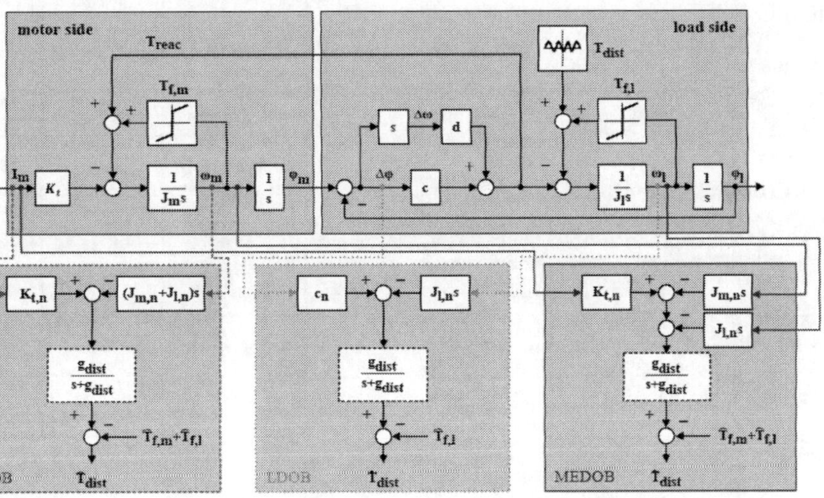

Fig. 1. Structure of the two-mass system including different disturbance observers and required input signals.

One major obstacle for practical application of the LDOB is the required knowledge of the spring constant between motor and load side. In particular, the rigidy of linear feed axes (e.g. ball-screw drives) depends on the current axis position and typically has non-linear behavior [12]. Furthermore, a direct position measuring system to determine the load position is necessary. Additional difficulties arise in the practical identification of the friction torque on the load side, which requires the disassembly of the mechanical coupling.

2.3 Multi-Encoder based disturbance observer (MEDOB)

As a third option, the multi-encoder based disturbance observer (figure 1, right) is considered, as already presented in [12, 14]. This approach is based again on the model concept of a two-mass system. The equations of motion of such a system are as follows:

$$J_m \frac{d\omega_m}{dt} = I_m K_t - T_{f,m} - T_{reac} \quad (4)$$

$$J_l \frac{d\omega_m}{dt} = T_{reac} - T_{f,l} - T_{dist} \quad (5)$$

The reaction torque T_{reac} is calculated as the torsional and damping forces transmitted via the elastic shaft:

$$T_{reac} = c(\varphi_m - \varphi_l) + d(\omega_m - \omega_l) \quad (6)$$

By combining equations (4), (5) and (6), the reaction torque is erased and a load side disturbance torque can be calculated according to the following estimation equation:

$$\hat{T}_{dist} \approx I_m K_{t,n} - J_{m,n} \frac{d\omega_m}{dt} - J_{l,n} \frac{d\omega_l}{dt} - \hat{T}_{f,m} - \hat{T}_{f,l} \quad (7)$$

The application of this structure also requires a encoder on the load side, but does not depend on the knowledge of the coupling parameters such as spring or damping constants. Compared to the DOB, this leads to increased estimation dynamics with simultaneously higher accuracy. With regard to the experimental determination of the influence of friction, suitable results may already be achieved with a total friction model in relation to the motor or load speed.

3 Simulation model

The investigation of the observer-based methods for the estimation of disturbance torques on mechatronic drive systems is performed in MATLAB® Simulink®. Based on a corresponding test rig with known parameters (figure 2), the modeling of the mechanical subsystem including a simulation of the friction conditions will be carried out. Subsequently, the control loops for speed and current are presented. For this paper, the simulation was performed with a cycle time of 62.5 µs.

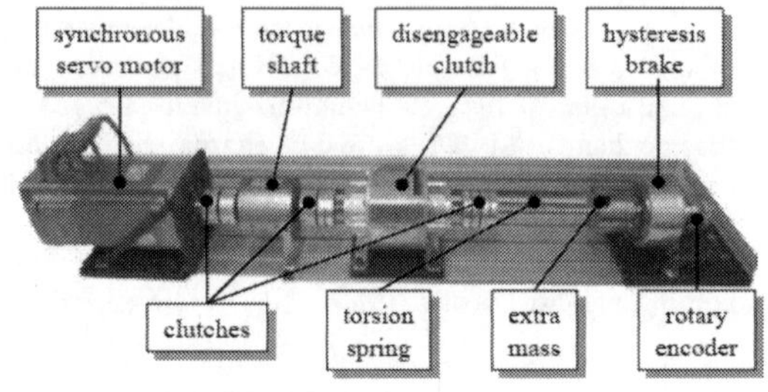

Fig. 2. Utilized drive system test rig.

3.1 Modeling of mechanics and friction

3.1.1 Modeling of the mechanical system

The transfer behavior of the mechanical subsystem of electromechanical axes is usually corresponding to a multi-mass mechanism. Depending on the mechanical complexity or the modeling quality, models of different complexity can be utilized to accurately reflect the real behavior.

A model widely used in control engineering for oscillating mechanics is the two-mass system, which was already mentioned in chapter 2. Furthermore, an additional load inertia $J_{l,2}$ can be coupled in the simulation. This leads to the model of a three-mass system. The elastic coupling of the additional load inertia is determined by the spring constant $c_{l1,l2}$ and the mechanical damping constant $d_{l1,l2}$. Higher-order models are rarely used in industrial practice or research. Reasons are the high manual modeling effort and the only slight increase in accuracy. The parameterization of the models can be found in table 1 and is based on [15]. A comparison of the two model approaches in the frequency response is shown in figure 3.

Table 1. Model parameters.

	Two-mass system	*Three-mass system*
$J_m\,(kg{\cdot}m^2)$	0.000869	0.000869
$J_{l,1}\,(kg{\cdot}m^2)$	0.000485	0.000485
$J_{l,2}\,(kg{\cdot}m^2)$	-	0.000685
$c_{m,l1}\,\left(\frac{Nm}{rad}\right)$	2150	2150
$c_{l1,l2}\,\left(\frac{Nm}{rad}\right)$	-	1800
$d_{m,l1}\,\left(\frac{Nms}{rad}\right)$	0.026	0.026
$d_{l1,l2}\,\left(\frac{Nms}{rad}\right)$	-	0.016

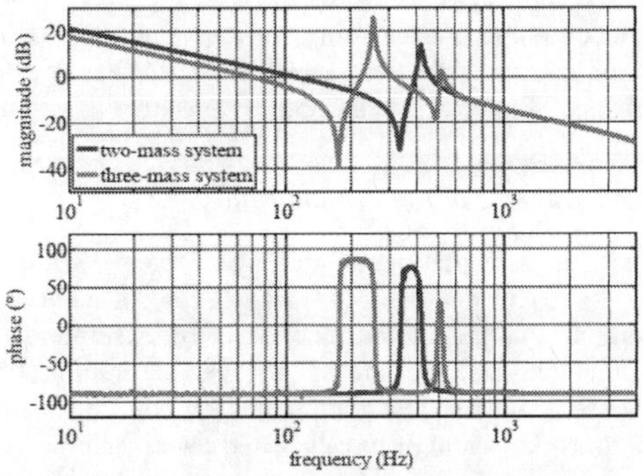

Fig. 3. Comparison of the models in the frequency response of the speed control plant.

3.1.2 Modeling of the friction conditions

In addition to the modeling of the transfer behavior of the mechanical system, the friction conditions at the test rig must be emulated. Therefore, experimental friction curves were recorded by repeated speed-controlled approaching of fixed speed points on the test rig. Subsequently, the average of these three curves was approximated utilizing a model approach as proposed in [16]. This approach represents the experimentally determined friction characteristic as a sum of four individual parts:

$$T_k = T_{f,c} + T_{f,lin} + T_{f,exp} + T_{f,log} \quad (8)$$

The first part consists of a constant, direction-dependent friction component T_c:

$$T_{f,c} = sgn(\omega) \cdot T_c \quad (9)$$

A positive linear part takes speed-dependent friction effects into account and extends the model by a factor σ to $T_{f,lin}$:

$$T_{f,lin} = \sigma \cdot \omega \quad (10)$$

The exponential friction torque component $T_{f,exp}$ as shown in equation (11) includes the transition between static and sliding friction in the low-speed range. Therefore, the parameter $\omega_{exp,0}$ defines the speed range in which the transition between static and sliding friction takes place. The form factor δ ultimately determines the exponential curve of the transition and can take positive and negative values for different friction surfaces. In addition, a static friction torque T_s must be taken into account, which describes the maximum frictional force at standstill.

$$T_{f,exp} = sgn(\omega) \cdot \left[(T_s - T_c) \cdot exp\left(- \left| \frac{\omega}{\omega_{exp,0}} \right|^{\delta} \right) \right] \quad (11)$$

The fourth friction component, which can be interpreted as rolling friction, is described as follows:

$$T_{f,log} = sgn(\omega) \cdot \left[T_{log} \cdot ln\left(\frac{|\omega|}{\omega_{log,0}} + 1 \right) \right] \quad (12)$$

Figure 4 shows the approximation of the measured friction conditions with the described model approach and its individual components.

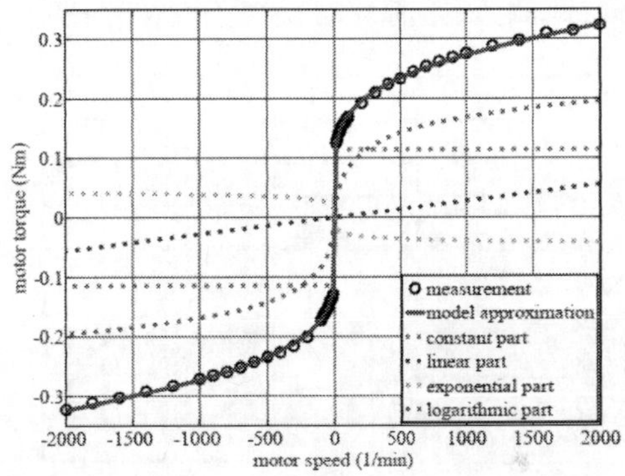

Fig. 4. Approximation of the experimental friction conditions and individual parts.

It can be stated that the approximation according to the proposed model approach simulates the friction conditions at the test rig very well and can therefore be utilized for the simulative investigations. Furthermore, by additional measurement of the motor side friction torque and its subsequent subtraction from the total friction torque, the friction influences on the load side for the two-mass system as well as the three-mass system can be mapped separately. This is of particular importance in the parameterization of the LDOB. The parameterization of the friction approximation for the mechanical models used in the simulation is given in table 2.

Table 2. Friction model parameters.

friction component	T_c *(Nm)*	$\sigma\left(\frac{Nms}{rad}\right)$	T_s *(Nm)*	$\omega_{exp,0}\left(\frac{rad}{s}\right)$	δ	T_{log} *(Nm)*	$\omega_{log,0}\left(\frac{rad}{s}\right)$
motor	0.1158	0.00026	0.0664	0.6560	-0.0098	0.0260	1.0900
$load_1$	-0.0042	0.000049	0.0014	1.000	-0.0062	0.0070	0.8813
$load_2$	-0.0042	0.000049	0.0014	1.000	-0.0062	0.0070	0.8813

3.2. Modeling of current and speed control

3.2.1 Supplementary model of the current control loop

The feedback control of mechatronic drive systems is usually implemented in a cascaded form. For this paper, a speed control consisting of the two control loops for current and rotational speed is assumed. In particular, the determination of the influence of the speed controller on the estimation of disturbance torques is one main goal of the paper. For this reason, an established supplementary model was used for simulating the current loop, as it can be found in [17]. Based on experimental measurements on the test rig, a second order model approach with additional dead time has proved to be a suitable approximation. In figure 5, the frequency response of the closed current control loop of the test rig and the utilized supplementary model is shown.

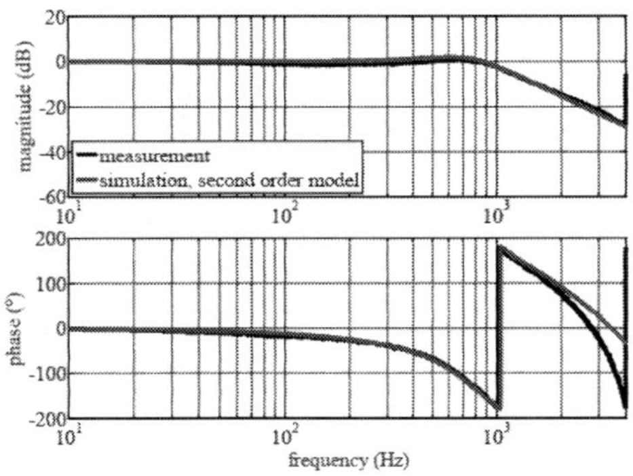

Fig. 5. Frequency response of the closed current loop and second-order model approximation.

As it can be seen, the model approach reflects the actual behaviour very well. Especially the course of the phase response is reproduced exactly over a wide frequency range. Furthermore, various additional current setpoint filters can be parameterized to attenuate possible resonance effects of the mechanical system.

3.2.2 Modeling of the speed control loop

In the course of this paper, the speed controller consists of a PI controller and the actual control values are provided by the motor encoder. Furthermore, it is possible to use an additional speed setpoint filter. Depending on the characteristics of the mechanical subsystem, different controller parameters (table 3) were examined, based on the investigations in [15]. The command and disturbance response in the frequency domain of all controller settings is shown in figure 6. Note that the disturbance in the right picture is acting as disturbance torque on the load side of the mechanical system.

Setting No. 1 was deliberately chosen conservatively in order to minimize the influence of the speed controller on the estimation results. In the command frequency response, the setting is characterized by a limited bandwidth. This is also reflected in the disturbance behavior. Disturbances acting on the load side are barely adjusted by the speed control over a wide frequency range. Setting No. 2 is based on the established adjustment rule according to the symmetrical optimum, which is characterized by a high bandwidth without noticeable magnitude amplification. This setting requires the utilization of an additional speed setpoint filter to avoid high overshoots in the time domain. Furthermore, a current setpoint filter was installed to reduce resonance effects. Its parameterization is based on automatic self-tuning procedures, which are commonly utilized by industrial distributed drive manufacturers. Concerning the disturbance behavior, setting No. 2 is capable of correcting disturbance influences over a relatively broad

Table 3. Speed Controller Settings.

No.	*Speed controller*		*Speed setpoint filter (SSPF)*	*Current setpoint filter (CSPF)*			
	K_p $\left(\frac{Nms}{rad}\right)$	T_n (s)	T_{SSPF} (s)	ω_{FCN} (Hz)	D_{FCN}	ω_{FCD} (Hz)	D_{FCD}
Two-mass system							
1	0.0802	0.00842	-	-	-	-	-
2	1.3532	0.002	0.002	418.18	0.0	418.18	0.25
3	2.0209	0.00552	-	418.18	0.2305	418.18	0.3982
Three-mass system							
4	1.6625	0.0025	0.0025	254.49	0.0	254.49	0.25
				520.18	0.0	520.18	0.25

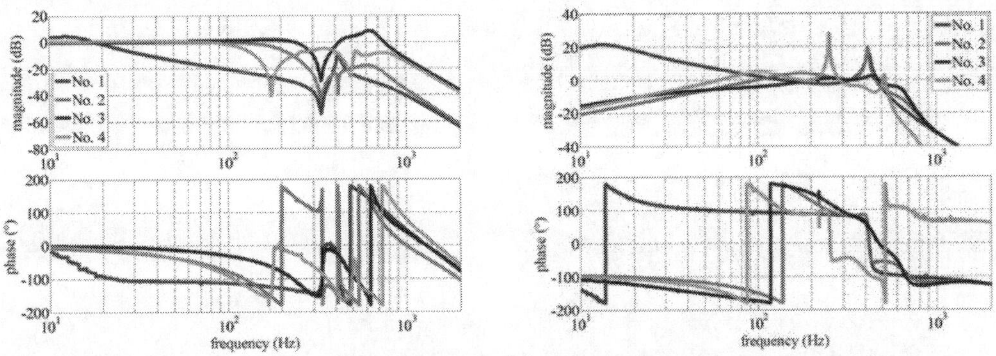

Fig. 6. Simulated command (left) and disturbance (right) frequency response for speed control settings No. 1-4.

frequency range up to approximately 100 Hz. However, in the region of the mechanical natural frequencies the disturbances are amplified. Setting No. 3 was tuned utilizing the simulation-based optimization presented in [15]. This method includes optimized parameters for the current setpoint filters. Therefore, in the command frequency response a significantly higher control dynamic can be achieved. For the disturbance frequency response, the advantages of the optimized controller parameters lead to no significant amplification in the resonant frequency range.

In order to allow comparability with the two-mass system, the symmetrical optimum was also utilized for the three-mass system (No. 4). Overall, the bandwidth of the speed control compared to the two-mass system as a result of the changed coupling parameters is significantly reduced.

4 Simulation results

In the following section, the three disturbance observers presented in chapter 2 are implemented in the simulation model and examined with regard to their performance in the time and frequency domain. For better comparability, the actual motor torque T_m,which is typically available within the drive-internal control system, is also considered as a benchmark. In addition to the variation of the speed controller setting, the influences of parameter changes of the mechanical system are investigated. This includes changes in load inertia and spring constant as well as the extension of the mechanical system to a three-mass system. Note that the low-pass filters of the observers mentioned in chapter 2 were not implemented.

4.1 Modification of the speed controller settings

At the beginning, the influence of changing speed controller settings on the disturbance estimation of a two-mass system with the parameters listed in table 1 is examined. For this purpose, the speed controller settings No. 1-3 shown in table 3 are utilized. The disturbance torque on the load side was implemented in the form of a pseudorandom bit stream (PRBS) with a magnitude of 2 Nm. To avoid static friction effects, a rotational speed offset of 50 1/min was specified on the setpoint side.

Figure 7 shows the frequency responses of the motor torque and the disturbance observers, respectively. In the left picture it can be seen that the application of the DOB already has advantages over a pure evaluation of the motor torque. Although a more dynamic controller setting like No. 3 contribute to an increase in the

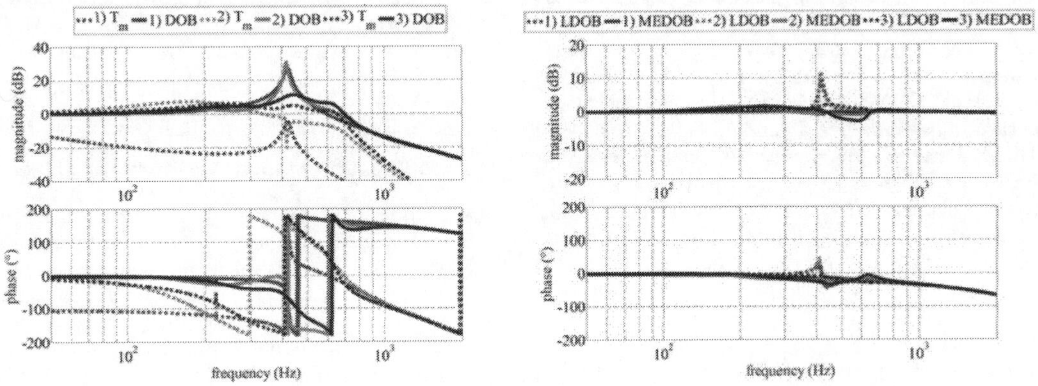

Fig. 7. Simulated frequency response for load side disturbance torque for speed controller settings No. 1-3.

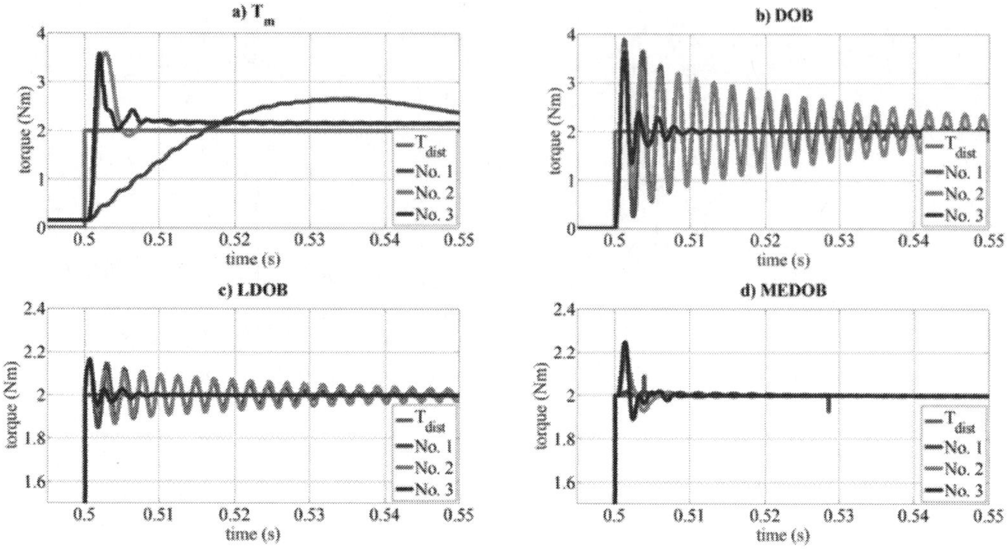

Fig. 8. Simulated step response for load-side disturbance torque for speed controller settings No. 1-3.

effective bandwidth of the motor torque, the existing deviations in the phase response are only partially compensated. In contrast, the DOB has no significant phase shift over a wider frequency range up to approximately 120 Hz for all controller settings. For the settings No. 1 and 2, the effects of the elastically coupled mechanics clearly appear in the magnitude characteristic. These effects can be partially compensated by the optimized speed controller settings (No. 3).

In the right picture, LDOB and MEDOB are compared. Note that due to the smaller deviations in the magnitude response its scaling was adjusted. Regardless of the controller parameterization, the estimation quality is enhanced for an even broader frequency range without significant phase shift for both observer types. In case of the LDOB, there is a noticeable increase in magnitude in the region of the natural frequency for runs No. 1 and 2. However, it is much lower compared to the DOB and can also be countered by an optimized controller setting as in run No. 3. For the MEDOB, the resonance point has no notable effects in the magnitude response even for controller settings No. 1 and 2. However, the more dynamic controller setting No. 3 leads to slight deviations in the range of higher frequencies, which is also expressed in the closed loop command frequency response of this setting (see figure 6).

The results in the frequency domain can be confirmed by applying a step disturbance torque and a subsequent examination in time domain. In figure 8 a-d, the simulation model is again operating with a speed offset of 50 1/min. At a simulation time of t_{sim} = 0.5 seconds, a step disturbance torque T_{dist} with a magnitude of 2 Nm was added on the load side.

The response of the motor torque in figure 8a shows significant differences depending on the controller setting. All three runs have in common that a constant deviation from the course of the disturbance torque due to unrecognized friction effects can be seen. In case of the DOB, the generally increased dynamics is also recognizable in the time domain. For the runs No. 1 and 2, the excitation of the natural frequency is clearly visible. The best results are achievable utilizing the simulation-based optimized setting No. 3. For the LDOB and MEDOB equally a significantly reduced overshoot and better damping of the oscillations occurs. While the optimized controller setting No. 3 leads to an improved disturbance estimation for the LDOB, a high estimation accuracy for setting No. 1 can be determined for the MEDOB.

In summary, an optimized setting of the speed control has certain advantages, in particular when utilizing the motor torque or the conventional DOB. However, improvements in the range of mechanical resonances can be achieved for the LDOB by means of optimized speed control settings. The MEDOB profits in particular from conservative or established speed control setting rules (e.g. symmetrical optimum) and overall appears much more robust than the remaining strategies. Note that the deviations of the LDOB and MEDOB result from the differentiation of the actual speed values. This effect would increase for higher sample times of the control loops.

4.2 Modification of load side moment of inertia

As mentioned in chapter 2, the nominal values for the moment of inertia are used for the parameterization of the observers. In practical applications, especially the values on the load side are not constant, for example due to changing workpieces or modifications of the mechanical configuration. In the frequency response of the mechanical system, this variation manifests in a vertical shift of the magnitude response in the range of frequencies smaller than the natural frequency and a

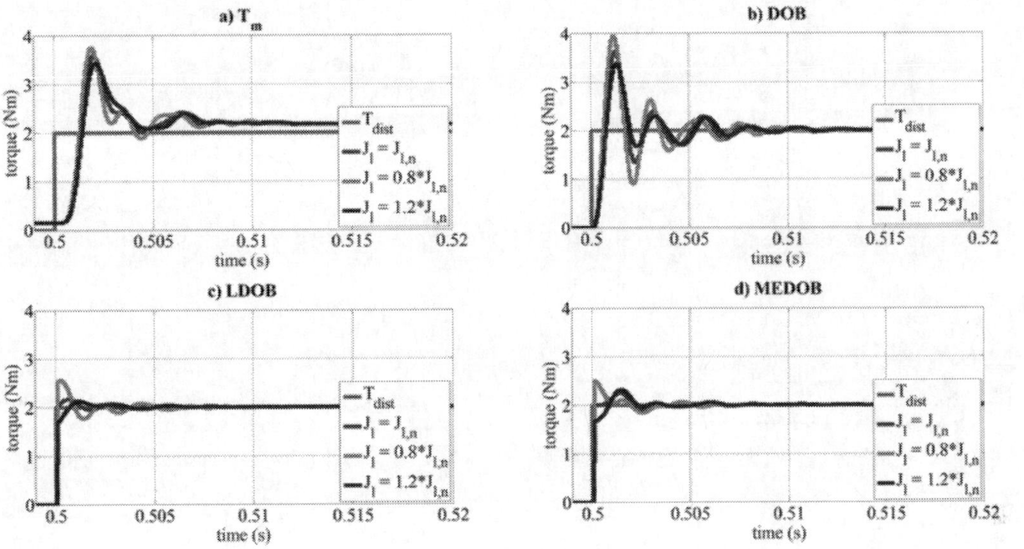

Fig. 9. Simulated step response for load-side disturbance torque with variation of load side inertia for speed controller settings No. 1-3.

horizontal displacement of the resonance points. Eventually, this has a direct impact on the quality of the drive control itself. Therefore, several methods for parameter identification [18] were developed in recent years which are able to detect these changes and adapt the controller parameters accordingly. However, these methods are currently rarely used in the industrial practice. For this reason, it is of particular interest how changes in the load side moment of inertia affect the estimation accuracy of the individual observers.

Figure 9 shows the step response of the observers as well as the motor torque with variation of the load side inertia to 80 or 120 percent of the nominal value. For all investigations, speed controller setting No. 3 was selected. To avoid disturbing static friction effects, again a speed offset of 50 1/min was specified.

For all the structures shown in figure 9 a-d, it can be stated that deviations in moment of inertia on the load side have only a minor influence on the estimation results. However, all methods have in common that a reduced load side moment of inertia (green curves) leads to higher overshoot and oscillations. This effect increases for higher deviations to the nominal values. In contrast, a higher moment of inertia (blue curves) leads to slightly reduced estimation dynamics, but smaller overshoots. Especially for the LDOB and MEDOB, there are only slight differences, which can be interpreted as improved robustness against variable load inertias. Note that larger deviations of the nominal inertia or a less robust setting of the speed controller parameters and current setpoint filters may have a negative impact on the estimation accuracy.

4.3 Modification of the spring constant

In addition to the process-related change of the load side inertia, it is possible that variations occur in the coupling parameters between motor and load side. For example, on linear feed axes (e.g. ball screw drive) changes in the load position or progressive wear of the mechanical elements (e.g. loss of pre-tension) affect the rigidity [19]. In the frequency response of the mechanical system, this variation manifest in a shift of the resonant and antiresonant frequency in the abscissa direction. Figure 10 a-d show the reaction of the motor torque and the observer structures again utilizing speed controller setting No. 3. The variation of the spring constant amounts 80 or 120 percent of the nominal value, respectively. As before, a step disturbance with a magnitude of 2 Nm at the simulation time t_{sim} = 0.5 seconds was applied on the load side. To neglect static friction effects, a speed offset of 50 1/min was predetermined.

For the course of the motor torque and DOB (Figure 10a and b) the variation of the spring constant ultimately leads to a modified estimation dynamic. A more rigid elastic coupling (blue curves) leads to a faster response of the disturbance magnitude and simultaneous only slight changes in overshoot. For the LDOB, a change in the rigidity directly affects the estimation results. Thereby, an increased spring constant generates static lower estimations, while their reduction leads to too high approximations. In contrast, the variation of the spring constant has no real impact on the MEDOB, which can be justified by reference to equations (4), (5), (6) and (7). Smaller deviations between the curves again result from the differentiation of the rotational speed of motor and load side.

4.4 Modification of the mechanical system

In industrial practice, the mechanical subsystems are characterized by several elastically coupled masses of

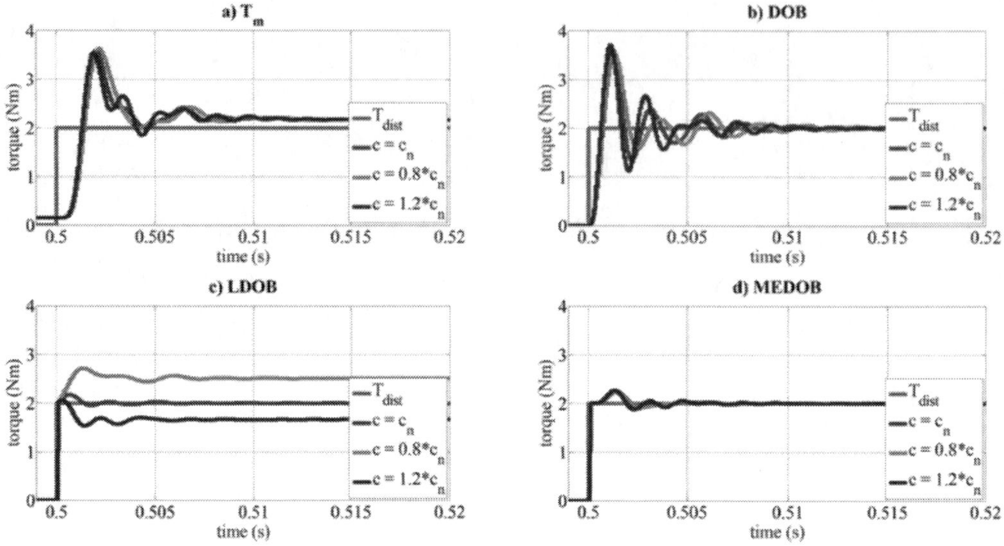

Fig. 10. Simulated step response for load-side disturbance torque with variation of the spring constant for speed controller settings No. 1-3.

inertia. Therefore, the simplification of the real system in the form of a two-mass is not always durable. In order to investigate the robustness of the observer structures more precisely, a second load inertia is implemented in the simulation model. The corresponding simulation parameters refer to table 1 and table 2, respectively. For the LDOB, the additional load side moment of inertia $J_{l,2}$ was utilized as nominal value $J_{l,n}$. The nominal spring constant c_n was calculated according to equation (13):

$$\frac{1}{c_n} = \frac{1}{c_{m,l1}} + \frac{1}{c_{l1,l2}} \tag{13}$$

For the MEDOB, it has proven to be beneficial to distribute the moment of inertia $J_{l,1}$ in equal parts on the motor and load side. Therefore, $J_{m,n}$ and $J_{l,n}$ are calculated as follows:

$$J_{m,n} = J_m + \frac{J_{l,1}}{2} \tag{14}$$

$$J_{l,n} = J_{l,2} + \frac{J_{l,1}}{2} \tag{15}$$

For the investigations in the time and frequency domain, the same disturbance parameters as for the two-mass system are applied. Note that the parametrization of the speed controller and current setpoint filters is limited to setting No. 4 (table 3). The corresponding response behavior in the frequency and time domain is shown in figure 11.

As before, the DOB has only a slight phase shift compared to the motor torque over a wider frequency range (up to about 70 Hz). However, the course of the magnitude is characterized by distinct peaks in the region of the mechanical resonance points. In comparison to the results of the two-mass system, slight deviations in magnitude are already recognizable for the DOB even in lower frequency ranges. In the time domain, the excitation of the mechanical natural

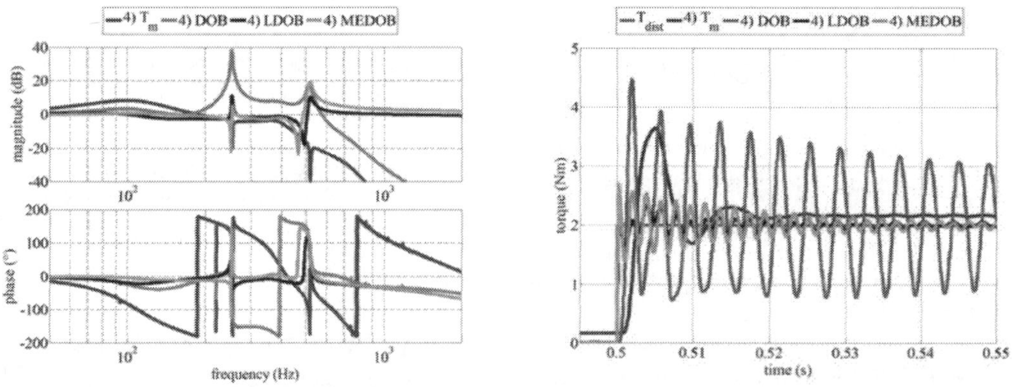

Fig. 11. Simulated response for load-side disturbance torque in frequency and time domain for speed controller setting No. 4.

frequencies can be clearly recognized for a step disturbance, which leads to distinct and slightly attenuated oscillations. This could be prevented by a suitable filtering as mentioned in chapter 2. However, this may have a negative impact on the dynamics of the estimation.

The smallest deviations in magnitude are obtained by the LDOB, which, however, start in lower frequency ranges at about 100 Hz. In the time domain, the oscillations are significantly less distinct compared to the DOB and decrease faster. When comparing the results with those of the two-mass system, the quality of the estimation is only slightly affected by the modified mechanical structure.

Again, the MEDOB achieves suitable estimation results compared to the other approaches. In the frequency domain, only slight deviations are detectable in the magnitude and phase response up to 200 Hz. The excitation of the mechanical resonance frequency at 518 Hz is also recognizable in the time domain by significant oscillations in case of a step disturbance excitation. However, these oscillations decline much faster compared to the DOB. Again, it is possible to further improve the quality of the estimation by an appropriate filtering as mentioned in chapter 2.

5 Conclusion & Outlook

The application of model-based approaches in the form of disturbance observers is suitable for estimating load side disturbances on speed-controlled mechatronic drive systems. Depending on the complexity of the observer structure and the characteristics of the drive system, results of different quality are achievable. To investigate the individual structures, a simulation model was created and parameterized based on experimental measurements on a drive test rig. It enables the emulation of mechanical systems with variable model order and coupling parameters as well as different settings of the speed controller.

For disturbance estimation, a solitary evaluation of the motor torque has turned out as the simplest option, but it particularly depends on the setting of the speed control and is not able to take friction effects into account. In contrast, the conventional disturbance observer has clear advantages in terms of estimation dynamics. For elastically coupled mechanical systems, however, its application is not recommended. The LDOB achieves very good estimation results regardless of the mechanical subsystem and speed controller settings. However, it requires the exact knowledge of the spring constant as well as the friction torque of the load side, which is difficult to identify especially in industrial applications. An appropriate compromise is provided by the MEDOB, which obtains a high estimation accuracy, especially for two-mass systems. Furthermore, the quality of the estimation is barely affected by changes in the coupling parameters. Merely mechanical systems of higher order have a limiting effect, whereby undesired oscillations should be prevented with a suitable signal filtering.

Future investigations should focus on the enhancement of the simulation model accuracy including consequential effects on the observer structures. For example, the detailed modeling of the current control or the implementation of quantized encoder signals should be considered. In addition, the influence of appropriate signal filters to avoid unwanted oscillations caused by the elastic coupling of the inertia masses have to be investigated. For example, a separate filtering of the input and output signals of the observers offers further improvements. Another research goal is the application on position-controlled, linear feed axes. An experimental verification, for example on a machine tool, is also necessary.

European Union
European Regional Development Fund
European Social Fund
Europe funds Saxony.
EFRE
ESF

Funded by the European Union (European Social Fund) and the Free State of Saxony.

References

1. B. Bindu, B. Vinod. Measurement of Cutting Forces in CNC Turning Centers: A Review, *International. Journal of Mechanical Engineering*, **3**(5): 77-87, (2015)
2. M. Rizal, J. Ghani, M. Nuawi, C. Haron. A Review of Sensor Systems and Application in Milling Processes for Tool Condition Monitoring, *Research Journal of Applied Sciences, Engineering and Technology*, **7**(10): 2083-2097 (2014)
3. D. Dimla. Sensor Signals for Tool-Wear Monitoring in Metal Cutting Operations – a Review of Methods, *International Journal of Machine Tools and Manufacturing*, **40**: 1073-1098 (2000)
4. J. Stein, K. Shin. Current Monitoring of Controlled DC Spindle Drives, *Journal of Dynamic Systems, Measurement and Control*, **108**: 189-295 (1986)
5. Y. Altintas. Prediction of Cutting Forces and Tool Breakage in Milling from Feed Drive Current Measurements, *Journal of Engineering for Industry*, **114**: 386- 392 (1992)
6. G. Kim, N. Chong. Indirect Cutting Force Measurement Considering Frictional Behaviour in a Machining Center using Feed Motor Current, *International Journal of Advanced Manufacturing Technology*, **15**: 478-484 (1999)
7. R. Sato, M: Hasegawa, K. Shirase. Cutting Force Monitoring based on the Frequency Analysis of Feed Motor Torques, Journal of SME Japan (2013)
8. T. Kim, J. Woo, D. Shin, J. Kim. Indirect Cutting Force Measurement in multi-axis simultaneous NC Milling Processes, *International Journal of Machine Tools and Manufacture*, **39**: 1717-1731 (1999)
9. G. Pritschow, J. Bretschneider, S. Fritz. Reconstruction of Process Forces within Digital

Servodrive Systems, *Annals of the German Academic Society for Production Engineering*, **5**(1): 73-78 (1999)

10. T. Murakami, F. Yu, K. Ohnishi. Torque Sensorless Control in Multidegree-of-Freedom Manipulator, *IEEE Transactions on Industrial Electronics*, **40**(2): 259-265 (1993)
11. Y. Kakinuma, Y. Sudo, T. Aoyama. Detection of chatter vibration in end milling applying disturbance observer, *CIRP Annals – Manufacturing Technology*, **60**: 109-112 (2011)
12. C. Mitsantisuk, M. Nandayapa, K. Ohnishi, S. Katsura. Design for Sensorless Force Control of Flexible Robot using Resonance Ratio Control based on Coefficient Diagram Method, *Journal for Control, Measurement, Electronics, Computing and Communications*, **54**(1): 62-73 (2013)
13. S. Katsura, J. Suzuki, K. Ohnishi. Pushing Operation by flexible Manipulator Taking Environmental Information into Account, *IEEE Transactions on Industrial Electronics*, **53**(5): 1688-1697 (2006)
14. Y. Yamada, Y. Kakinuma. Sensorless Cutting force estimation for full-closed controlled ball-screw-driven stage, *International Journal of Advanced Manufacturing Technology*, **87**: 3337-3348 (2016)
15. K. Hipp. Einsatz hybrider Optimierungsverfahren zur Inbetriebnahme elektromechanischer Systeme, *Berichte aus dem IWU*, **98** (2017)
16. M. Kunc. Identifikation und Modellierung von nichtlinearen Dämpfungseffekten in Vorschubachsen für Werkzeugmaschinen, *Ergebnisse aus der Produktionstechnik*, **11** (2013)
17. S. Hofmann. Identifikation parametrischer Modelle für geregelte, elektromechanische Achsen mit modifizierter, sukzessiver Polkompensation, *Berichte aus dem IWU*, **69** (2012)
18. A. Hellmich, K. Hipp, H. Schlegel, R. Neugebauer. Parameter identification of NC-axes during regular operation of a machine tool, *Advanced Materials Research*, **1018**: 419-426 (2014)
19. D. Maier. Sensorlose online Zustandserfassung von Vorschubantriebskomponenten in Werkzeugmaschinen, *Berichte aus dem institute für Maschinenelemente*, **157** (2015)

Vibration and cooling performances of piezoelectric cooling fan: numerical and experimental investigations

Thomas Jin-Chee Liu*, *Yu-Shen* Chen, *Hsi-Yang* Ho, and *Jyun-Ting* Liu

Department of Mechanical Engineering, Ming Chi University of Technology, Taishan, New Taipei City, Taiwan

Abstract. In this paper, the vibration and cooling performances of the piezoelectric cooling fan are studied. The finite element and experimental methods are adopted for the analyses. The natural frequency, mode shape, flapping amplitude, and cooling performance are discussed for the primary design. The numerical results have good agreement with the experimental measurements. For the cooling purpose, the piezoelectric cooling fan has to work under the natural frequency. The aspect ratio 2:3 is the optimal geometry of the fan blade.

1 Introduction

As shown in Fig. 1, the piezoelectric cooling fan uses the flapping blade to make the air flow for the electronic cooling. It is quite different from the traditional rotating cooling fan. Using the piezoelectric actuator, the piezoelectric effect makes the cyclic deformation so that the blade can vibrate and flap. Under the natural frequency, the piezoelectric cooling fan can perform the maximum flapping displacement and air flow.

In the past references [1-6], it is concluded that the piezoelectric cooling fan consumes lower electric power. Also, it has good cooling performance for the small-size electronic devices. Undoubtedly, the piezoelectric cooling fan is a good cooling solution for many electronic products.

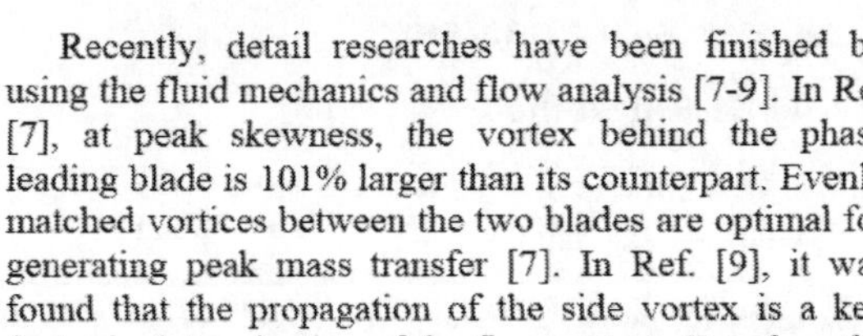

Fig. 1. Working condition of piezoelectric cooling fan.

Recently, detail researches have been finished by using the fluid mechanics and flow analysis [7-9]. In Ref. [7], at peak skewness, the vortex behind the phase leading blade is 101% larger than its counterpart. Evenly matched vortices between the two blades are optimal for generating peak mass transfer [7]. In Ref. [9], it was found that the propagation of the side vortex is a key factor in determination of the flow pattern. Based on the numerical results, it was found that whether the side vortex propagates or not is an important factor in determining three-dimensional flow pattern around the piezoelectric fan [9].

In this paper, the vibration and cooling performances of the piezoelectric cooling fan will be studied. The numerical and experimental methods are adopted for the analyses. The natural frequency, mode shape, flapping amplitude, and cooling performance will be discussed for the primary design. Based on the vibration mechanics and temperature reduction, the present study will find the optimal geometry of the fan blade and associated conditions.

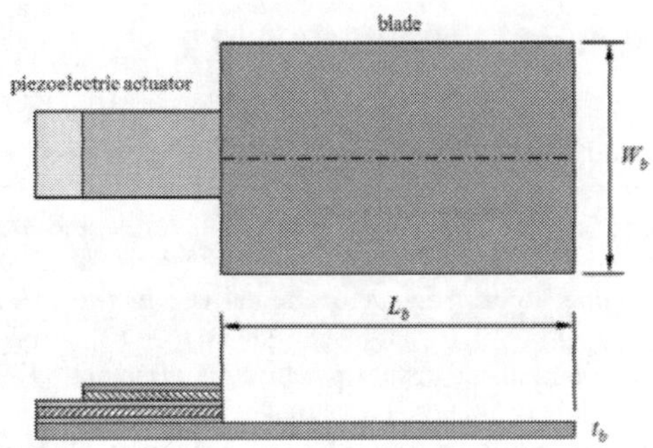

Fig. 2. Geometry of piezoelectric cooling fan.

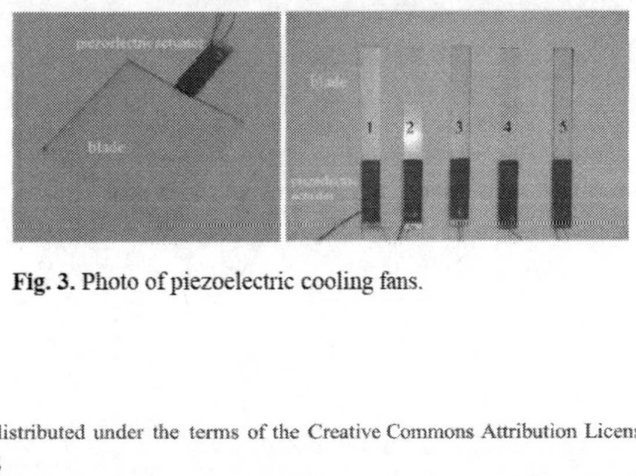

Fig. 3. Photo of piezoelectric cooling fans.

---
* Corresponding author: jinchee@mail.mcut.edu.tw

2 Definition of research

The geometry and photo of the piezoelectric cooling fan are shown in Fig. 2 and 3. It contains two main parts: the blade and piezoelectric actuator. The dimensions of the rectangular blade (W_b, L_b, t_b) are variables and will be studied for the vibration and cooling performances.

In Fig. 4, the detail geometry of the bimorph piezoelectric actuator is illustrated. All dimensions are constant values. The piezoelectric material type is PZT-5H, i.e. a kind of the piezoelectric ceramic. The *P*-arrows define the poling directions. Three electrodes are attached to the piezoelectric material. Due to relatively thin value, the thickness of the thin-film electrode is ignored in the numerical analysis.

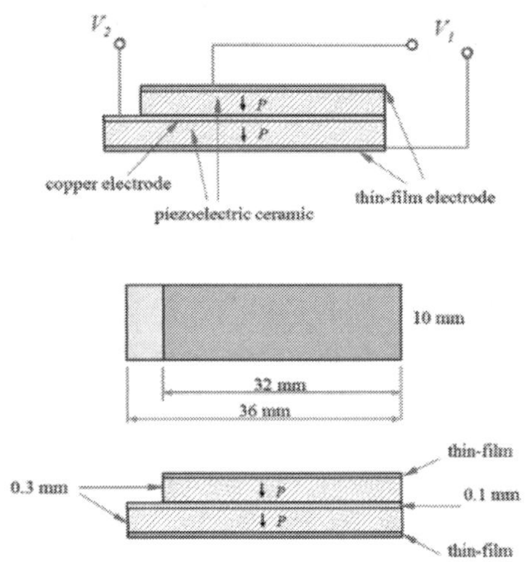

Fig. 4. Geometry of piezoelectric actuator.

As shown in Fig. 4, the piezoelectric material is actuated by the electric field between two electrodes. When the voltage difference (V_2-V_1) is applied on both electrodes, the piezoelectric actuator deforms and bends due to the electric field and poling direction. For making the cyclic deformation, the alternative current (AC) is applied on two thin-film electrodes as $V_1 = V_{amp}\sin(2\pi ft)$ V. In addition, the copper electrode is grounded ($V_2 = 0$ V). In above equation, the parameters f and t are the AC frequency and time, respectively. Thr AC amplitude is V_{amp}=100 V.

The electro-mechanical piezoelectric behavior is ruled by the constitutive equation under the Cartesian coordinate x_1-x_2-x_3 as follows [10]:

$$\begin{Bmatrix} T_{11} \\ T_{22} \\ T_{33} \\ T_{23} \\ T_{13} \\ T_{12} \\ D_1 \\ D_2 \\ D_3 \end{Bmatrix} = \begin{bmatrix} c_{11} & c_{12} & c_{13} & 0 & 0 & 0 & 0 & 0 & e_{31} \\ & c_{11} & c_{13} & 0 & 0 & 0 & 0 & 0 & e_{31} \\ & & c_{33} & 0 & 0 & 0 & 0 & 0 & e_{33} \\ & & & c_{44} & 0 & 0 & 0 & e_{15} & 0 \\ & & & & c_{44} & 0 & e_{15} & 0 & 0 \\ & & & & & c_{66} & 0 & 0 & 0 \\ & & sym. & & & & -\varepsilon_{11} & 0 & 0 \\ & & & & & & & -\varepsilon_{11} & 0 \\ & & & & & & & & -\varepsilon_{33} \end{bmatrix} \begin{Bmatrix} S_{11} \\ S_{22} \\ S_{33} \\ 2S_{23} \\ 2S_{13} \\ 2S_{12} \\ -E_1 \\ -E_2 \\ -E_3 \end{Bmatrix} \quad (1a)$$

$$c_{66} = \frac{1}{2}(c_{11} - c_{12}) \quad (1b)$$

where T, S, D, E, c, e and ε are the stress, strain, electric displacement, electric field, elastic constant, piezoelectric constant, and permittivity, respectively. In Eq. (1), the poling direction of the piezoelectric material is along the x_3-axis.

The piezoelectric material is PZT-5H. Two materials, PET and PC plastics, are used to make the blade. Table 1 and 2 list all material data for the numerical simulations.

Table 1. Material data of PZT-5H [10].

Material : PZT-5H		
Stiffness constants ($\times 10^{10}$ N/m^2)	c_{11}	12.6
	c_{12}	7.91
	c_{13}	8.39
	c_{33}	11.7
	c_{44}	2.3
	c_{66}	2.35
Piezoelectric constants (C/m^2)	e_{31}	-6.5
	e_{33}	23.3
	e_{15}	17
Permittivity ($\times 10^{-8}$ C/ Vm)	ε_{11}	1.505
	ε_{33}	1.302
Density (kg/m^3)	ρ	7500

Table 2. Material data of isotropic materials.

Material	Young's modulus E (GPa)	Poisson's ratio v	Density ρ (kg/m^3)
Copper	110	0.33	8800
PC plastics	2.803	0.32	1205
PET plastics	5.384	0.242	1420

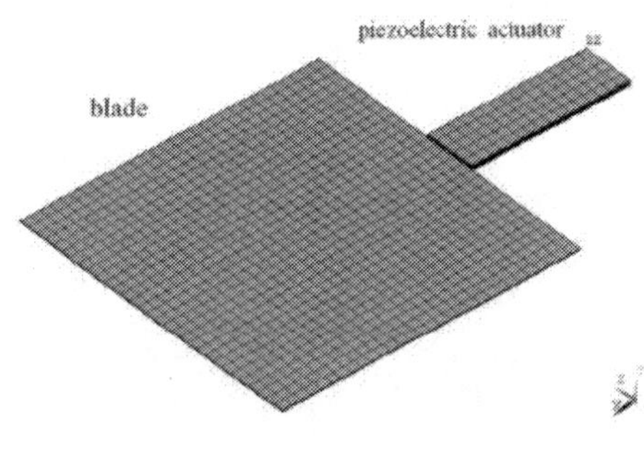

Fig. 5. Finite element model of piezoelectric cooling fan.

3 Methods of analyses

3.1 Numerical method

In this study, the numerical method is the finite element method. Also, the software ANSYS is employed. In Fig. 5, it shows a typical finite element model of the piezoelectric cooling fan. The model composes of SOLID226 and SOLID95 elements. The SOLID226 elements have the coupled-field analysis so that the piezoelectric effect can be simulated. In ANSYS, the modal analysis is used to calculate the natural frequencies and mode shapes of the vibration bahaviors.

3.2 Experimental method

In Figs. 6-8, the experimental equipments for the flapping amplitude and cooling performance of the piezoelectric cooling fan are shown. The AC power drives the piezoelectric cooling fan so that the air flow can be produced for the cooling process. The laser displacement sensor and thermo-meter are used to obtain the data of the flapping amplitude (δ_t) and cooling performance (temperature reduction of the heatsink). The distance between the blade and heatsink is 10 mm.

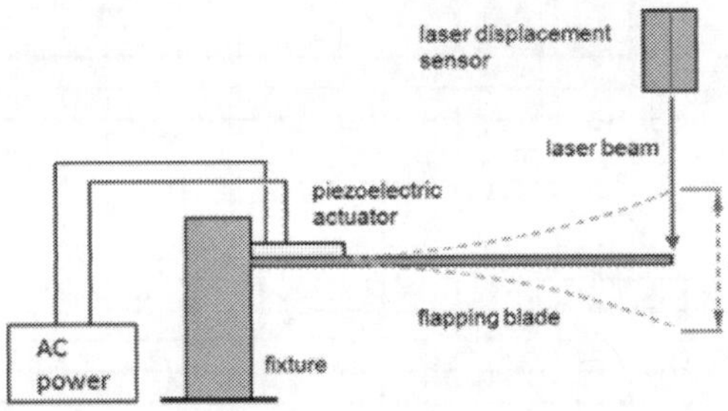

Fig. 6. Experimental equipments for flapping amplitude.

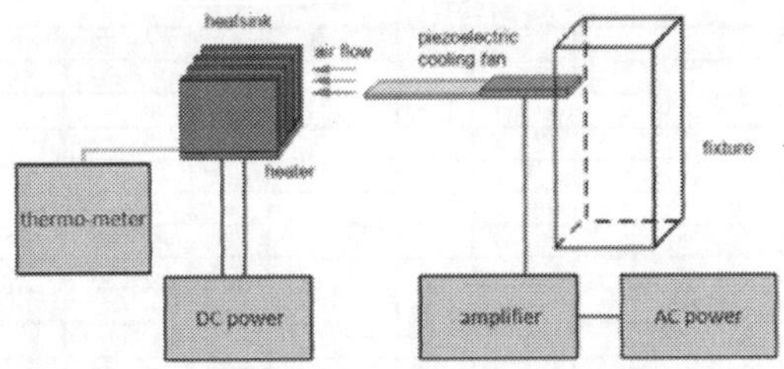

Fig. 7. Experimental equipments for cooling performance.

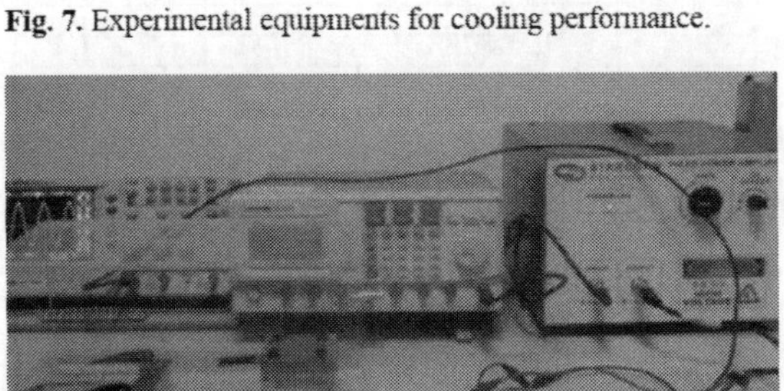

Fig. 8. Photo of experimental equipments.

4 Results and discussions

4.1 Vibration behaviours

In this section, the dimensions of the PET blade are W_b=10 mm, L_b=60 mm, and t_b=0.25 mm. The natural frequency is determined according to numerical modal analysis and experimental maximum flapping amplitude. In Fig. 9 and 10, the finite element and experimental results show the mode shapes and natural frequencies of the piezoelectric cooling fan. Both numerical and experimental values are almost the same. It confirms the good accuracy of the research methods. Also, the finite element method can predict the vibration behaviour of the piezoelectric cooling fan before the real cooling system is established.

The maximum flapping amplitude occurs under the natural frequency, i.e. the resonance condition. The larger flapping amplitude makes larger air flow and better cooling performance.

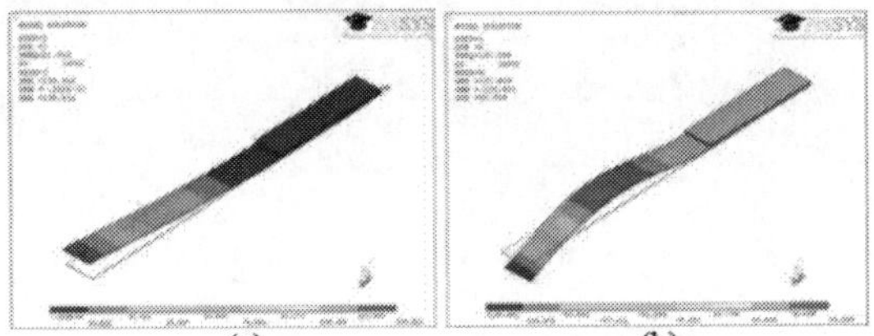

(a) (b)

Fig.9. Numerical mode shapes. (a) first-mode natural frequency = 21.853 Hz, (b) second-mode natural frequency= 135.094 Hz.

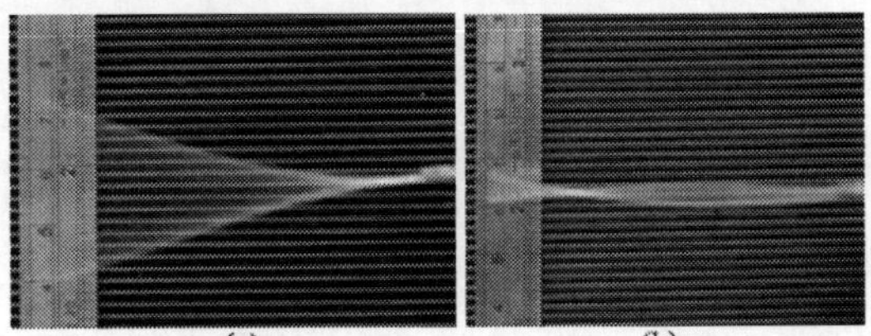

(a) (b)

Fig. 10. Experimental mode shapes. (a) first-mode natural frequency= 21.9 Hz (b) second-mode natural frequency= 134.8 Hz. (V_{amp}=100 V)

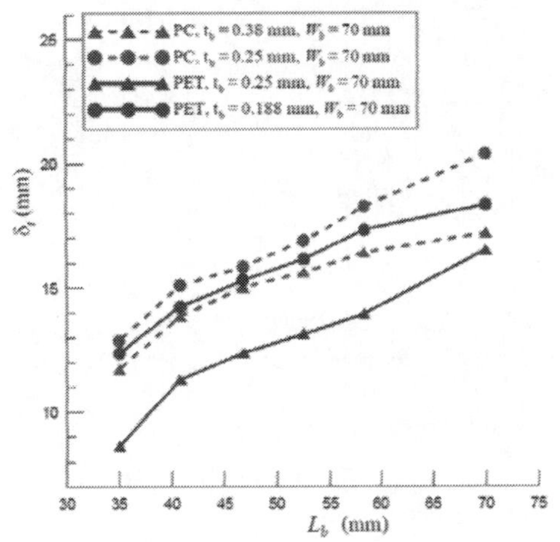

Fig. 11. Variation of δ_t with different geometric properties.

4.2 Flapping amplitude

According to Fig. 10, the flapping amplitude of the first-mode vibration is larger than that of the second-mode. In this study, the first-mode flapping amplitude δ_t (as shown in Fig. 6 and 10a) is adopted for the electronic cooling. Also, the amplitude δ_t is obtained under the first-mode natural frequency and V_{amp}=100 V.

The variations of δ_t with different geometric properties are shown in Fig. 11 and 12. The wider, shorter and thicker blades have smaller δ_t. The larger flapping amplitude makes larger air flow and better cooling performance. The amplitude δ_t is an important value for the piezoelectric cooling fan design.

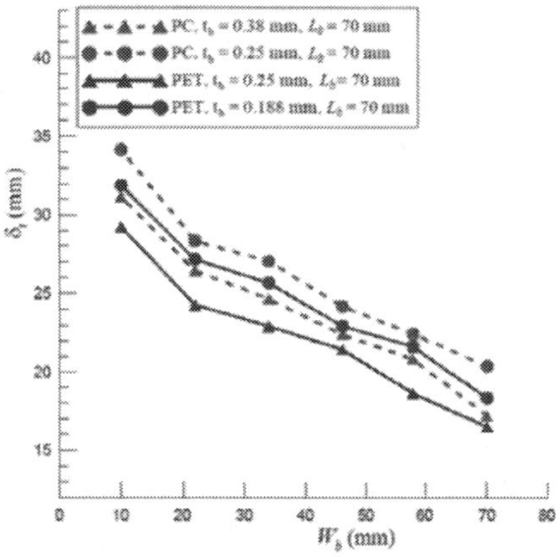

Fig. 12. Variation of δ_t with different geometric properties.

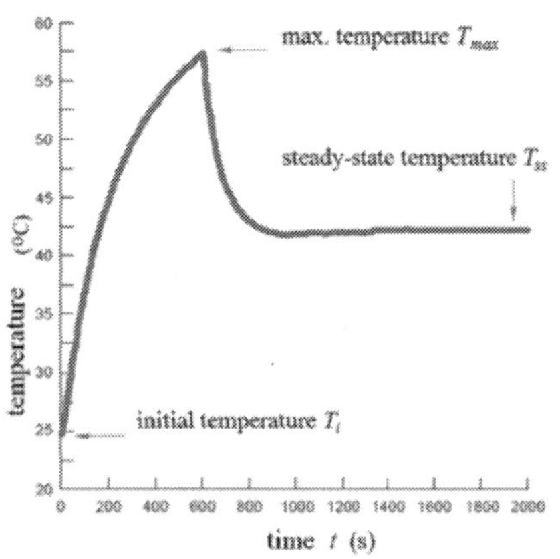

Fig. 13. Typical temperature history.

4.3 Cooling performance

In Fig. 7 and 8, it shows the experimental configuration for the cooling performance. To prevent from environmental disturbance, the transparent structure covers the heat source and piezoelectric cooling fan. The temperature at the heatsink bottom is measured by the thermo-meter. Figure 13 shows the typical temperature history. During t= 0~600 s, the DC power and heater provide the thermal energy and temperature rise from T_i to T_{max}. At t=600 s, the piezoelectric cooling fan begins to work under the first-mode natural frequency. The steady-state temperature T_{ss} is obtained at t=2000 s. The temperature reduction ΔT of the heatsink is defined as T_{max}−T_{ss}. The larger value of ΔT expresses the better cooling performance.

In Table 3~6, the ΔT values of different blades are listed. Four cases: 46.8×70 mm, 40×60 mm, 33.4×50 mm, and 27×40 mm have largest ΔT values, i.e. best cooling performance. These optimal cases have the same geometric property: $L_b/W_b \approx 2/3$.

Table 3. Cooling performances.

PET, t_b = 0.25 mm							
$L_b \times W_b$	ΔT	$L_b \times W_b$	ΔT	$L_b \times W_b$	ΔT	$L_b \times W_b$	ΔT
(mm)	(°C)	(mm)	(°C)	(mm)	(°C)	(mm)	(°C)
70 × 70	14.1	60 × 60	16.3	50 × 50	17.4	40 × 40	19.4
70 × 58	13.6	60 × 50	15.2	50 × 42	15.8	40 × 34	18.4
70 × 46	12.8	60 × 40	14.4	50 × 34	14.6	40 × 28	17.8
70 × 34	12.4	60 × 30	12.6	50 × 26	13.4	40 × 22	16.7
70 × 22	9.7	60 × 20	11.3	50 × 18	12.9	40 × 16	15.9
70 × 10	6.8	60 × 10	9.1	50 × 10	10.1	40 × 10	14.8
70 × 70	14.1	60 × 60	16.3	50 × 50	17.4	40 × 40	19.4
58.4 × 70	15.5	50 × 60	18.2	41.7 × 50	18.3	33.5 × 40	20.2
52.5 × 70	16.6	45 × 60	18.5	37.5 × 50	20.2	30 × 40	22.3
46.8 × 70	18.4	**40 × 60**	20.1	**33.4 × 50**	21.6	**27 × 40**	23.6
40.8 × 70	17.5	35 × 60	19.7	29.2 × 50	20.3	23.3 × 40	21.1
35 × 70	16.1	30 × 60	18	25 × 50	18.7	20 × 40	19.2

Table 4. Cooling performances.

PET, t_b = 0.188 mm							
$L_b \times W_b$	ΔT	$L_b \times W_b$	ΔT	$L_b \times W_b$	ΔT	$L_b \times W_b$	ΔT
(mm)	(°C)	(mm)	(°C)	(mm)	(°C)	(mm)	(°C)
70 × 70	9.6	60 × 60	12.1	50 × 50	14.7	40 × 40	17.1
70 × 58	8.8	60 × 50	11.6	50 × 42	13.2	40 × 34	16.6
70 × 46	8.2	60 × 40	11.1	50 × 34	12.8	40 × 28	15.7
70 × 34	7.3	60 × 30	10.5	50 × 26	12.2	40 × 22	13.6
70 × 22	6.6	60 × 20	9.3	50 × 18	11.4	40 × 16	12.3
70 × 10	5.3	60 × 10	8.8	50 × 10	9.8	40 × 10	11.8
70 × 70	9.6	60 × 60	12.1	50 × 50	14.7	40 × 40	17.1
58.4 × 70	12	50 × 60	13.9	41.7 × 50	16.2	33.5 × 40	18
52.5 × 70	13.9	45 × 60	15.8	37.5 × 50	17.9	30 × 40	19.2
46.8 × 70	16.4	**40 × 60**	17.8	**33.4 × 50**	18.8	**27 × 40**	21.4
40.8 × 70	15.5	35 × 60	16.6	29.2 × 50	18.1	23.3 × 40	20.2
35 × 70	14.7	30 × 60	16.1	25 × 50	17.4	20 × 40	18.3

Table 5. Cooling performances.

PC, t_b = 0.25 mm							
$L_b \times W_b$	ΔT	$L_b \times W_b$	ΔT	$L_b \times W_b$	ΔT	$L_b \times W_b$	ΔT
(mm)	(°C)	(mm)	(°C)	(mm)	(°C)	(mm)	(°C)
70 × 70	10.8	60 × 60	11.9	50 × 50	14.7	40 × 40	18.1
70 × 58	10.1	60 × 50	11.1	50 × 42	12.8	40 × 34	16.8
70 × 46	9.6	60 × 40	10.7	50 × 34	11.9	40 × 28	16.1
70 × 34	8.8	60 × 30	9.8	50 × 26	11.1	40 × 22	15.3
70 × 22	7.3	60 × 20	8.3	50 × 18	10.2	40 × 16	13.1
70 × 10	5.6	60 × 10	6.2	50 × 10	8.6	40 × 10	11.9
70 × 70	10.8	60 × 60	11.9	50 × 50	14.7	40 × 40	18.1
58.4 × 70	11.9	50 × 60	14.6	41.7 × 50	18	33.5 × 40	18.9
52.5 × 70	14.8	45 × 60	17.9	37.5 × 50	18.7	30 × 40	20.5
46.8 × 70	18.4	**40 × 60**	19.4	**33.4 × 50**	19.9	**27 × 40**	22
40.8 × 70	16	35 × 60	18.1	29.2 × 50	19.2	23.3 × 40	20.1
35 × 70	15.4	30 × 60	17.3	25 × 50	16.4	20 × 40	17.9

Table 6. Cooling performances.

PC, t_b = 0.38 mm							
$L_b \times W_b$	ΔT	$L_b \times W_b$	ΔT	$L_b \times W_b$	ΔT	$L_b \times W_b$	ΔT
(mm)	(°C)	(mm)	(°C)	(mm)	(°C)	(mm)	(°C)

70 × 70	14.9	60 × 60	17	50 × 50	18.1	40 × 40	19.6
70 × 58	13.8	60 × 50	16.4	50 × 42	16.8	40 × 34	18.6
70 × 46	13.2	60 × 40	15.5	50 × 34	15.6	40 × 28	17.9
70 × 34	12.4	60 × 30	14.3	50 × 26	14.7	40 × 22	16.9
70 × 22	10.9	60 × 20	13.1	50 × 18	13.3	40 × 16	16.1
70 × 10	9.2	60 × 10	11	50 × 10	11.9	40 × 10	15.3
70 × 70	14.9	60 × 60	17	50 × 50	18.1	40 × 40	19.6
58.4 × 70	17	50 × 60	18.6	41.7 × 50	21.6	33.5 × 40	20.8
52.5 × 70	19.3	45 × 60	19.7	37.5 × 50	22.9	30 × 40	23.2
46.8 × 70	20.4	40 × 60	20.9	33.4 × 50	24.8	27 × 40	25.6
40.8 × 70	19.4	35 × 60	20.1	29.2 × 50	22.6	23.3 × 40	23.5
35 × 70	16.7	30 × 60	18.9	25 × 50	21.2	20 × 40	20.1

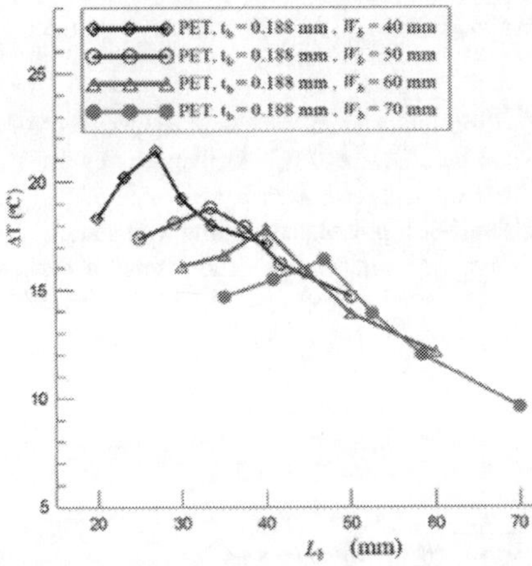

Fig. 14. Variation of ΔT with different geometric properties.

In Fig. 14, four curves of ΔT variations are plotted according to the data in Table 4. When $L_b/W_b \approx 2/3$, it shows the largest ΔT value in each curve. The wider blade makes larger air flow. Shorter, thicker and stiffer blades have higher natural frequency. The above reasons dominate the cooling performance.

5 Conclusions

The vibration and cooling performances of the piezoelectric cooling fan have been studied in this paper. The natural frequency, mode shape, flapping amplitude, and temperature reduction have been obtained. The numerical results have good agreement with the experimental measurements. As a result, the piezoelectric cooling fan has to work under the natural frequency. The aspect ratio 2:3 ($L_b/W_b \approx 2/3$) is the optimal geometry of the fan blade for the cooling purpose.

The authors are grateful to the Ming Chi University of Technology for the financial support under contract number "100-Start-Academic-Research-E-01". The authors are also grateful to the Ministry of Economic Affairs of R.O.C. Government in Taiwan for the financial support under contract number 103-EC-17-A-31-S1-207.

References

1. M. Toda, S. Osaka, Proceedings of the IEEE **67**, 1171 (1979)
2. J.H. Yoo, J.I. Hong, W. Cao, Sensors and Actuators **79**, 8 (2000)
3. T. Acikalin, S.V. Garimella, A. Raman, J. Petroski, International Journal of Heat and Fluid Flow **28**, 806 (2007)
4. M. Kimber, S.V. Garimella, International Journal of Heat and Mass Transfer **52**, 4470 (2009)
5. W.J. Sheu, R.T. Huang, C.C. Wang, Sensors and Actuators A: Physical **148**, 115 (2008)
6. H.K. Ma, H.C. Su, C.L. Liu, W.H. Ho, International Communications in Heat and Mass Transfer **39**, 603 (2012)
7. A. Hales, X. Jiang, International Journal of Heat and Mass Transfer **139**, 362 (2019)
8. A. Hales, X. Jiang, International Journal of Heat and Mass Transfer **137**, 52 (2019)
9. M.H. Oh, J. Seo, Y.H. Kim, M. Choi, European Journal of Mechanics / B Fluids **75**, 339 (2019)
10. J. Yang, *The mechanics of piezoelectric structures* (World Scientific Publishing, New Jersey, 2006)

Design and Testing of a Valveless Piezoelectric Pump with Conical Vortex Diodes

Lei Zou, *Kai* Li, and *Jun* Huang*

National Research Center of Pumps, Jiangsu University, Zhenjiang, P.R. China

Abstract. The valveless piezoelectric pump delivers fluid based on the difference of flow resistance of internal tube. In this paper, a vortex diode with conical tangential tube is proposed, which possesses the great characteristic of reverse cut-off, and the analysis of the forward and reverse flow resistance of the conical vortex diode is verified by finite element analysis. Then, a valveless piezoelectric pump with conical vortex diodes as the internal channel is designed, and the prototype is manufactured. The results of the output performance experiment show that the maximum output flow rate of the valveless piezoelectric pump is 10.44 g/min at the drive frequency of 45Hz, and the maximum output pressure is 560 Pa at the drive frequency of 35Hz. The proposal of the valveless piezoelectric pump provides a good reference for more new types of valveless piezoelectric pumps.

1 Introduction

Valveless piezoelectric pumps driven by the inverse piezoelectric effect of piezoelectric materials has the advantages of simple structure, high reliability and miniaturization, which is widely applied in cell transmission, fuel cell and chip cooling.[1-3] Valveless piezoelectric pumps mostly utilize flow resistance differential tube to replace the valve. The common flow resistance differential tube is diffuser/nozzle, Tesla tube, and Y-shaped tube [4-7].

As a kind of flow resistance differential tube, the vortex diode has no mechanical moving parts and is adopted for fluid transmission. In the reverse flow of the vortex diode, there is a strong vortex motion, which exhibits a great function of reverse cut-off. Due to the characteristic of unequal flow resistances in forward and reverse, the vortex diode is utilized in the transportation of nuclear industry, chemical industry and pharmaceutical industries [8-9].

In this paper, applying the flow resistance characteristics of the diffuser/nozzle tube to the design of a vortex diode, the new vortex diodes with conical tangential tubes are proposed. And the forward and reverse flow resistances of the conical vortex diode are calculated by finite element simulation. Subsequently, a valveless piezoelectric pump with conical vortex diodes as the inner channel is designed, and the prototype is manufactured. Finally, the output performance experiment of the valveless piezoelectric pump with conical vortex diodes are carried out to verify the effectiveness of the design scheme.

2 Working principle and simulation

2.1 Design and simulation

In the previous study, the unequal flow resistance in different directions of the diffuser/nozzle had been proved, and the structure optimization of the diffuser/nozzle tube was carried out by Olssion [10]. The results demonstrate that the diffuser/nozzle tube possesses the maximum flow resistance difference with the diffusion angle is 7 °. Moreover, the fluid in the tangential tube of the vortex diode flows in a straight line. Combining with the flow resistance characteristics of the diffuser/nozzle tube, the tangential tube of the vortex diode is structurally changed. A new vortex diode (conical vortex diodes) is designed, and the tangential tube is designed to be a diffuser and the diffusion angle is 7°, illustrated in Fig.1.

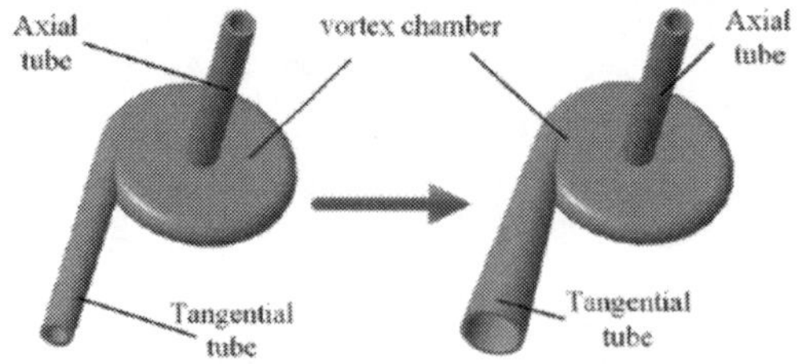

Fig. 1. Structural design of the conic vortex diode.

The process of the fluid entering the vortex chamber from the axial tube and flowing out of the tangential tube, it is defined as a forward flow. On the contrary, the process of the fluid entering the vortex chamber from the tangential tube and flowing out of the axial tube, it is defined as reverse flow. For the conical vortex diode, the resistance coefficient ξ is adopted to describe the flow characteristics and defined in equations (1). The

* Corresponding author: huangjun@ujs.edu.cn

https://doi.org/10.1051/matecconf/202030604003

impedance ratio ε is adopted to define the performance of the unidirectional transport fluid of a conic vortex diode and defined in equations (2).

$$\xi = \frac{2\Delta P}{\rho v^2} \tag{1}$$

$$\varepsilon = \frac{\xi_F}{\xi_R} \tag{2}$$

Where, ρ (1000kg/m^3) is the density of the fluid, ΔP (Pa) is the pressure drops between the axial tube and the tangential tube of the conical vortex diode, and v (m/s) is the velocity of the fluid flowing into the conical vortex diode. ξ_R is the resistance coefficient of the reverse flow, ξ_F is the resistance coefficient of the forward flow.

The fluid flows in the pump chamber in different forms during the forward and reverse flow, which results in unequal flow resistances. The characteristic of the different flow resistances can be applied to design the internal flow channels of a valveless piezoelectric pump.

2.2 Numerical simulation

In order to obtain the condition of fluid flowing in the conical vortex diode, the ANSYS CFX software is employed to perform the finite element analysis. The structural parameters of the conical vortex diode are shown in Table 1. The characteristics of the forward and reverse flow in conical vortex diode are evaluated by pressure drops between inlet and outlet. The flow field inside the conical vortex diode is set as the fluid domain, the material is the water. Fig. 2 is the finite element model of a conical vortex diode. In forward flow, the axial tube is defined as the inlet, the tangential tube is defined as the outlet. And in reverse flow, the tangential tube is defined as the inlet, the axial tube is defined as the outlet. The remaining wall is set as the no slip wall. The inlet boundary is determined as the pressure boundary, which is 200,500, 800, 1000, 1500, 2000, 3000, and 5000Pa. The outlet boundary is set as the open boundary (One atmosphere). In the forward flow, the fluid flowing in the conical vortex diode is not complex, and the k-ε model is selected for calculation. However, in the reverse flow, there is a strong vortex motion in the vortex chamber, the RNG k-ε model is selected for calculation.

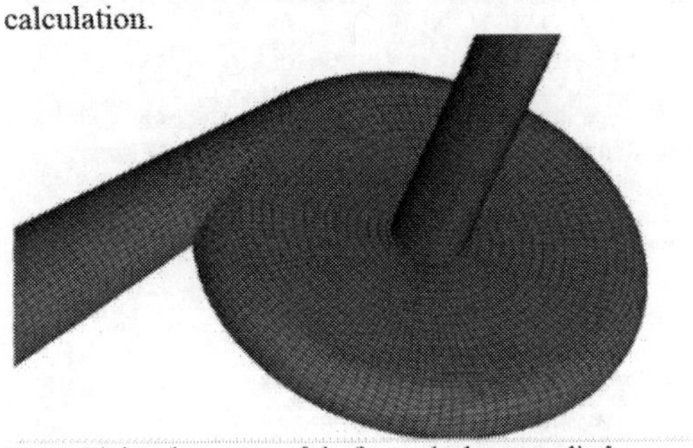

Fig.2. Finite element model of a conical vortex diode

Table 1. Structural parameters of the conical vortex diode.

Structure	Parameter
Axial tube diameter *Da*	1 mm
Vortex chamber height *h*	1 mm
Vortex chamber diameter D_1	6 mm
Angle of tangential tube	7°

Fig.3 illustrates a velocity streamline diagram of a conical vortex diode at 500Pa. In the forward flow, the fluid enters the vortex chamber from the axial tube and forms a small spiral motion in the pump chamber. And then the fluid flows out from the tangential tube. In the reverse flow, the fluid enters the vortex chamber from the tangential tube, and moves along the circular wall surface. The fluid forms a vortex motion in the vortex chamber under the influence of the circular wall surface, and the vortex motion exists in the whole vortex chamber .Then the fluid flows out from the axial tube in a spiral form.

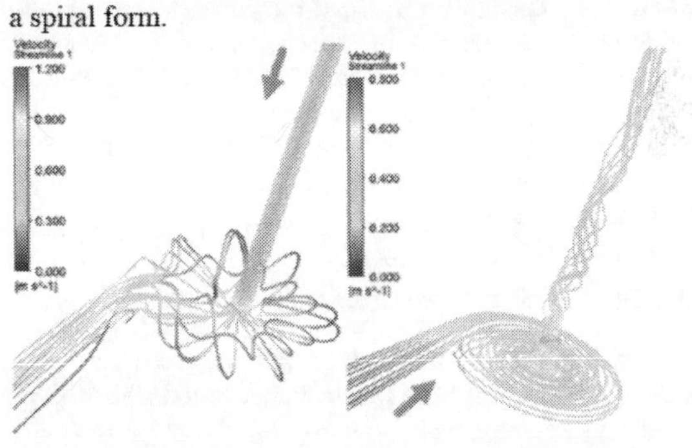

Fig.3. Streamline diagram of a conical vortex diode (ΔP=500Pa)

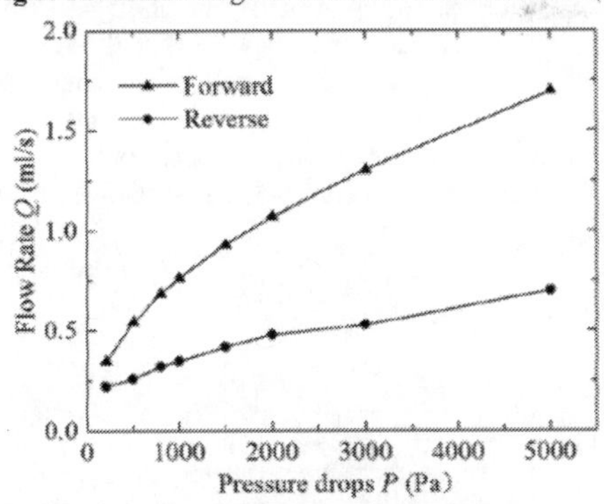

(a) Flow rate of simulation

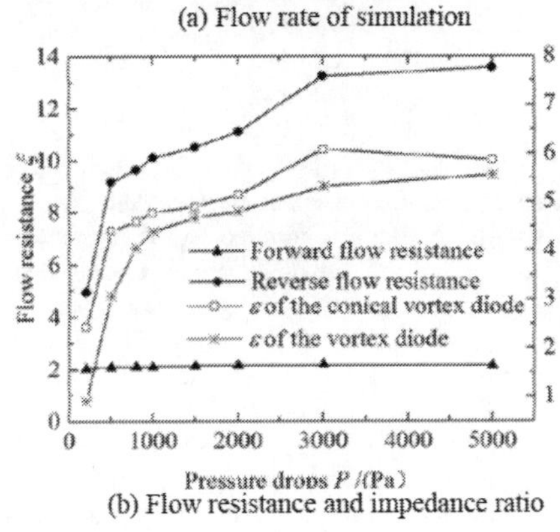

(b) Flow resistance and impedance ratio

Fig. 4. Results of simulation calculation.

Fig. 4 illustrates the simulation results of the conical vortex diode with under various pressure drops. The flow rate in the reverse flow is always higher than that of the forward flow. Then substituting the results into equation (1), (2), the resistance coefficient ξ and the impedance ratio ε is obtained, shown in Fig.4 (b). As the pressure drops increases, the forward and reverse flow resistance coefficients ξ increase, and the reverse flow resistance coefficient ξ_R remains greater than that of the forward flow.

Combining Fig. 3 and 4, although there is a spiral motion in the pump chamber during the forward flow, the spiral motion is small, which results in lower flow resistance. However, there is a strong vortex motion in the vortex chamber of the reverse flow, the flow resistance becomes larger. Comparing the impedance ratios ε of different vortex diode under the same pressure drops, the impedance ratio ε of the conical vortex diode is always higher than that of the vortex diode. The result of the comparison confirms that the design of the conical vortex diode can enhance characteristics of the unequal flow resistance. Moreover, the conical vortex diode can be applied for a valveless piezoelectric pump for one-way transmission.

2.3 Design of the piezoelectric pump

Fig.5 illustrates the schematic diagram of the piezoelectric pump. The piezoelectric vibrator is utilized as the power source, and the sealing ring isolates the pump chamber from the piezoelectric vibrator and keeps the pump chamber sealed, the dimensions of the seal ring is 27×20×0.2 mm. A pair of conical vortex diodes are placed vertically and placed under the pump chamber, which acts as an internal "valve". The dimensions of the valveless piezoelectric pump with conical vortex diode are 40×40×18mm. The diameter of pump chamber is 20 mm and the depth is 1 mm.

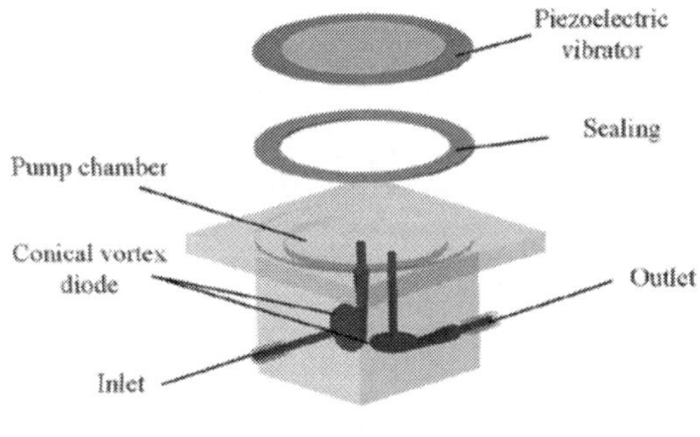

Fig.5. Schematic diagram of the piezoelectric pump

An alternating voltage is applied to the piezoelectric vibrator. Under the function of the inverse piezoelectric effect, the piezoelectric vibrator generates vibration, which causes the variation in the volume of the pump chamber. The difference in flow resistance of the conical vortex diode leads to the unequal flow under different working conditions (Up-down vibration of piezoelectric vibrator) of the piezoelectric pump. Theoretically the piezoelectric pump can realize the one-way transmission for the fluid.

3. Experiment and results

3.1. Fabrication and experimention

The prototype of the valveless piezoelectric pump with conical vortex diodes is manufactured by Stereo lithography Apparatus (SLA) technology. The processing precision of the SLA technology is 0.1mm, and the processing material is ABS photosensitive resin. The base material of the piezoelectric vibrator is copper alloy and the parameters of the piezoelectric vibrator are shown in Tab 2.

Fig.6 illustrates the schematic diagram of the output performance experiment. The deionized water is adopted as the fluid medium in the experiment. The signal generator provides a sinusoidal voltage, and the drive voltage is amplified by the power amplifier to 200V (peak-to-peak value). An electronic balance (resolution: 0.01g) is utilized to measure the flow rate of the piezoelectric pump per minute. The output back pressure of the piezoelectric pump is evaluated by measuring the height of the fluid at the outlet tube. By changing the range of driving frequencies from 10Hz to 110Hz, the output performance of the valveless piezoelectric pump is carried out. And the amplitude of the center position of the piezoelectric vibrator is measured by a CCD Laser displacement sensor (Keyence LK-G35, resolution: 0.001mm).

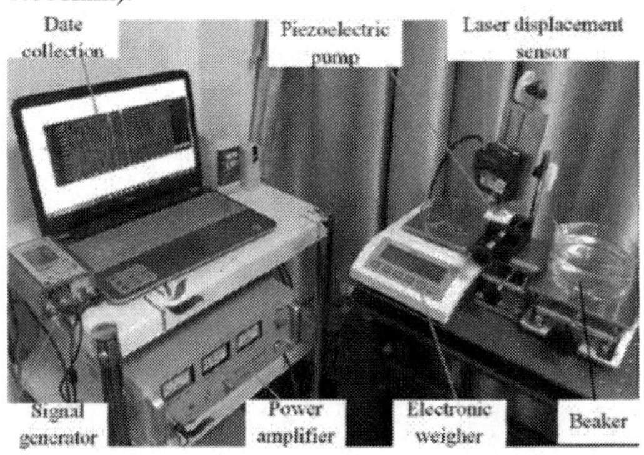

Fig. 6. Photograph of experiment.

Table 2. Structural parameters of the piezoelectric vibrator.

Structure	Parameter
External diameter	27±0.1mm
Base thickness	0.15mm
Ceramic diameter	19±0.2mm
Ceramic thickness	0.22mm
Resonant Frequency	4.1±0.5KHz
Resonant Impedance	⩽300 Ω
Static Capacitance	28000(PF) ±30%

3.2 Results and discussion

Fig.7 (a) illustrates the output flow rate and amplitude curves varying with frequency of the valveless piezoelectric pump with conical vortex diodes. At the frequencies range of 10Hz to 110Hz, the piezoelectric pump exhibits obvious output flow rate. The maximum amplitude (single peak) of the piezoelectric vibrator is 39.7μm at the drive frequency of 50Hz. The variation trend of the output flow rate is consistent with that of the piezoelectric vibrator. The output flow rate gradually increases with a range of 10Hz to 45Hz, the maximum output flow rate is 10.44 ml/min at the drive frequency of 45Hz. Fig.7 (b) illustrates the output back pressure and amplitude curves varying with frequencies of the valveless piezoelectric pumps with conical vortex diodes. With the change of frequencies, the vibrator amplitude curve is in the form of a parabola, and the maximum amplitude (single peak) of the piezoelectric vibrator is 44.2 μm at the drive frequency of 33Hz. The variation of amplitude of piezoelectric vibrator affects the output back pressure of piezoelectric pump, the maximum output pressure is 560 Pa at the drive frequency of 35Hz.

The experimental results verify the reliability and effectiveness of the design scheme for the valveless piezoelectric pump with conical vortex diodes.

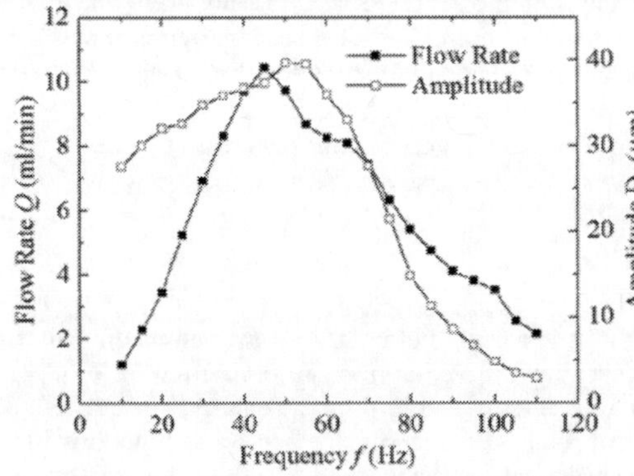

(a) Curves of output the flow rate and amplitude

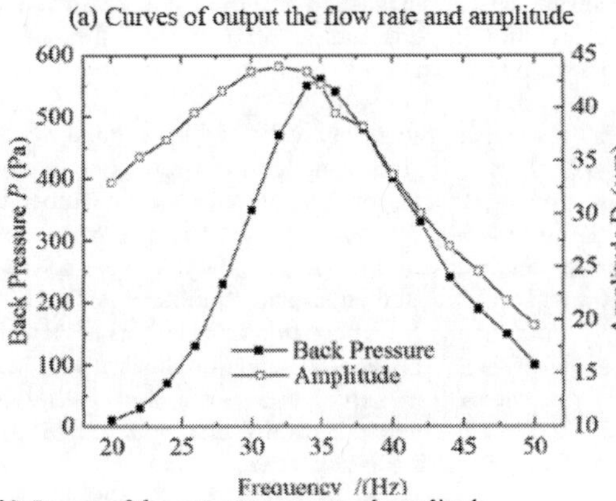

(b) Curves of the output pressure and amplitude

Fig. 7. Results of performance experimental.

4 Conclusion

A valveless piezoelectric pump with conical vortex diodes is designed with vortex diode as the internal flow channel. The finite element method is adopted to simulate the flow resistance of the conical vortex diode. The simulation results demonstrated that the unidirectional cut-off characteristics of the conical vortex diode. Then, a valveless piezoelectric pump with the conical vortex diode is designed, and a prototype of the valveless piezoelectric pump is fabricated. Moreover, the output performance experiment of the piezoelectric pump is carried out, and the experiment results demonstrate that the maximum flow rate is 10.44 ml/min at the drive frequency of 45Hz, and the maximum output pressure is 560 Pa at the drive frequency of 35Hz.

The authors would like to thank the National Natural Science Foundation of China (Grant No. 51605200) and the Senior Talent Start-up Foundation of Jiangsu University (Grant No. 14JDG145) for the financial support.

References

1. X. F. Leng, J. H. Zhang, Y. Jiang. Theory and experimental verification of spiral flow tube-type valveless piezoelectric pump with gyroscopic effect *Sensors & Actuators A Physical*, **195(6):**1-6 (2013).
2. J. Huang, J. H. Zhang, S. Wang. Analysis of the Flow Rate Characteristics of Valveless Piezoelectric Pump with Fractal-like Y-shape Branching Tubes *Chinese Journal of Mechanical Engineering*, **27(3):** 628-634 (2014).
3. Q. H. Xiao, L. G. Chen, F. R. Sun. Constructal entransy dissipation rate and flow-resistance minimizations for cooling channels *Science in China Series E: Technological Sciences*, **53(9):**2458-2468, (2010).
4. E. Stemme and G. Stemme. A valveless diffuser/nozzle based fluid pump Sensors & Actuators A **39** :159–167 (1993)
5. R. Gamboa, C. J. Morris, F. K. Forster. Improvement Fixed Valve Micropump Performance Through Shape Optimization of Valves *Journal of Fluids Engineering*, **127**(2):339, (2005).
6. J. Huang, J. Zhang, W. Shi. 3D FEM analyses on flow field characteristics of the valveless piezoelectric pump. *Chinese Journal of Mechanical Engineering*, **29**(4):825-831 (2016)
7. S. S. Wang, X. Y. Huang, C, Yang. Valveless micropump with acoustically featured pumping chamber *Microfluidics & Nanofluidics*, **8**(4):549-555 (2010).
8. S. X. Qu, Y. H. Wu, Z, Z He. Surrogate Fluid Experimental Study and CFD Simulation on the Hydraulic Characteristics of Vortex Diode *Nuclear Science and Engineering*, 1-8, (2017).
9. Jain P, Bhandari V M, Balapure K. Hydrodynamic cavitation using vortex diode: An efficient approach for elimination of pathogenic bacteria from water *Journal of environmental management*, **242**: 210-219, (2019).
10. Olsson A. Olsson A. Valveless diffuser micropumps *Signaler, sensorer och system*, (1998).

Anti-seismic characteristics of safety valve and its spring failure prediction analysis

Fanrong Meng, *Huibing* Zhang, *Ye* Dai*, *Hanbo* Zhang, *Wenqiang* Wei and *Rui* Liu

Harbin University of Science and Technology, School of Mechanical and Power Engineering, Heilongjiang Harbin 150080, China

Abstract. Safety valve is an important guarantee for nuclear power plant system. Its working environment is harsh, and medium is high-temperature corrosive. It is important to study the dynamic characteristics of nuclear safety valve under unsteady condition to improve the stability and safety of nuclear power plant piping system. Using the finite element method, mass simplification and Rayleigh method, the frequency response analysis of the safety valve and the spring is predicted to predict the seismic capacity under the earthquake load. Based on the verification of the dynamic characteristics of the valve body, the stress and strain analysis of the spring is carried out. To explore the failure condition of the safety valve spring, the lateral deviation linearity of the spring was tested and analyzed by setting up the offset test bench. The results show that the safety valve with the above analysis method can meet the working requirements under the shaking condition. The research results provide an important theoretical basis for the design and analysis of nuclear safety valves.

1 Preface

The safety valve is one of the most basic components in the nuclear power plant piping system and is mainly used to control the pressure, flow and flow of the fluid [1, 2]. Safety valve is an important guarantee for nuclear power plant safety system. Its working environment is harsh, and its medium has high temperature and corrosiveness. It is important to study the dynamic characteristics of nuclear safety valve under unsteady condition to improve the stability and safety of nuclear power plant piping system significance.

With the frequent occurrence of earthquake disasters, if the designed safety valve does not meet the seismic requirements, the earthquake will lead to nuclear valve nuclear leakage, which not only causes environmental pollution but also causes huge losses to human beings. Therefore, it is very necessary to analyze the seismic characteristics of the safety valve. Design safety valves not only for seismic analysis of safety valves, but also for failure prediction analysis of key spring components. This paper analyzes the spring failure prediction of key components of safety valves. The spring is a key component of the safety valve. The spring plays an important role in the stability and reliability of the safety valve. Due to the special working environment of the safety valve, the spring is required to have high temperature resistance and corrosion resistance as well as good fatigue resistance. Under the normal working condition of the safety valve, it is in a compressed state for a long time, so it is necessary to predict the failure before designing the spring, and then optimize the design. At present, spring-loaded safety valves are widely used in the industrial field, and both petrochemical and nuclear power plants play an important role. Spring functions mainly include storage energy, buffer pressure, control return, balance device and other functions [3-5]. The spring acts as the elastic part of the safety valve, and the calculation of the natural frequency of the safety valve spring is an important factor for the natural frequency of the safety valve [6-9].

In this paper, the safety valve will be studied and analyzed from the following two aspects: one is the seismic characteristics of the safety valve; the other is the spring failure prediction analysis. Perform frequency response analysis on the entire safety valve and key component springs, and stress and strain analysis on key spring components. By constructing a spring lateral offset test bench, the lateral deviation linearity of the spring in the safety valve is detected and analyzed. Analyze the experimental results and summarize the seismic capacity of the safety valve.

2 Safety valve natural frequency analysis

The research object of this paper is a spring direct load type safety valve. Its working principle uses the load to be applied to the valve disc. The main parts include valve body, valve seat, spring, valve flap, etc. The structure diagram is shown in Fig. 1. The valve two-dimensional model is shown in Fig. 2.

* Corresponding author: daiye312@163.com

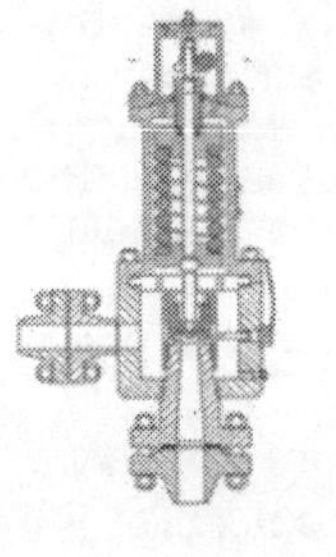

Fig. 1. Schematic diagram of safety valve structure.

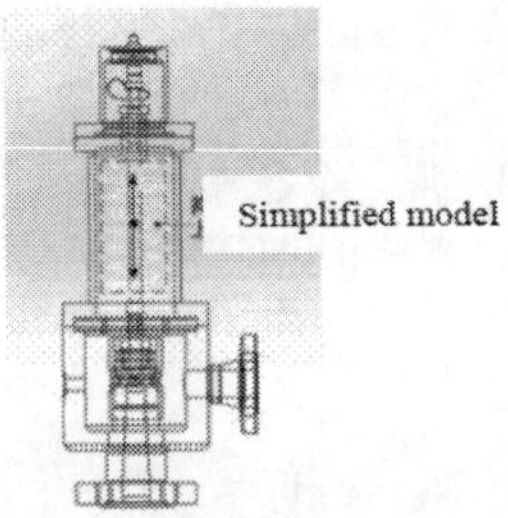

Fig. 2. Two-dimensional model of safety valve

The purpose of safety valve stress analysis is to verify its structure to meet the needs of production work. Apply a certain load on the valve body and other key components to analysis of the valve body deformation and stress conditions[2]. In the main parts of the safety valve, the allowable stress value of the valve body is the minimum[3]. Taking the valve body as an example, the structural stress analysis process is analyzed, and the other parts and components are analyzed emphatically.

2.1 Establish a safety valve model

The finite element and mass simplification method is used to analyze the seismic characteristics of the safety valve. The seismic performance of the safety valve is generally related to its own natural frequency. The first-order natural frequency value of the designed safety valve is required to be greater than 33 Hz. To improve the analysis efficiency of the safety valve under the premise of ensuring the calculation accuracy, the model can be simplified as appropriate, and the spring, the upper spring seat and the lower spring seat are regarded as a whole, and the removed mass is added to the valve seat, and the spring is applied to the upper and lower spring seats. The force can be seen as the force of a particle on the spring seat, which makes the calculated natural frequency value more accurate.

2.2 Safety valve modal analysis pretreatment

Create a three-dimensional model of the safety valve, including the valve body assembly, valve seat, valve flap and other structures.

The safety valve model does not include the spring, the upper spring seat, and the lower spring seat. Instead, the mass equivalent of this part of the removal is added to the adjustment screw to ensure that the overall quality of the safety valve remains unchanged. This not only reduces the analysis difficulty of the safety valve, but also improves the analysis efficiency of the safety valve.

2.3 Safety valve natural frequency calculation result

The whole safety valve is imported into the finite element in x-t format, and parameters are set, meshed and restrained. Finally, the first four-order modal analysis results of the safety valve are obtained, as shown in Fig. 3-Fig. 6.

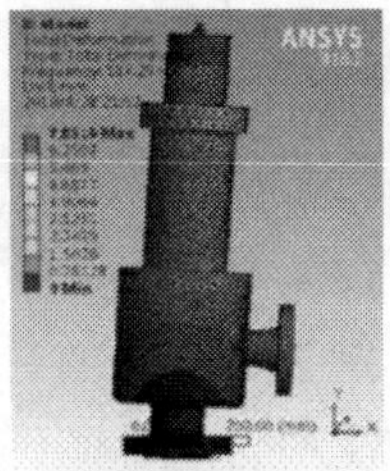

Fig.3. First-order mode shape

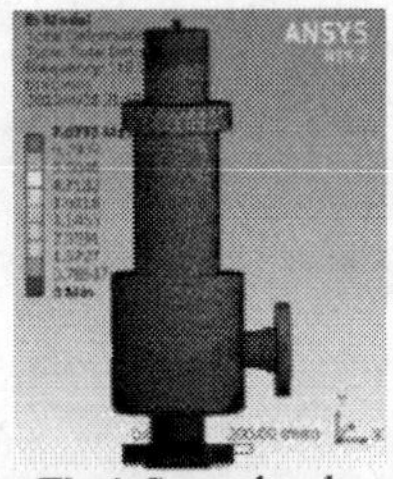

Fig.4. Second-order mode shape

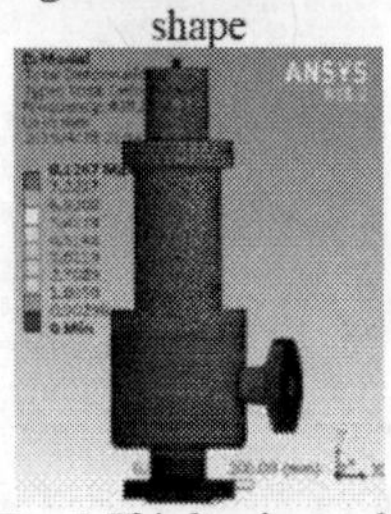

Fig.5. Third-order mode shape

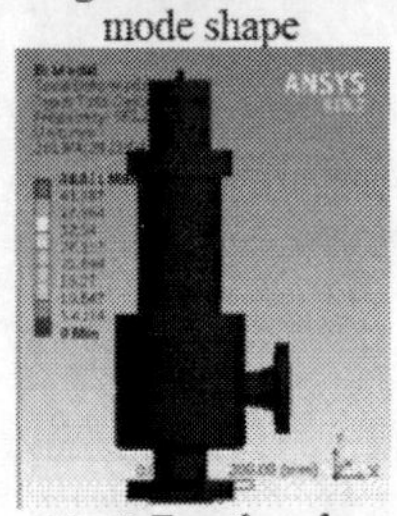

Fig.6. Fourth-order mode shape

It is shown in the above figures that the first-order mode shape of the safety valve belongs to the bending mode. From **Fig. 3**, the first-order natural frequency of the safety valve is 117.29 Hz. The first-order natural frequency of the safety valve is designed to be greater than 33 Hz, which meets the design conditions and therefore meets the safety requirements and valve seismic requirements.

3 Calculation and analysis of natural frequency of spring

In this paper, the Rayleigh method is used to calculate the kinetic energy of the spring's distributed mass into the total kinetic energy of the system, and the single-degree-of-freedom system is used to find the natural frequency [10]. A mathematical model is established for the spring, as shown in **Fig. 7**, assuming that each mass of the spring is fixed to one end of a straight elastic rod, and the other end is the same as the static deformation of each section under the axial force.

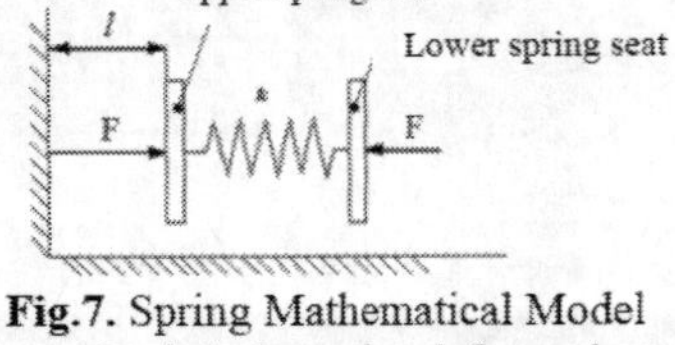

Fig.7. Spring Mathematical Model

According to Hooke's law, the deformation of each section of the spring is linearly related to the distance of the fixed end.

The kinetic energy of the entire system of the spring model is:

$$T_S = \frac{1}{2} m_e x^2 \quad (1)$$

The displacement $\frac{\xi}{l} x$ from the left end is ξ, and the kinetic energy of the $d\xi$ spring is

$$dT_S = \frac{1}{2} \frac{m_s}{l} d\zeta \left(\frac{\zeta}{l}\right)^2 \quad (2)$$

The total kinetic energy of the spring is

$$T_S = \int_0^1 dT_S = \frac{1}{2}\frac{m_s}{3}x^2 \quad (3)$$

By combining the above formulas, you can get:

$$m_e = \frac{1}{3}m_s \quad (4)$$

It can be seen that the equivalent mass of the spring is one-third of the mass of the effective number of springs, and the total kinetic energy of the system is :

$$\mathrm{T} = \frac{1}{2}m\dot{x}^2 = \frac{1}{2}\left(m + \frac{m_s}{3}\right)\dot{x}^2 \quad (5)$$

The potential energy of the system is

$$V = \frac{1}{2}kx^2 \quad (6)$$

Let $x = A\cos(wt - \varphi)$ be calculated from the conservation of energy: $T_{max} = T_{min}$

$$\omega = \sqrt{\frac{k}{m+\frac{m_s}{3}}} \quad (7)$$

$$f = \frac{\omega}{2\pi} = \frac{1}{2\pi}\sqrt{\frac{k}{m+\frac{m_s}{3}}} \quad (8)$$

Where:m_e is the equivalent mass of the spring; m_s is the mass of the effective number of springs; m is the mass of the moving parts; for the safety valve, including the mass of the valve disc, the valve stem, the lower spring seat, and the number of invalid turns of the spring; The free length of the effective number of turns of the spring.

Modal analysis of the safety valve spring is carried out. The modal analysis is a numerical technique for calculating the vibration characteristics of the structure. The structural vibration characteristics include the natural frequency and the vibration mode [11-12].

In this paper, the first-order natural frequency value of the safety valve is obtained, and the result meets the seismic requirements. Then the natural frequency of the spring is analyzed to see if it meets the seismic requirements. The results are shown in Fig. 8-Fig. 11.

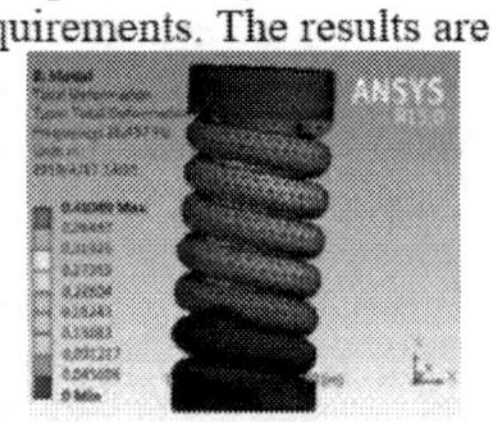

Fig.8. Spring first-order mode shape

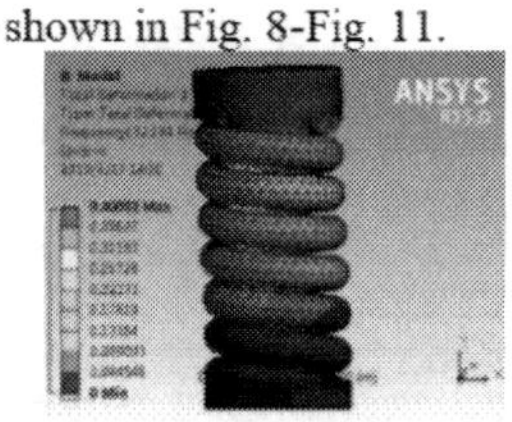

Fig.9. Spring second-order mode shape

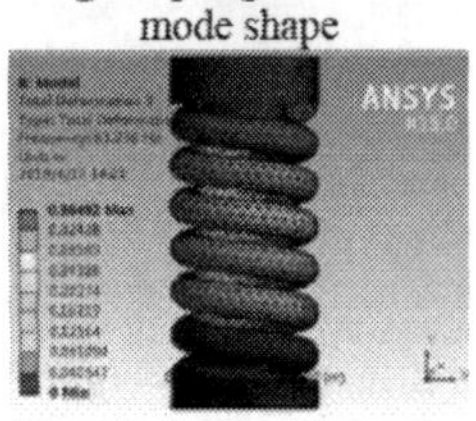

Fig. 10. Spring third-order mode shape

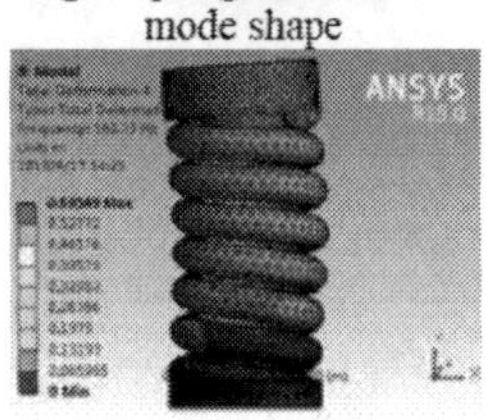

Fig. 11. Spring fourth-order mode shape

From the above simulation results, the following conclusions can be drawn: the first-order mode mode and the second-order mode shape of the spring and the fourth-order mode shape belong to the bending mode, and the spring array is orthogonal.

4 Safety valve spring stress analysis

4.1 Spring strain analysis

Due to the working principle of the safety valve, the two sides of the spring are subjected to the set pressure, and the spring is subjected to the axial pressure, assuming that the spring cross section is in the same plane as the spring axis. The internal forces of the spring section include the shear force Q and the torque MT, which are obtained according to the equilibrium relationship:

$$\begin{cases} Q = F \\ M_T = \frac{1}{2}FD \end{cases} \quad (9)$$

Where: D is the diameter of the coil and F is the axial pressure to which the spring is subjected.

It is assumed that the shear stress caused by the shear force Q is evenly distributed in the cross section, as shown in Fig. 12, the shear stress caused by the shear force Q is :

$$\tau_Q = \frac{4F}{\pi d^2} \quad (10)$$

According to the linear elastic circular axis torsion theory, the distribution of shear stress caused by the torque T in the linear elastic range is shown in the figure. The maximum shear stress is:

$$\tau_{max} = \frac{8FD}{\pi d^3} \quad (11)$$

$$\frac{\tau_Q}{\tau_{max}} = \frac{d}{2D} \quad (12)$$

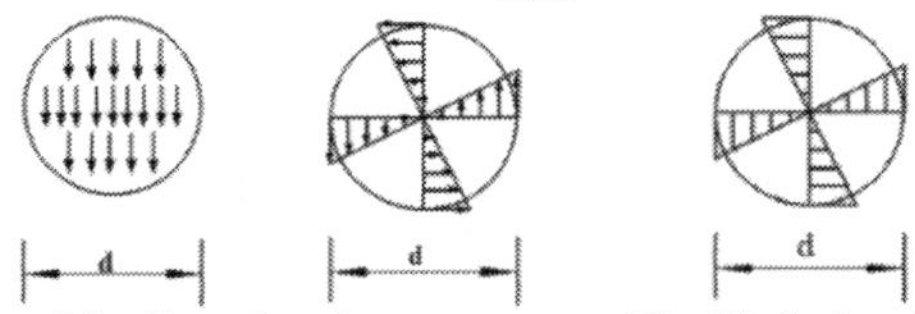

Fig. 12. spring shear stress distribution **Fig. 13.** Spring shear strain distribution

It can be obtained from the above formula that when the spring cross-sectional diameter d is much smaller than the average diameter D of the coil, the shear strain distribution on the cross-section is negligible, and the strain distribution on the spring cross-section is as shown in Fig.12. and Fig.13. According to the principle of variable performance, the relationship between the axial compression amount λ of the spring and the axial pressure F can be obtained by using the formula.

$$\mathrm{F} = \frac{Gd^4}{8D^3N}\lambda \quad (13)$$

Where N is the total number of turns of the spring. Through the above theoretical analysis, the shape variable and the shear stress are inversely proportional to the spring diameter under the axial or radial load of the safety valve spring.

4.2 Spring strain analysis

A mathematical model is established for the safety valve spring, the upper spring seat and the lower spring seat, and the stress and strain analysis is performed by using the finite element, and the results are shown in Fig. 14-Fig. 17.

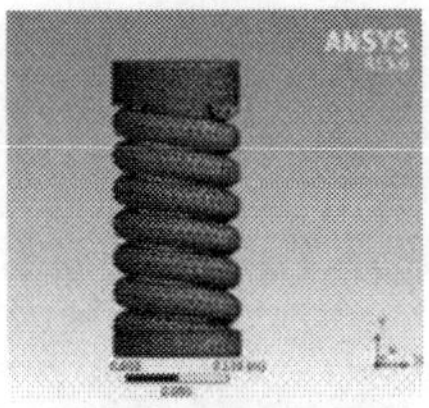

Fig. 14. Spring finite element model

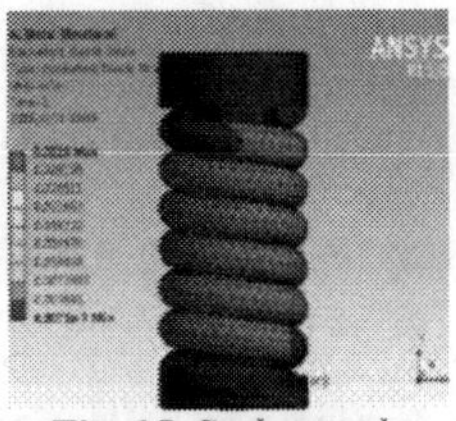

Fig. 15. Spring strain analysis results

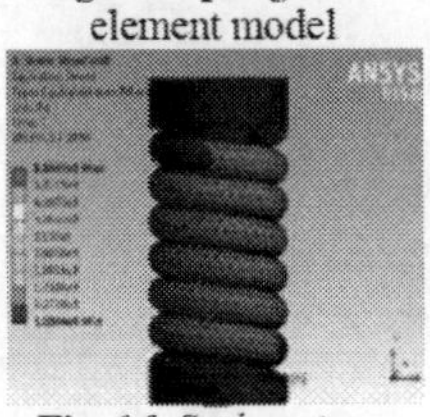

Fig. 16. Spring stress analysis results

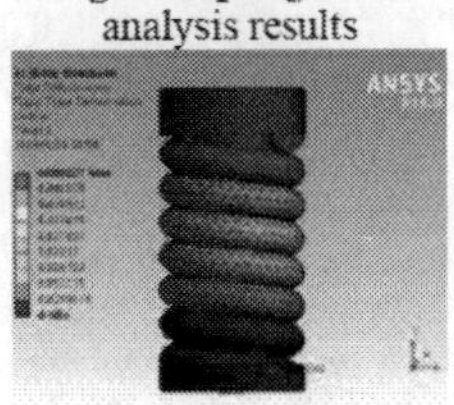

Fig. 17. Total spring deformation

From the simulation analysis results, it can be seen that the total deformation of the spring strain analysis is 0.05mm, and the stress analysis and strain analysis amount are relatively small, negligible, and meet the spring design requirements.

5 Comparison of theoretical analysis and simulation analysis results

Comparing the results of theoretical stress analysis and finite element stress analysis of spring, under the applied load, it can be seen that the amount of spring deformation is small and meets the requirements of seismic characteristics.

Comparing the natural frequency modal analysis and finite element analysis results of the spring, the interface definition safety valve spring setting material is analyzed on the finite element: the density is 7850Kg/m^3, and the elastic modulus is $2.0392 \times 10^{11} Pa$, Poisson's ratio is 0.33. The result f = 30 Hz can be calculated from the above formula. The analysis result is 28.46 Hz, and the theoretical calculation result is 30 Hz. Deviation calculation:

$$\delta = \frac{30 - 28.46}{30} \times 100\% = 4.6\%$$

6 Safety valve spring lateral displacement experiment

Considering the test requirements comprehensively, it is proposed to use the hammer method for modal test, the piezoelectric acceleration sensor is used for data acquisition of the safety valve, and the piezoelectric hammer is used for excitation. The one-way sensor measuring point layout is shown in Fig. 18, and three measuring points are arranged in each direction, for a total of six measuring points. Among them, measuring points 1 can test the natural frequency and bending mode of the Y direction, and the measuring points 2 can test the natural frequency and bending mode of the X direction. The two measuring points correspond to the first six channels of the dynamic signal processing system, and the third channel is the hammer channel.

The specific test scheme of artificial random excitation is: single input, multiple output, analysis frequency 500Hz, sampling frequency 1280Hz, points/frame: 8192, collecting 20 frames of data, cumulative sampling time 128 seconds. Random tapping, hard disk acquisition, a total of 4 tests, respectively, different directions for different directions of excitation, of which one-way sensor layout program 2 times.

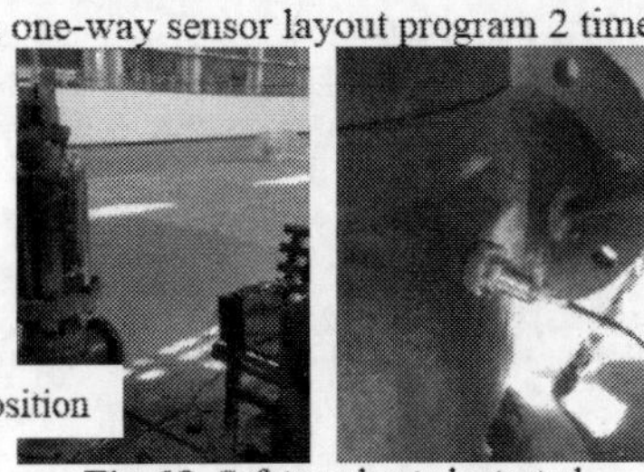

Fig. 18. Safety valve to be tested

The time domain curve measured by the hammer excitation at the position of the measuring point 3 is as shown in Fig. 19. According to the exciting direction, the natural frequency and the bending mode of the structure along the Y direction can be measured. According to the power spectral density function, the natural frequency of the structure along the Y direction can be identified. According to the imaginary part of the frequency response function (component analysis method), the natural frequency and the bending mode of the structure along the Y direction can be identified.

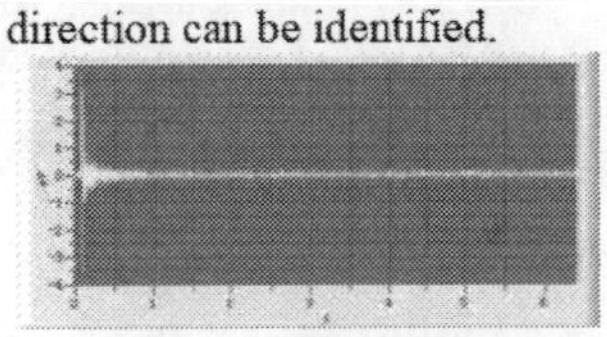

(a)

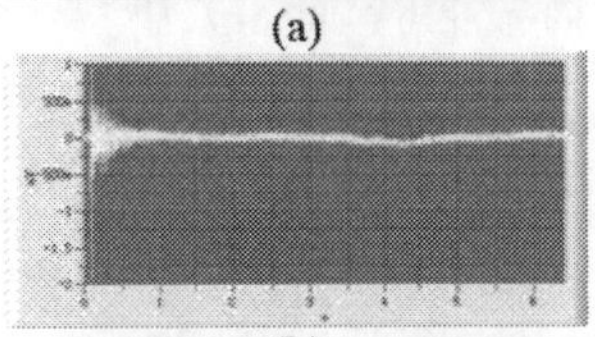

(b)

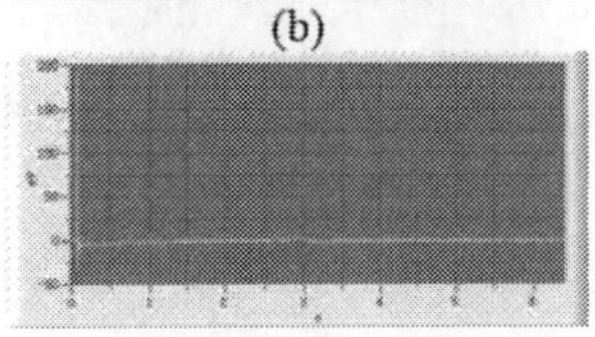

(c)

Fig. 19. Time domain waveform

The natural frequency of the safety valve measured in this test has little difference between different excitation modes and different modal parameter identification methods, and the results are more accurate. The Y-direction and the X-direction natural frequency are relatively close, mainly because the structure is a quasi-circumferential symmetrical structure. Compared with the finite element calculation results, the deviation is within 10%, which is acceptable within the allowable range of engineering error. Comparing the corresponding mode shape results with the finite element results, the test test mode is consistent with the finite element calculation mode.

7 Conclusion

1) Propose a method of quality simplification. The method ensures that the overall quality of the safety valve is unchanged, and the finite element analysis mode diagram is obtained, and the natural frequency value is greater than 33 Hz, which satisfies the requirements of seismic working conditions.
2) The Rayleigh method is used to calculate the natural frequency of the spring. And combining theory and analysis to obtain the fourth-order natural frequency value of the spring, the first-order modal shape and the second-order modal shape and the fourth-order modal shape spring shape are obtained.
3) Perform stress and strain analysis on the key components of the spring to obtain the condition of failure of the safety valve spring. Under the axial or radial load, the deformation and shear stress are inversely proportional to the spring diameter.
The research is supported by National Natural Science Foundation of China (Grant No. E51505109), Natural Science Foundation of Heilongjiang Province of China (Grant No. LH2019E062), University Nursing Program for Young Scholars with Creative Talents in Heilongjiang Province (Grant No. UNPYSCT-2017077), Fundamental Research Foundation for Universities of Heilongjiang Province (LGYC2018JC040), China Postdoctoral Science Foundation (Grant No. 2016M591539), Heilongjiang Postdoctoral Science Foundation (Grant No. LBH-Z15102).

References

1. Chen Junjie, Yin Zhihong, He Jianghua, Shang Guanwen. Study on Modeling and Dynamic Characteristics of Air Spring System with Throttle Damping Hole and Additional Air Chamber[J]. Journal of Mechanical Engineering, 2017, 53(08): 166-174.
2. Godunov S K. Finite-difference method for numerical computation of discontinuous solutions of the equations of fluid dynamics [J]. 2017, 34:232-235.
3. Chen Junjie, Yin Zhihong, He Jianghua, Shang Guanwen. Study on Modeling and Dynamic Characteristics of Air Spring System with Throttle Damping Hole and Additional Air Chamber[J]. Journal of Mechanical Engineering, 2017, 53(08): 166-174.
4. Shen Wei, Zhang Qiangsheng, Chen Tianmin, *et al.* Seismic analysis and stress assessment of nuclear-grade main steam safety valve[J]. Journal of Power Engineering, 2016, 36(2): 162-167.
5. Shen Wei, Zhang Qiangsheng, Chen Tianmin, *et al.* Seismic analysis and stress assessment of nuclear-grade main steam safety valve[J]. Journal of Power Engineering, 2016, 36(2): 162-167.
6. Galbally D, García G, Hernando J, *et al.* Analysis of pressure oscillations and safety relief valve vibrations in the main steam system of a Boiling Water Reactor[J]. Nuclear Engineering & Design, 2015, 293(3): 258-271 .
7. DAI Ye, LAI Yi-nan, ZHANG Yuan, REN Sheng-le, WANG Jian. A rapid simplification method based on rules and fuzzy judgments[J]. Journal of Harbin University of Science and Technology, 2013, 18(04): 37-41.
8. Li L Q, Ju N P, Zhang S, *et al.* Correction to: Seismic wave propagation characteristic and its effects on the failure of steep jointed anti-dip rock slope[J]. Landslides, 2018:1-2.
9. Dai Ye , Wu Nai-Qun* , Liu Rui , Meng Fan-Rong , Li Yuan. Integrated Design Method and Experimental Study for the Safety Performance Analysis of Nuclear Power Valve [J]. International Journal of Control and Automation, 2016, 9(10) : 201~208.
10. Song Yixin. Research on the Natural Frequency Analysis Method of ANSYS Safety Valve[J]. Science and Technology Innovation, 2017(26): 77-78.
11. Zhou Bo, Wang Zhiyong, Xue Shifeng. Mechanical model of shape memory alloy superelastic coil spring[J/OL]. Journal of Mechanical Engineering: 1-9[2019-05-06].
12. Gao Hongxing, Chi Maoru, Zhu Wei, Liu Degang, Lin Jun. Study on Air Spring Model[J]. Chinese Journal of Mechanical Engineering, 2015, 51(04): 108-115.

Minimization for current peak-to-peak value in switched reluctance motor

Chaoyi Shang[1], *Aide* Xu[2,*], *Qianni* Li[2], and *Lepeng* Huang[1]

[1]College of Electrical Engineering of Ships, Dalian Maritime University, 116026 Dalian, China
[2]College of Information and Science Technology, Dalian Maritime University, 116026 Dalian, China

Abstract. For switched reluctance motor (SRM), due to the excessive peak-to-peak current occurs at the initial stage under the direct instantaneous torque control (DITC) algorithm, the motor efficiency has been affected to some extent. For this propose, a DITC optimization algorithm combined with high torque per ampere (HTPA) is introduced in this paper. Based on the traditional maximum torque-to-current ratio (MTPA) principle, the changing points of the inductance slope can be calculated. These points are also refer to the maximum torque-to-current ratio. Taking these points as the basis for sector division of the mechanical angle of each phase, the operation rule is optimized in the newly divided interval to select the corresponding voltage vector. The optimal voltage vector is obtained by the method of segmented variable duty cycle modulation. The new algorithm was validated on a three-phase 12/8 switched reluctance motor. Compared to conventional DITC, HTPA-DITC reduces peak current evidently. At the same time, the three-phase stator current is reduced under the same load conditions, thereby increasing the torque-to-current ratio of the motor.

1 Introduction

Switched reluctance motor (SRM) is a new type of variable speed motor. Compared with asynchronous motor and permanent magnet synchronous motor (PMSM) , it has a series of advantages, such as simple structure, low cost , reliable power circuit , and high efficiency. However, due to the double salient structure of the body and the high non-linearity of the electromagnetic relationship, there are several difficulties in efficiency maximization and current value minimization under the constant torque [1].

As one of the field-weakening control methods, maximum torque per ampere (MTPA) has been used to optimize the efficiency of traditional motors. MTPA control strategy can generate the minimum current corresponding to the given torque, for the optimal scheme of torque per ampere (TPA).

Based on the rotating coordinate system and current vector control, the MTPA control for PMSM is proposed in [2]. As a new decoupling controller, MTPA has always been developed on current optimization [3-4] and loss analysis by researchers. Literature [5] proposes a prediction control based on pulse width modulation (PWM) and MTPA current optimization. Literature [6] analyzes the loss under different field-weakening controls based on the deadbeat prediction scheme in detail. Among them, the performance of MTPA is better than other field-weakening control, such as maximum torque per voltage control.

Due to the advantages of the structure and working principle of PMSM, both the rotor and accurate torque expressions are accurate. However SRM is not qualified to construct rotating coordinate system because of the loss rotor flux. In addition, the high non-linearity of SRM will increase complexity, especially the current decoupling control, the application of traditional MTPA in SRM is limited. In order to solve these issues, adopting high torque per ampere (HTPA) to analyze the relationship between torque and current is practical. HTPA is based on the idea of MTPA to find the maximum points of TPA corresponding to different torques for SRM. These points are also the maximum torque -current ratio under the given torque. The ideal of HTPA can be introduced into smooth torque control as a current optimization module.

As an effective method to solve SRM smooth torque control, direct instantaneous torque control (DITC) formulates different hysteresis control rules for single-phase operation region and overlap region of two-phase simultaneous operation according to real-time rotor position, and directly selects switch vector output to power converter through hysteresis control rules. The hysteresis control rule directly selects the switch vector output to the power converter [7-8]. However, DITC requires complex operation rules to generate smooth torque in the overlap region, the current peak is too high in the initial stage of current rise so does the torque. This performance leads to the lower value of TPA. This phenomenon can lead to excessive peak-to-peak current and low motor efficiency.

* Corresponding author: aidexu@dlmu.edu.cn

For the excessive peak-to-peak current value at the initial stage caused by DITC algorithm of SRM, a new current optimization scheme of DITC combined with HTPA control strategy is proposed, that is HTPA-DITC. Through the ideal of HTPA, the largest torque-current ratio point in one electrical cycle can be obtained. In addition, the mechanical angle corresponding to the point can also be obtained. Using the special mechanical angle and piecewise variable duty cycle modulation, overlap region of DITC is sector-divided and operation rule is optimized as well. The optimization control strategy is verified in simulation and experiment.

2 Principle of the proposed HTPA

2.1. Torque calculation of switched reluctance motor

According to kirchhoff's voltage law, the single-phase instantaneous voltage of SRM is expressed as

$$U_k = i_k R + \frac{d\psi_k}{dt} = i_k R + \frac{\partial \psi_k(\theta,i_k)}{\partial i_k}\frac{di_k}{dt} + \frac{\partial \psi_k(\theta,i_k)}{\partial \theta}\frac{d\theta}{dt} \quad (1)$$

In the formula, k represents phase number, $k=1,2,3$, U_k represents phase voltage, i_k represents phase current, R represents the winding impedance of the motor, $\psi_k(\theta,i_k)$ represents phase flux linkage.

In SRM, torque is usually calculated according to the magnetic common energy, and the general formula is shown as follows

$$T_k = \frac{\partial}{\partial \theta} W'(\theta,i_k) \mid i_k = const \quad (2)$$

Where $W'(\theta,i_k)$ represents the magnetic co-energy, which is expressed by equation (3)

$$W'(\theta,i_k) = \int_0^{i_k} \psi_k(\theta,i_k) di_k = \int_0^{i_k} L_k(\theta,i_k) i_k di_k \quad (3)$$

Formula (3) shows that the magnetic common energy is a function of position Angle and phase current, $L_k(\theta,i_k)$ is the phase inductance. When the magnetic saturation effect is ignored, $L_k(\theta,i_k)$ can be reduced to a function independent of phase current $L_k(\theta)$. In the linear inductance model, the phase torque and composite torque can be expressed as equations (4) and (5):

$$T_k = \frac{1}{2} i_k^2 \frac{dL_k(\theta)}{d\theta} \quad (4)$$

$$T_{total} = \sum_{k=1}^{3} T_k(\theta,i_k) \quad (5)$$

Captions should be typed in 9-point Times. They should be centred above the tables and flush left beneath the figures.

2.2 Analysis of HTPA point

When the winding pressure drop is ignored, formula (1) can be rewritten as

$$U_k = L_k \frac{di_k}{dt} + i_k \frac{dL_k}{dt} = L_k \frac{di_k}{dt} + i_k \frac{dL_k}{d\theta}\omega_r \quad (6)$$

From (6), (7) is available

$$\frac{U_k}{\omega_r} = L_k \frac{di_k}{d\theta} + i_k \frac{dL_k}{d\theta} \quad (7)$$

where ω_r is rotor speed, in the linear inductance model, The change of inductance slope is called θ_{lu}. From the opening angle θ_{on} to θ_{lu}, inductance is remained constant as $L = L_{\min}$, $i(\theta_{on}) = 0$ then

$$i_k(\theta) = \frac{U_k}{L_{\min}} \cdot \frac{\theta - \theta_{on}}{\omega_r} \quad (8)$$

The rate of current change is

$$\frac{di_k(\theta)}{d\theta} = \frac{U_k}{\omega_r L_{\min}} \quad (9)$$

According to equations (8) and (9), the current reaches its maximum at θ_{lu}, and the rate of current changing is inversely proportional to the speed. When the speed rises, the phase current change rate decreases, and the current tracking ability decreases [9]. As the load is increased, in order to provide sufficient torque, the peak current value has to increase to obtain a sufficiently large current.

To solve the problem of excessive peak-to-peak current, the torque-current relationship can be established based on the ideal of HTPA. HTPA means the torque ratio generated by stator phase current reaches the maximum value. Under the constant output torque situation, if the motor consumes less stator current, the efficiency of the motor can be improved.

The average torque of an electric period, root mean square value of current and the definition of TPA can be expressed by equation (10), (11), (12) respectively:

$$T_{av} = \frac{\int_0^{\frac{2\pi}{N_r}} \frac{1}{2} i_k^2 \frac{dL_k(\theta)}{d\theta} d\theta}{2\pi/N_r} \quad (10)$$

$$I_{rms} = \sqrt{\frac{\int_0^{\frac{2\pi}{N_r}} i_k^2 d\theta}{2\pi/N_r}} \quad (11)$$

$$\mu = \frac{T_{av}}{I_{rms}} = \sqrt{\frac{N_r}{8\pi}} \frac{\int_0^{\frac{2\pi}{N_r}} i_k^2 \frac{dL_k(\theta)}{d\theta} d\theta}{\sqrt{\int_0^{\frac{2\pi}{N_r}} i_k^2 d\theta}} \quad (12)$$

N_r is the logarithm of rotor poles. In the inductance linear region, $dL_k(\theta)/d\theta$ is a constant value, and the inductance slope in the nonlinear region is zero. From equations (9) and (12), θ_{lu} is the point with the maximum torque current ratio.

3 Implementation of HTPA-DITC

3.1 Re-division the sectors

Although θ_{lu} is the point with the maximum torque current ratio, it can also be called as critical overlap. The inductance is very small before this point, the current rises very quickly, then the inductance enters the linear region, the slope of the inductance is the largest, and the current is quickly consumed. As the large TPA represents less copper consumption, to a certain extent, it means higher efficiency. In this paper, the mechanical angle of each phase is divided into five regions with θ_{lu} as the dividing point and the nonlinear inductance model, as shown in figure 1. Taking phase A as an example, A_{on} is the opening angle, A_{off} is the relevant breaking Angle, and A_{lu} is the change point of inductance slope of phase A.

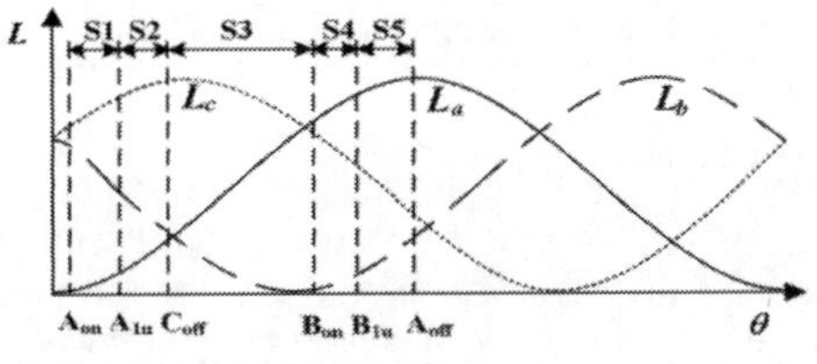

Fig.1. Re-division the sectors

3.2 Operation rule of HTPA-DITC

If the three-phase winding working order of SRM is C→A→B→C, hysteresis band is . Different and variable duty cycles are applied to the overlapping region and single-phase conducting region respectively. Among them, the phase A can be taken as an example:

S1: Phase A acts as the excitation phase. When A is turned on, the inductance of phase A is small, the rate of change of inductance is almost zero, and the torque that can be supplied is small. In this region, the torque is mainly supplied by phase C. When the torque error is less than the lower bound of the hysteresis loop, $\Delta T \leq -T_H$, switch off A-phase. When the torque error is larger than the upper limit, turn it on.

S2: At this time, the A-phase inductance is still small, but the inductance change rate is increasing, and the torque variation in this sector is relatively severe. Before the C phase switch off, the torque still needs to be provided by C phase. When the error is within the torque hysteresis and over zero, phase A should under the free-wheeling state.

S3: The sector is a single-phase operation region. When the torque error is over zero, the phase A is turned on, and the duty ratio is determined according to the duty ratio formula. When the torque error is less than zero, phase A continues to flow. In this sector phase C is completely shut down.

S4: At this time, the A phase will be transformed into the demagnetization phase, and phase B will face the same situation when it is just turned on. The torque is mainly provided by phase A.

S5: Before phase A is completely turned off and the torque of phase B has not been established yet, it is still necessary to excite Phase A. After the torque of phase B is established, the phase A is brought into the freewheeling state, and after the turn-off angle, the back pressure is applied.

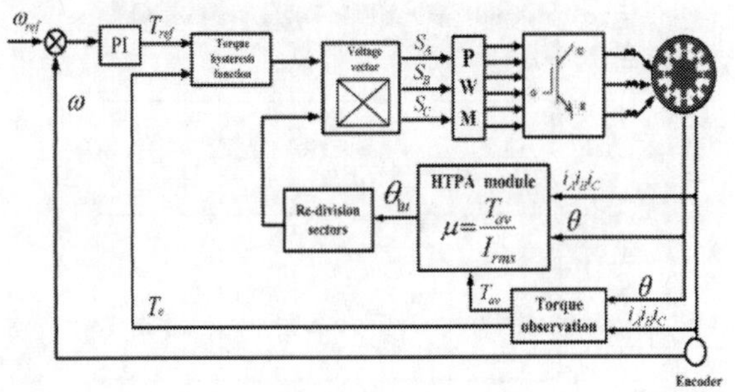

Fig.2 Control block diagram of HTPA-DITC

4 Simulation analysis

In this paper, the DITC and HTPA-DITC are simulated and verified in MATLAB from a three-phase 12/8-pole SRM. At the same turn-on turn-off angle, 650 rpm, 5 N·m, the comparison of HTPA-DITC and DITC is shown in figure 3. In addition, 1000 rpm, 5 N·m, the comparison of HTPA-DITC and DITC is shown in figure 4.

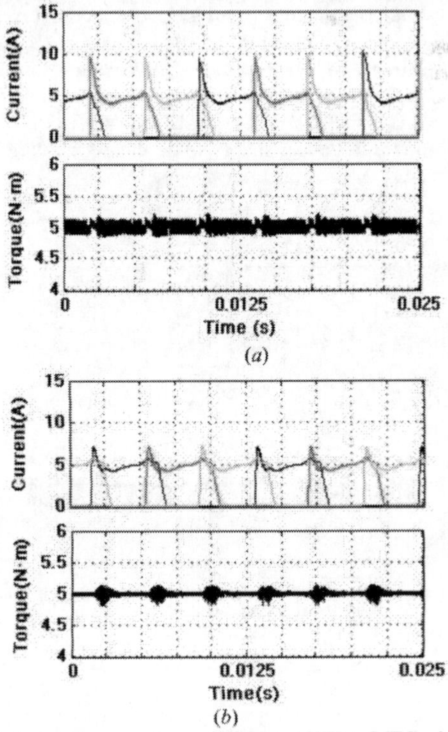

Fig.3 Simulation waveform of DITC and HTPA-DITC with 650 rpm, 5 N·m (*a*)DITC(*b*)HTPA-DITC

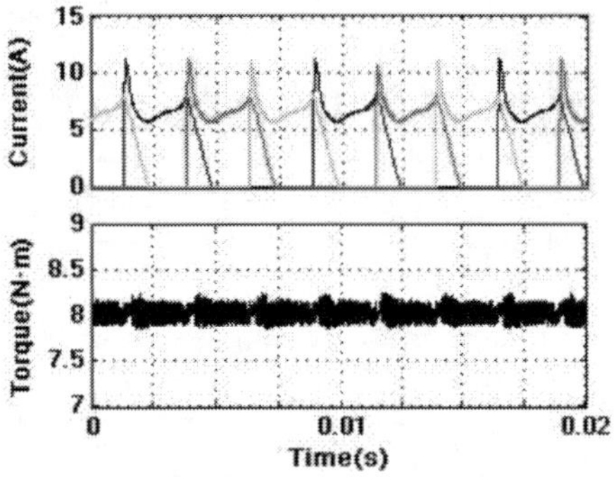

(*a*)

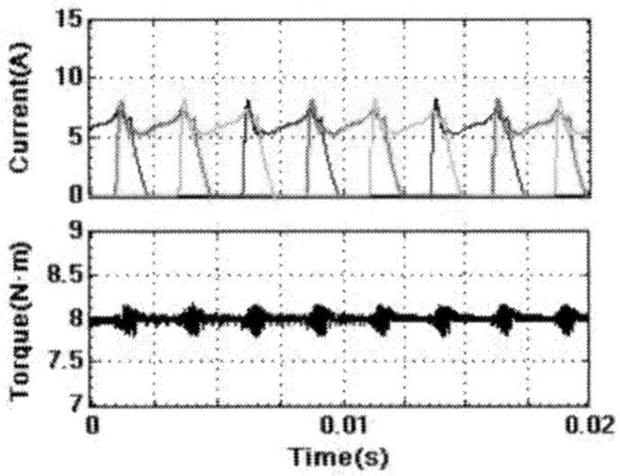

(*b*)

Fig.4 Simulation waveform of DITC and HTPA-DITC with 1000 rpm, 8 N·m (*a*)DITC (*b*)HTPA-DITC

As can be seen from figure 3 and figure 4, by HTPA control and segmenting the duty cycle modulation, the voltage vector is applied to the re-divided sectors, which can effectively reduce the current peak value about 2.1A and 3.7A. Figure 5 is a comparison of the flux linkage of DITC and HTPA-DITC. the DITC algorithm's flux diagram is a three-leaf shape, while the optimized DITC flux is more even at the tip of the blade.

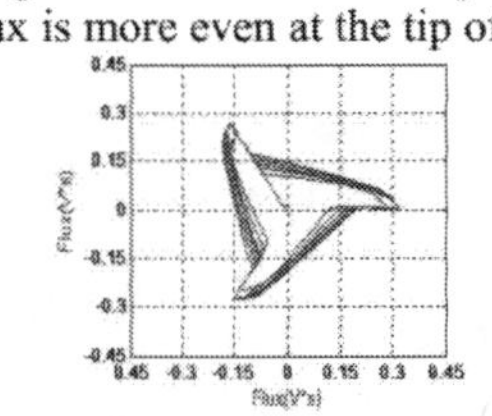

(*a*)

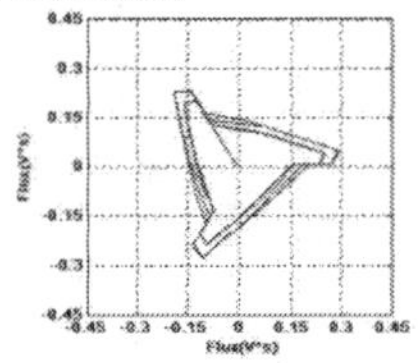

(*b*)

Fig.5 Flux-linkage circle of DITC and HTPA-DITC(*a*)DITC (*b*)HTPA-DITC

Figure 6 shows TPA line chart and current peak value of DITC and HTPA-DITC under different loads and the different speeds.

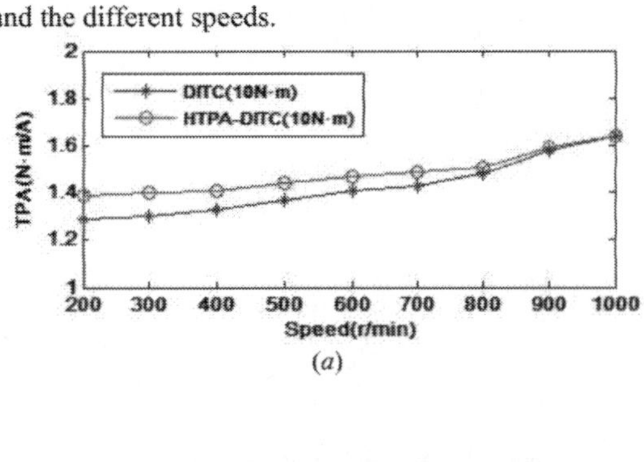

(*a*)

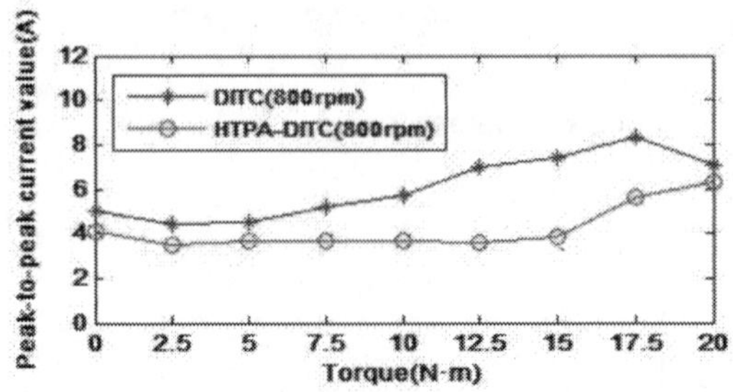

(*b*)

Fig.6. Comparison of DITC and HTPA-DITC with different situations (*a*)DITC (*b*)HTPA-DITC

According to figure 5, the DITC algorithm's flux diagram is a three-leaf shape, while the optimized DITC flux is more even at the tip of the blade. It can be seen from figure. 6 that the torque current ratio of HTPA-DITC is higher than DITC in a large range, and the peak-to-peak current value of HTPA-DITC is lower than DITC, especially to the medium speed with light loading.

5 Conclusion

This paper points out many problems in traditional DITC, such as high peak-to-peak current and low torque-to-current ratio at the early stage of current rising. By combining the ideal of MTPA, the points of HTPA can be obtained. These points make the interval of HTPA-DITC redivide and conduction rule change. HTPA-DITC also optimizes the conduction rule, and modulates the optimal voltage vector by using the segmentation variable duty ratio. Therefore, PWM-DITC not only improves the current waveform, but also reduces the peak-to-peak current value, and further optimizes the torque-to-current ratio.

This work is supported by: Special Innovation Projects for the Double-First Class Construction under grant no.CXXM2019SS005, Dalian Science Technology Innovation Fund Project under grant no.2018J12GX039 and National Natural Science Foundation of China under grant no.51407021.

References

1. 1.A. Xu, C. Shang, J. Chen, J. Zhu and L. Han, A New Control Method Based on DTC and MPC to Reduce Torque Ripple in SRM. IEEE Access, **7**, 68584-68593(2019).
2. S. Morimoto, M. Sanada and Y. Takeda, *Wide-speed operation of interior permanent magnet synchronous motors with high-performance current regulator*. IEEE Tran on Industry Applications, **30**, 920-926(1994).
3. T. Inoue, Y. Inoue, S. Morimoto and M. Sanada, *Maximum Torque Per Ampere Control of a Direct Torque-Controlled PMSM in a Stator Flux Linkage Synchronous Frame*. IEEE Trans on Industry Applications, **52**, 2360-2367(2016).

4. K. Li and Y. Wang, *Maximum Torque Per Ampere (MTPA) Control for IPMSM Drives Based on a Variable-Equivalent-Parameter MTPA Control Law.* IEEE Trans on Power Electronics, **34**, 7092-7102(2019).
5. R. S. Dastjerdi, M. A. Abbasian, H. Saghafi and M. H. Vafaie, *Performance Improvement of Permanent-Magnet Synchronous Motor Using a New Deadbeat-Direct Current Controller*. IEEE Trans on Power Electronics, **34**, 3530-3543(2019).
6. M. Saur, F. Ramos, A. Perez, D. Gerling and R. D. Lorenz, *Implementation of deadbeat-direct torque and flux control for synchronous reluctance machines to minimize loss each switching period.* APEC, 215-220(2016).
7. C. Shang, A. Xu, L.Huang and J. Chen, *Flux linkage optimization for direct torque control of switched reluctance motor based on model predictive control.* IEEJ Tran on Electrical and Electronic Engineering, **14**,1105-1113(2019).
8. Zhu Yesheng, Zhang Guobao, Huang Yongming. *PWM -Based direct Instantaneous torque control of switched reluctance Machine.* Trans of China Elect Society,**32**, 31-39(2017).
9. YE J, BILGIN B, EMADI A. *An extended-speed low-ripple torque control of switched reluctance motor drives.* IEEE Trans on Power Electronics, **30**, 1457-1470(2015).

MATEC Web of Conferences **306**, 04006 (2020) https://doi.org/10.1051/matecconf/202030604006
ICMME 2019

Vibration Energy Harvest for Low Frequency using Double-piezoelectric Cantilever Beam

Chuttipong Tuma* and *Danai* Phaoharuhansa

Department of Mechanical Engineering, Faculty of Engineering, King Mongkut's University of Technology Thonburi, Bangkok, Thailand

Abstract. This study introduces design methods of natural frequency for double-piezoelectric cantilever beams in order to absorb electricity from mechanical vibration. The natural frequencies of double cantilever beams are designed by quadratic equation such as 6.45 and 18.14 Hz. On the other hands, FEA is used to recheck the result. FEA simulation presents the natural frequencies such as 4.4614, 8.4793, 13.8020 and 25.1050 Hz. It works in low frequency and large bandwidth. The experimental results show the peak electric power at 5 and 22 Hz. The error may occur by stiffness of piezoelectric and effective of mounting position.

1 Introduction

Piezoelectric is used to convert vibration energy to electricity. It is usually attached on mechanism, which absorb some force or pressure. Stress and strain with occur in the structure. Then, piezoelectric generate electricity relative to amount stress and strain. So that it obtains good performance of high frequency.

The most piezoelectric cantilever beam is designed only one frequency, it has a natural frequency related to ambient vibration sources [1]. The single piezoelectric cantilever beam gives narrow bandwidth. Performance of single cantilever beam decreases when ambient vibration sources changed frequency.

Several papers [2-5] present the multiple cantilever beam arrays. It increases bandwidth and electric power for scavenge vibration energy ambient. So, it advantages more than single cantilever beam. The most multiple cantilever beams were designed natural frequency between 22 to 245 Hz. Multiple cantilever beams stacked to increase the bandwidth of natural frequencies. It is well known as piezoelectric cantilever beam arrays. The bandwidth of piezoelectric cantilever beam arrays is tuning mass or dimension beam.

Frequency of double-piezoelectric cantilever beams can tune by changing tip mass or stiffness. Therefore, this paper deals quadratic equation and finite element method (FEM) for tuning natural frequency of the double-piezoelectric cantilever beams. FEA is descriptive detail of geometry, is a powerful for mechanical engineering problems. The solutions of FEA are verified condition of the problem [6-7].

This paper aims to design the double cantilever beams to harvest the energy, frequency of which should lower than 22 Hz. The low frequency will be suitable for ambient energy.

2 Piezoelectric energy harvest

Piezoelectric material can generate electric when it receives some force. Stress and strain occur in material as shows in the Fig. 1. It calls direct piezoelectric phenomenal.

The relationship of stress, strain and electric field of piezoelectric [8], it is given as

$$T_p = E_p S_p - e_p E_p \tag{1}$$

where, T_p and S_p are stress and strain in piezoelectric plate. E_p is young's modulus of piezoelectric, e_p is piezoelectric coefficient and E_p is electric field of piezoelectric.

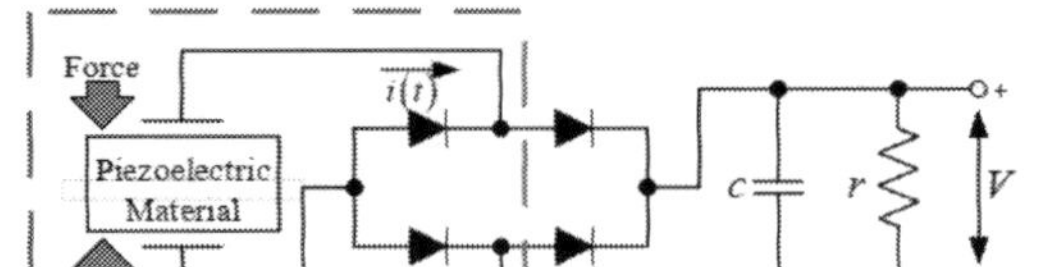

Fig. 1. Phenomenal direct piezoelectric.

3 Double mass spring system

The double-piezoelectric cantilever beams as shown in Fig. 2, they consist of two springs and two tip masses. The clamped area is attached to moving base. The base excitation has been moving sinusoidal. The mathematical model of double-piezoelectric cantilever beams under base excitation is given as

* Corresponding author: chuttipong.tuma@gmail.com

$$y(t) = y_0 \sin(\omega t) \quad (2)$$

$$m\ddot{x} + kx = U \quad (3)$$

where, $m = \begin{bmatrix} m_1 & 0 \\ 0 & m_2 \end{bmatrix}$, $x = \begin{bmatrix} x_1 \\ x_2 \end{bmatrix}$, $\ddot{x} = \begin{bmatrix} \ddot{x}_1 \\ \ddot{x}_2 \end{bmatrix}$, $U = \begin{bmatrix} k_1 y(t) \\ 0 \end{bmatrix}$ and $k = \begin{bmatrix} k_1 + k_2 & -k_2 \\ -k_2 & k_2 \end{bmatrix}$, m_1 is first tip mass and m_2 is second tip mass, k_1 is first spring stiffness constant and the k_2 is second spring stiffness constant, y_0 and ω are the amplitude of base excitation and frequency of base excitation, respectively. $\ddot{x}_1$ and $\ddot{x}_2$ are the acceleration of m_1 and m_2, respectively. x_1 and x_2 are the displacement of m_1 and m_2, respectively.

The natural frequency of the two-degree freedom system can find from underdamped vibration equation, it is written as

$$m\ddot{x} + kx = 0 \quad (4)$$

Two possible natural frequencies obtained from quadratic equation [9] as

$$\det\left(-\omega^2 \mathrm{M} + \mathrm{K}\right) = 0 \quad (5)$$

In this paper, double cantilever beams are designed two natural frequency, such as 6.45 and 18.41 Hz.

4 Mechanical of single cantilever beam

Fig. 3 shows the single cantilever beam with the tip mass system. The cantilever beam has behaviour as spring. The spring stiffness is calculated by deflection of cantilever beam [10], it is given as

$$\delta_{max} = \frac{mgl^3}{3EI} \quad (6)$$

where, m is the mass of tip mass, E is the young's modulus of material, l is length of beam and I is the moment of inertia of cantilever beam.

The spring stiffness of cantilever beam is expressed as

$$k = \frac{F}{x} = \frac{mg}{\delta_{max}} = \frac{3EI}{l^3} \quad (7)$$

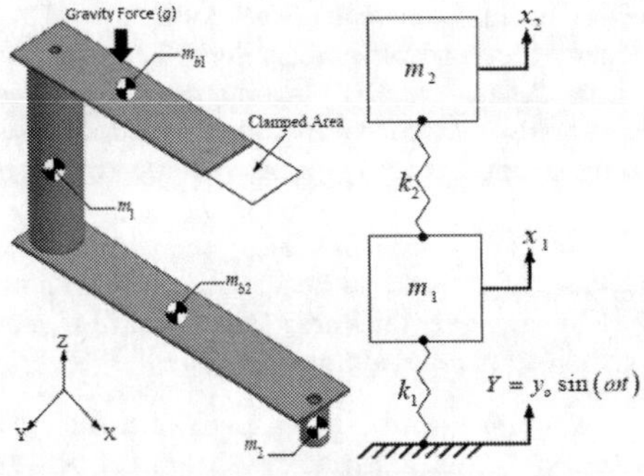

Fig. 2. Geometric of double mass spring.

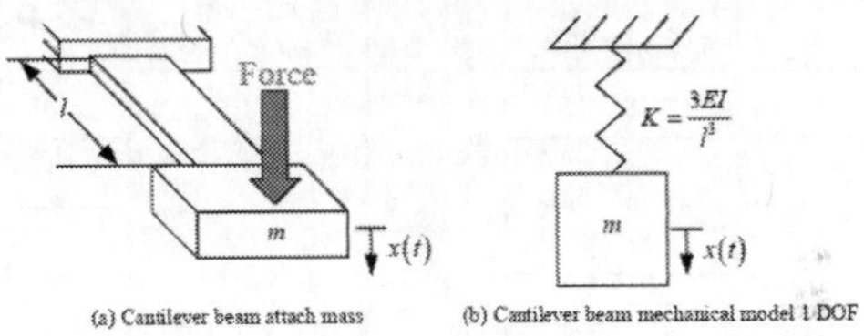

Fig. 3. Single cantilever beam with tip mass.

5 Finite element analysis of double-piezoelectric cantilever beam design

Finite element analysis (FEA) can derive the mathematical model of engineering problem.

The structure of double cantilever beams was calculated maximum stress because the double cantilever beams are received by gravity force as shown in Fig. 2. When a cantilever beam has a stress nearby yield strength of mechanical material properties which provide higher electrical power.

From the shape function of beam element, beam element stiffness matrix can be derived from Galerkin's method [11]. Stiffness of the beam is defined that

$$K = \frac{EI}{L^3}\begin{bmatrix} 12 & 6L & -12 & 6L \\ 6L & 4L^2 & -6L & 2L^2 \\ -12 & -6L & 12 & -6L \\ 6L & 2L^2 & -6L & 4L^2 \end{bmatrix} \quad (15)$$

Lumped-mass element matrix in FEA is defined that

$$M = \frac{\rho AL}{420}\begin{bmatrix} 156 & 22L & 54 & -13L \\ 22L & 4L^2 & 13L & -13L^2 \\ 54 & 13L & 156 & -22L \\ -13L & -13L^2 & -22L & 4L^2 \end{bmatrix} \quad (12)$$

5.1 Static Structure analysis of double cantilever beams

The stress on the structure is calculated by static structural analysis. In this paper, cantilever beams are created by aluminum material. Spring stiffness can calculate from Eq. (7) and natural frequency of double cantilever beams can calculate from Eq. (5). The proprieties of top and bottom cantilever beams shown as Table 1 and 2, respectively. Maximum stress is occurring on top cantilever beam because it received all masses. The comparison between spring stiffness with stress, which relate to increasing tip mass on cantilever beams is shown in Fig. 4. The indicates comparison between the thickness of a cantilever beam with stress is shown in Fig. 5. The cantilever thickness 1 mm beam has a stress nearby yield strength of aluminum properties.

Table 1. Proprieties of top cantilever beam

Thickness (mm)	Stiffness of Top cantilever beam (N/m²)	Width (mm)	Length of Top cantilever Beam (mm)	Tip mass (kg)
0.5	65.625			0.0038
1.0	525.0			0.305
1.5	1771.87	30	100	1.03
2.0	4200.0			2.4402
2.5	8203.13			4.766

Table 2. Proprieties of bottom cantilever beam

Thickness (mm)	Stiffness of Bottom cantilever beam (N/m²)	Width (mm)	Length of bottom cantilever Beam (mm)	Tip mass (kg)
0.5	19.44			0.0015
1.0	155.6			0.0122
1.5	525	30	150	0.041
2.0	1244.4			0.0972
2.5	2430.56			0.1898

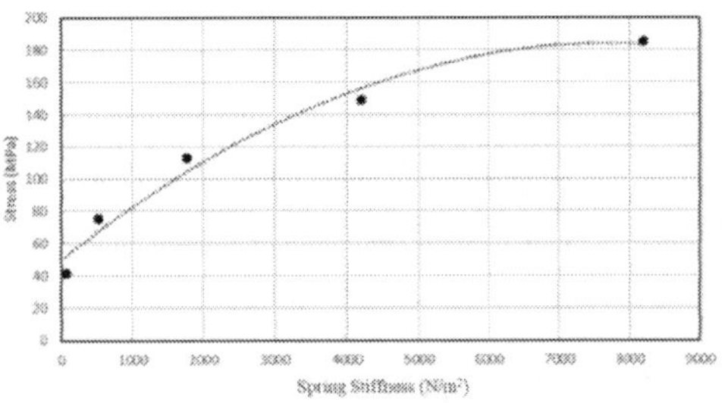

Fig. 4. Spring stiffness versus stress.

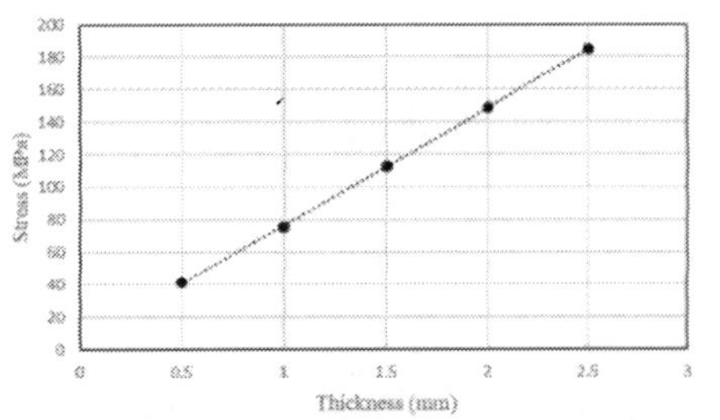

Fig. 5. Thickness stiffness versus stress.

5.2 Modal analysis of double cantilever beams

According to the double spring mass system, it can use modal analysis to solve the dynamics model. The modal analysis can derive the natural frequency and mode shape of structure. The boundary condition is shown in Fig. 2 and the simulation is neglected gravity force. Fig. 6 presents the modal analysis result and it consists of 4 modes shapes such as bending and swaying of the both beams.

The first mode shape is that the bending mode of the top plate cantilever beam and the natural frequency equals 4.4614 Hz as shown in Fig. 6(a). The second mode is that the bending mode of the bottom plate cantilever beam and the natural frequency equals 13.802 Hz as shown in Fig. 6(b). The third mode shape is that bending mode of both plate cantilever beam and natural frequency equals 25.105 Hz as shown in Fig. 6(c). The fourth mode shape is that the swaying mode of the bottom plate cantilever beam and the natural frequency equals 8.4793 Hz as shown in Fig. 6(d).

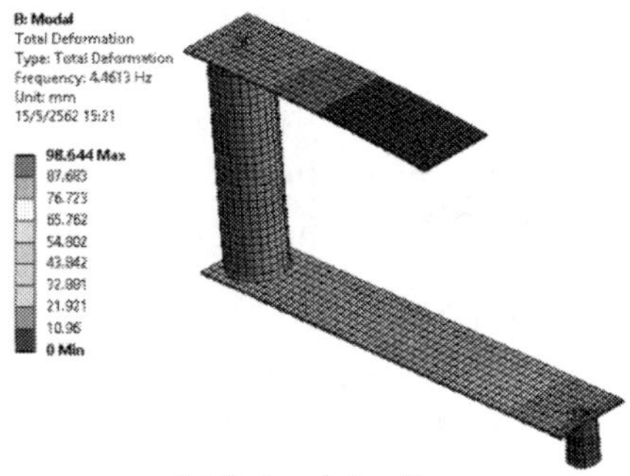

(a) first mode bending

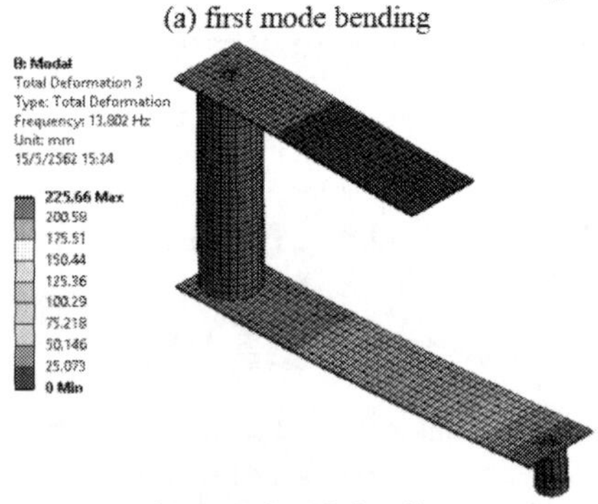

(b) second mode bending

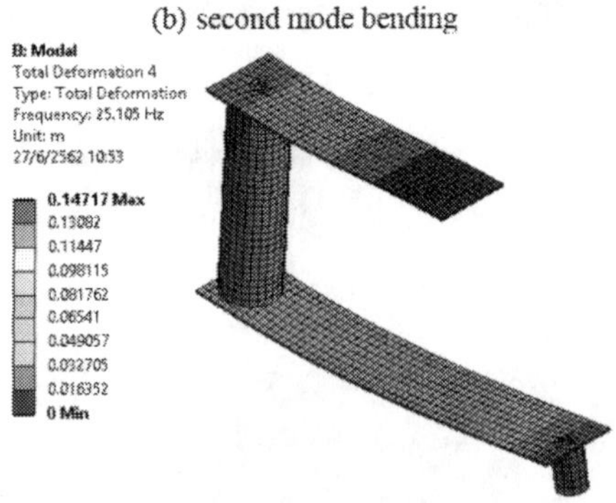

(c) third mode bending

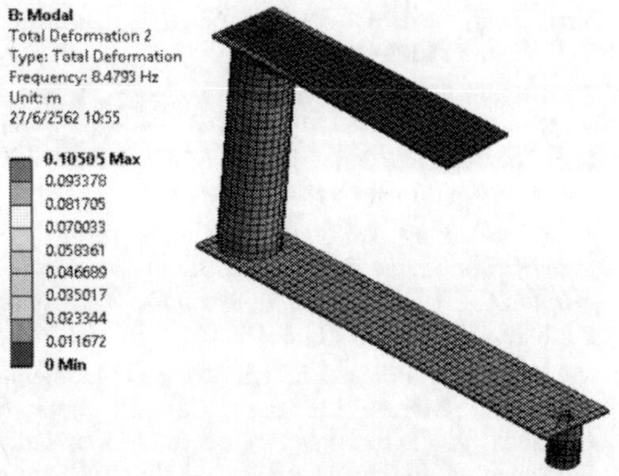

(d) fourth mode swaying

Fig. 6. Natural frequency and mode shape of double cantilever beam arrays system (a) bending mode of top plate cantilever beam (b) bending mode of bottom plate cantilever beam (c) bending mode of both plate cantilever beam (d) swaying mode of bottom plate cantilever beam.

6 Experiment and result

To experiment the double-piezoelectric cantilever beams in Fig. 2, base excitation experiment kit has developed as shown in Fig. 7. It consists of roller guide, moving base, stepping motor and encoder. The beam is clamped at clamp area with a moving base as shown in Fig. 2 and the moving base has been moved on the roller guide by stepping motors. The frequency of moving base is between 0.15-30 Hz.

PPA-1001 is a piezoelectric plate that it used to harvest energy in the experiment, and the properties are exhibited in Table 3 maximum power obtain at 70 kΩ. The power of double-piezoelectric cantilever beams is defined as

$$P_V = \frac{V^2}{r} \tag{16}$$

where, V is the voltage across the electronic load resistance and r is the electronic load resistance.

Frequency (Hz)	22	22
Maximum Power (mW)	1.4	4.4
Maximum Voltage (V)	9	17.3
Maximum Current (mA)	0.1	0.3

Table 3. Properties of PPA-1001 piezoelectric plate

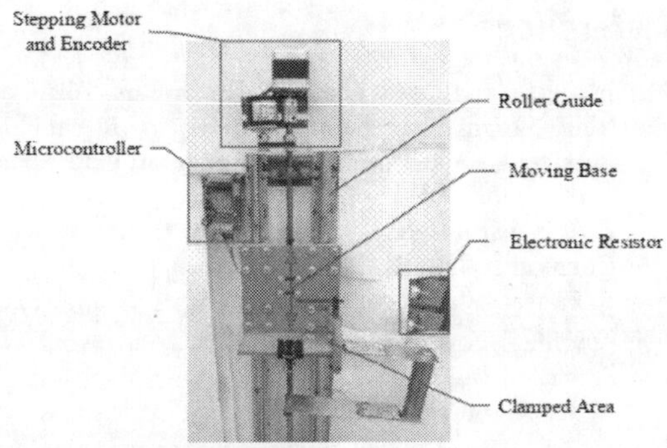

Fig. 7. Base excitation prototype.

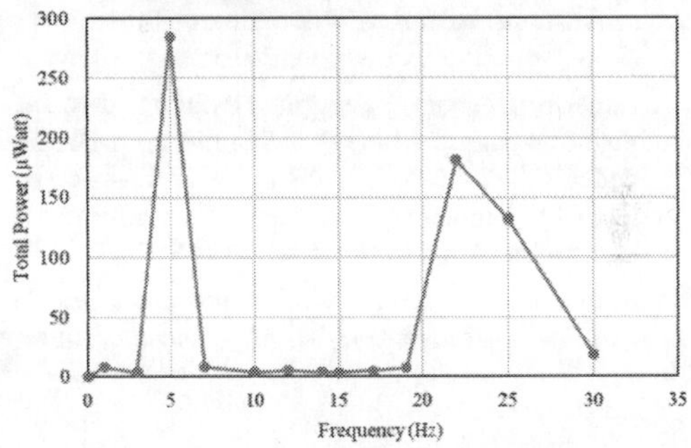

Fig. 8. Total power output versus frequency at 1-30Hz.

Power of double-piezoelectric cantilever beams at 0.15-0.2 Hz is an increase in a linear trend-line equation. Power of double-piezoelectric cantilever beams are equal, 0.2995 μW at 0.15 Hz, 0.4159 μW at 0.175 Hz and 0.2 μW at 0.2 Hz.

Fig. 8 exhibits compare between total power of double-piezoelectric cantilever beams in the frequency range 1-30 Hz, the first peak total power is 284.4 μWatt at 5 Hz, and the second peak total power is 181.78 μWatt at 22 Hz.

7 Conclusion

Double-piezoelectric cantilever beams in this study are designed for low frequency. The natural frequencies of beams are designed at 6.45 and 18.41 Hz, respectively. FEA simulation presents the natural frequencies such as 4.4614, 8.4793, 13.8020 and 25.1050 Hz, respectively. The error of FEA simulation occurs by calculating mass of cantilever beams. The peak power of experiment is 284.4 μWatt at 5 Hz and 181.78 μWatt at 22 Hz, respectively. The frequencies at peak power are bending mode of top plate and bending mode of both plates. Therefore, the natural frequencies of the experiment have a slight error from mathematics model and FEA simulation. The error may occur by the clamp position and the stiffness of piezoelectric plate.

References

1. Eli S. Leland, Jessy Baker, Eric Carleton, Elizabeth Reilly, Elaine Lai, Brian Otis, Jan M. Rabaey, V. Sundarajan and Pual K. Wright, Improving power output for vibration-Based energy scavengers, Energy harvesting & conservation, IEEE Pervasive Computing **4**, pp.28-36 (2005)
2. In-Ho Kim, Hyung-jo Jung, Bo Mi Lee and Seon-jun Jang, Broadband energy harvesting using two degree-of-freedom vibration body, Applied Physics Letter **98**, pp.1-3 (2011)
3. Yu-Yin Chen and Dejan Vasic, *Energy harvesting of two cantilever beams structure: Inter facing circuit discussion*, SPIE Smart Structures and Nondestructive Evaluation (2015)
4. Eduard Dechant, Feodor Dedulov, Leonid Y. Fetisov and Mikhail Shamonin, Bandwidth widening of piezoelectric cantilever beam arrays by mass-tip tuning for low-frequency vibration energy harvesting, Applied Science **7**, 12(2017)
5. S. C. Lin, B.S. Lee, W.J. Wu and C.K. Lee, *Multi-cantilever piezoelectric MEMS generator in energy harvesting*, IEEE International Ultrasonics Symposium (2009)
6. C. Birleanu, S. Talu, *Machine elements. Designing and computer assisted graphical representations*, Cluj-Napoca, Romania, Victor Melenti Publishing house, (2001). ISBM 973-99539-6-4.
7. T. Nitulescu, S. Talu, *Applications of descriptive geometry and computer aided design in engineering graphics*, Cluj-Napoca, Romania, Risoprint Publishing house, (2001). ISBN 973-656-102-X.
8. Renato Calio, Udaya Bhaskar Rongala, Domenico Camboni, Mario Milazzo, Cesare Stefanini, Gianluca de Petris and Calogero Maria Oddo, Piezoelectric Energy Harvesting Solution, Sensors **14**, pp.4755-4790 (2014). ISSN 1424-8220.
9. S. Graham Kell, *Fundamentals of Mechanical Vibrations*, McGraw-Hill Companies **2**, pp.296 (2000)
10. Singiresu S. Rao, *Mechanical Vibrations*, Prentice Hall **5**, pp.28 (2011)
11. Xiaolin Chen and Yijun Liu, *Finite Element Modeling and Simulation with ANSYS Workbench*, CRC Press **1**, pp. 63-65, 228-230 (2015)

https://doi.org/10.1051/matecconf/202030604007

Minimization torque ripple for SRM based on flux linkage partition in DB-DTFC

Qianni Li[1], *Aide* Xu[1,*], *Chaoyi* Shang[2], and *Lepeng* Huang[2]

[1]College of Information and Science Technology, Dalian Maritime University, 116026 Dalian, China
[2]College of Electrical Engineering of Ships, Dalian Maritime University, 116026 Dalian, China

Abstract. This paper proposes a novel deadbeat torque and flux control (DB-DTFC) to reduce torque ripple for switched reluctance motor (SRM). DB-DTFC combines the advantages of direct torque control (DTC) and space-vector modulation (SVM). DB-DTFC leads current vector control into DTC in order to find the equation between torque and current through deadbeat prediction theory i.e. a beat reaches a given point. In addition, the deadbeat calculation module here is similar to that of permanent magnet synchronous motor. Based on *dq*0 reference frame of SRM, the most suitable *dq*0 axis current of next moment corresponding to different torque errors is calculated and predicted. According to the calculated *dq*0 axis current, the optimal space voltage vectors can be selected to reduce torque ripple. In order to verify the effectiveness and correctness of the proposed scheme, DB-DTFC is verified and compared with the DTC-SVM by simulation.

1 Introduction

As a new type of reluctance motor, switched reluctance motor (SRM) has attracted much attention because of its outstanding advantages, such as simple and robust structure, wide speed range, good reliability and high efficiency [1]. However, double salient structure and the principle of minimum reluctance lead to SRM's nonlinear strong coupling characteristic. This characteristic is an important reason for large noise and torque ripple of SRM [2].

Traditional current chopper control and angle control can not make speed regulation and torque ripple reduction compatible perfectly, then space-vector modulation (SVM) and direct torque control (DTC) are proposed. With the development of DTC, many DTC-based control method has been explored. [3] proposes direct instantaneous torque control (DITC) based on double hysteresis control. [4] proposed the direct torque and predictive flux control method based on model prediction of SRM.

Current vector control (CVC) is an important part of vector control. The traditional CVC is only closed loop control of stator current in the *d-q* coordinate system without torque closed loop. When the idea of CVC is introduced into DTC, the relationship between torque and current can be found. Deadbeat direct toque flux control (DB-DTFC) of induction machine (IM) [5] is proposed. DB-DTFC not only has the advantages of DTC and SVM, but also can calculate the *d-q* axis current through torque error, so as to the selection of optimal voltage vectors. Because of these advantages, DB-DTFC has been applied interior permanent magnet synchronous machine (IPMSM) [6] and synchronous reluctance motor [7].

Since the promotion of DB-DTFC, many scholars have made contributions on improving DB-DTFC for IPMSM. In [8] deadbeat direct current control is proposed to reduce computation load of DB-DTFC . Although there is no torque closed loop in this method, the author use the idea of maximum torque per ampere to combine the torque and current. [9] proposes the loss minimization DB-DTFC from the point of loss and proved the validity of the method. Based on DTC-SVM , reference [10] proposes an optimal deadbeat control method which combined adjacent voltage vectors with optimal voltage vectors. Although DB-DTFC is widely used and developed in other motors, the application of DB-DTFC in SRM has been greatly limited. The reasons of the restriction are uncertain torque expression and the absence of rotor flux linkage.

According to [11], the mathematical model of *dq*0 coordinate system makes it possible for DB-DTFC to be applied on SRM. [12] proposes a new current vector control method based on the *dq*0 coordinate system and emphasizes the effect of zero-phase current on the torque control. Another vector control about speed-sensorless SRM drive is also proposed based on the *dq*0 coordinate system [13].

Since DB-DTFC have both advantages of DTC and SVM, this paper proposes a novel DB-DTFC of SRM based on *dq*0 coordinate system and deadbeat prediction theory. The *dq*0 coordinate system is established to find the equation between *dq*0 axis current and torque. As for the deadbeat prediction, it is used to find the optimal current and voltage under different torque errors. Due to

* Corresponding author: aidexu@dlmu.edu.cn

the difference between SRM and IPMSM, DB-DTFC of SRM is different from that of IPMSM, especially in the case of zero phase current optimization. The proposed method uses torque flux decoupling control and finds the optimal $dq0$ axis current. DB-DTFC is verified the validity and correctness by simulation. It is also compared with DTC-SVM for the performance of torque ripple under different conditions.

2 Principle of DB-DTFC scheme

2.1. Mathematical model of static coordinate

According to Kirchhoff's Law of Voltage, the k-phase voltage equation of SRM is:

$$u_k = R_k i_k + \frac{d\psi_k}{dt} \tag{1}$$

Expression (1) can also be written in matrix form as

$$\begin{bmatrix} u_a \\ u_b \\ u_c \end{bmatrix} = R\begin{bmatrix} i_a \\ i_b \\ i_c \end{bmatrix} + p(\begin{bmatrix} L_a & 0 & 0 \\ 0 & L_b & 0 \\ 0 & 0 & L_c \end{bmatrix}\begin{bmatrix} i_a \\ i_b \\ i_c \end{bmatrix}) \tag{2}$$

where R is the stator resistance; i_a, i_b, i_c are the phase currents of A, B, C phase, respectively; L_a, L_b, L_c are self-inductance of the corresponding phase; p is differential operator.

Among the matrix neglecting the mutual inductance, the self-inductance part is expressed as:

$$L_k(\theta_r) = L_{dc} + L_{ac}\cos(2\theta_r - (\mathrm{k}-1)\frac{2\pi}{N}) \tag{3}$$

where L_{dc} and L_{ac} are DC self-inductance and DC self-induction amplitude, respectively; θ_r is rotor position angle and N is phase number, $k = 1, 2, ..., N$,.

The instantaneous torque expression of SRM can be defined as:

$$T_e = \frac{P}{2}(i_a^2\frac{\partial L_a}{\partial\theta_r} + i_b^2\frac{\partial L_b}{\partial\theta_r} + i_c^2\frac{\partial L_c}{\partial\theta_r}) \tag{4}$$

where T_e is the instantaneous torque; P is magnetic pole number.

2.2 Mathematical model of rotating coordinate

According to literature [11], "rotor flux" generated by the DC current, when DC current is applied to each circuit. The rotating coordinate system is established by rotating the "rotor flux". In the synchronous rotating coordinate, current can be expressed as:

$$\begin{bmatrix} i_d \\ i_q \\ i_0 \end{bmatrix} = \sqrt{\frac{2}{3}}\begin{bmatrix} \cos 2\theta & \cos(2\theta - \frac{2\pi}{3}) & \cos(2\theta + \frac{2\pi}{3}) \\ -\sin 2\theta & -\sin(2\theta - \frac{2\pi}{3}) & -\sin(2\theta + \frac{2\pi}{3}) \\ \frac{1}{\sqrt{2}} & \frac{1}{\sqrt{2}} & \frac{1}{\sqrt{2}} \end{bmatrix}\begin{bmatrix} i_a \\ i_b \\ i_c \end{bmatrix} \tag{5}$$

where θ is the electrical angle; i_q and i_d are the q-axis and d-axis components of i_a, i_b and i_c, i_0 is zero-phase current.

Both substituting (3), (5) into (2) and ignoring asynchronous components can get the new matrix:

$$\begin{bmatrix} u_d \\ u_q \\ u_0 \end{bmatrix} = R\begin{bmatrix} i_d \\ i_q \\ i_0 \end{bmatrix} + \begin{bmatrix} L_{dc} & 0 & \frac{L_{ac}}{\sqrt{2}} \\ 0 & L_{dc} & 0 \\ \frac{L_{ac}}{\sqrt{2}} & 0 & L_{dc} \end{bmatrix} p(\begin{bmatrix} i_d \\ i_q \\ i_0 \end{bmatrix}) + 2\omega_r\begin{bmatrix} 0 & -L_{dc} & 0 \\ L_{dc} & 0 & \frac{L_{ac}}{\sqrt{2}} \\ 0 & 0 & 0 \end{bmatrix}\begin{bmatrix} i_d \\ i_q \\ i_0 \end{bmatrix} \tag{6}$$

2.3 Derivation of deadbeat control algorithms

Considering (4) as the torque expression in deadbeat control, when the speed in constant, $\partial L_a/\partial\theta_r$, $\partial L_b/\partial\theta_r$ and $\partial L_c/\partial\theta_r$ are determined by motor parameters. They can be regarded as constants. Therefore when the sampling time is small enough, only the three-phase current needs to be adjusted to control the torque increase and decrease .

Substituting (5) into (4), i_a, i_b and i_c can be replaced by i_q, i_d and i_0 to achieve the same control effect of torque as three-phase stator current. According to the decoupling control of DB-DTFC based on IPMSM [6], flux and torque can be decoupled into two independent systems. Torque is affected both i_q and i_0 [12], flux is only controlled by i_d.

In the torque control, considering i_q, i_0 as two different variables and considering torque as a constant, one quadratic equation about i_q and torque can be obtained, the other quadratic equation about i_0 and torque can also be obtained.

For the first quadratic equation about i_q and torque, the smaller value $i_q^*(k)$ can be acquired by adopting the quadratic formula. $i_q^*(k)$ can be expressed as:

$$i_q^*(k) = \frac{E_1}{D_1} \tag{7}$$

$$\begin{cases} D_1 = 4a\sin^2 2\theta + 4c\cos^2(\frac{\pi}{6}+2\theta) + 4b\sin^2(\frac{\pi}{3}+2\theta) \\ E_1 = 2i_d(a\sin 4\theta + b\sin(\frac{2\pi}{3}+4\theta) - c\sin(\frac{\pi}{3}+4\theta)) \\ \quad - \left\{ \begin{aligned} &(8a\sin^2 2\theta + 8c\cos^2(\frac{\pi}{6}+2\theta) + 8b\sin^2(\frac{\pi}{3}+2\theta))* \\ &[\frac{6\Delta T_e}{P} - 2i_d^2(a\cos^2 2\theta + b\cos^2(\frac{\pi}{3}+2\theta) + c\sin^2(\frac{\pi}{6}+2\theta)) - \\ &i_0^2(a+b+c) + 2\sqrt{2}i_0 i_d(b\cos(\frac{\pi}{3}+2\theta) + c\sin(\frac{\pi}{6}+2\theta) - a\cos 2\theta)] \\ &+[2i_d(a\sin 4\theta + b\sin(\frac{2\pi}{3}+4\theta) - c\sin(\frac{\pi}{3}+4\theta)) \\ &+2\sqrt{2}i_0(a\sin 2\theta + c\cos(\frac{\pi}{6}+2\theta) - b\sin(\frac{\pi}{3}+2\theta))]^2 \end{aligned} \right\}^{\frac{1}{2}} \\ \quad + 2\sqrt{2}i_0(a\sin 2\theta + c\cos(\frac{\pi}{6}+2\theta) - b\sin(\frac{\pi}{3}+2\theta)) \end{cases}$$

where $a = \partial L_a/\partial\theta_r$, $b = \partial L_b/\partial\theta_r$ and $c = \partial L_c/\partial\theta_r$; T_e is replaced by ΔT_e for reflecting the change of torque and $\Delta T_e = T_e^* - T_e(k)$. $a = \partial L_a/\partial\theta_r$, $b = \partial L_b/\partial\theta_r$ and $c = \partial L_c/\partial\theta_r$.

As for the other quadratic equation about i_0 and torque, the method of solving the minimum value $i_0^*(k)$ is similar to that of $i_q^*(k)$, $i_0^*(k)$ can be expressed as:

$$i_0^*(k) = \frac{E_2}{D_2} \quad (8)$$

$$\begin{cases} D_2 = 2(a+b+c) \\ E_2 = 2i_d(a\sin 4\theta + b\sin(\frac{2\pi}{3}+4\theta) - c\sin(\frac{\pi}{3}+4\theta)) \\ \quad - \left\{ \begin{aligned} &[2\sqrt{2}i_d(b\cos(\frac{\pi}{3}+2\theta) + c\sin(\frac{\pi}{6}+2\theta) - a\cos 2\theta) \\ &2\sqrt{2}i_q(a\sin 2\theta + c\cos(\frac{\pi}{6}+2\theta) - b\sin(\frac{\pi}{3}+2\theta)]^2 \\ &4(a+b+c)*[\frac{6\Delta T_e}{P} - 2i_d^2(a\cos^2 2\theta + b\cos^2(\frac{\pi}{3}+2\theta) + c\sin^2(\frac{\pi}{6}+2\theta)) \\ &-2i_q^2(a\sin^2 2\theta + c\cos^2(\frac{\pi}{6}+2\theta) + b\sin^2(\frac{\pi}{3}+2\theta)) \\ &+2i_d i_q(a\sin 4\theta + b\sin(\frac{2\pi}{3}+4\theta) - c\sin(\frac{\pi}{3}+4\theta))] \end{aligned} \right\}^{\frac{1}{2}} \\ \quad + 2\sqrt{2}i_q(a\sin 2\theta + c\cos(\frac{\pi}{6}+2\theta) - b\sin(\frac{\pi}{3}+2\theta)) \end{cases}$$

The two factors i_q and i_0 affecting the torque need to be set to $i_q(k+1)=i_q^*(k)$, $i_0(k+1)=i_0^*(k)$ and make sure that the values of the next moment are the smallest current values.

Due to the short sampling time, current can be discretized. In the discrete time, substituting $i_q^*(k)$ and $i_0^*(k)$ into (6), q-axis voltage $u_q^*(k)$ for torque control can be obtained as:

$$u_{sq}^*(k) = R_s i_q^*(k) + L_{dc}\frac{i_q^*(k) - i_q(k)}{T_s} + 2\omega_r(L_{dc} i_d(k) + \frac{L_{dc}}{\sqrt{2}} i_0^*(k)) \quad (9)$$

As for the flux control, i_d controls flux directly and d-axis voltage $u_d^*(k)$ for flux can also be obtained from:

$$\psi_s(k+1) = \psi_s^* = u_{qds}(k)T_s + \psi_s(k) \quad (10)$$

In order to simplify the calculation, the motor model is oriented in the direction of stator and the flux linkage coincides with the d-axis, so as to make ψ_{sq} equal to zero. Then discretizing the flux linkage can get the flux control expression about d-axis voltage $u_d^*(k)$ to make the next moment flux reach the given value.

$$u_{sd}^*(k) = \frac{|\psi_s^* - \psi_{sd}(k)|}{T_s} + R_s i_{sd}(k) \quad (11)$$

$u_q^*(k)$ and $u_d^*(k)$ are the set of optimal voltage vectors for the kth-control cycle. They satisfy the given conditions for flux and torque. In this way, the DTC torque flux hysteresis loop can be replaced and the deadbeat regulation of the switched reluctance motor can be achieved successfully.

2.4 Space-vector modulation

$u_q^*(k)$ and $u_d^*(k)$ can be transformed into $u_\alpha(k)$ and $u_\beta(k)$ by inverse rotation coordinate system which can be directly used in space-vector control module. Based on voltage vector principle of SRM, the basic voltage vectors of different sectors are redefined.

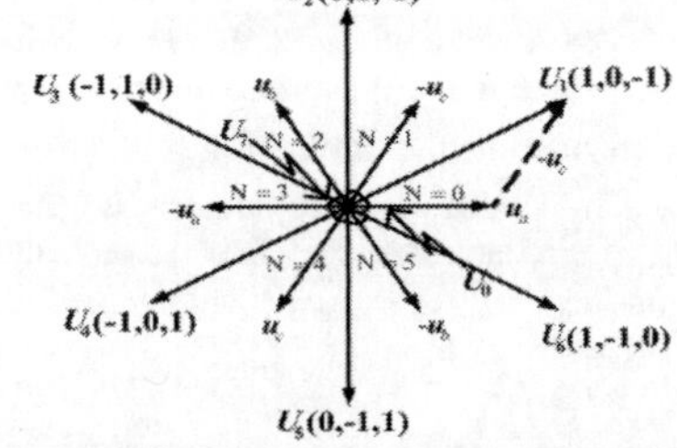

Fig. 1 Basic space voltage vectors of SVM

According to volt-second balance principle, if the voltage vector is calculated and needed in the first sector, $u_\alpha(k)$ and $u_\beta(k)$ can be expressed as:

$$\begin{cases} U_\alpha = \frac{T_1}{T_s}|U_1|\cos(\frac{\pi}{6}) + \frac{T_2}{T_s}|U_2|\cos(\frac{\pi}{2}) \\ U_\beta = \frac{T_1}{T_s}|U_1|\sin(\frac{\pi}{6}) - \frac{T_2}{T_s}|U_2|\sin(\frac{\pi}{2}) \end{cases} \quad (12)$$

Since the unique three-phase asymmetric bridge power converter of SRM, the maximum amplitude of SRM space voltage vector is $\sqrt{3}U_{dc}/2$ and $|U_1|=|U_2|=\sqrt{3}U_{dc}/2$. Among them, U_{dc} is the bus voltage. The time of different sector vectors can be calculated as table.1.

Tab. 1 Voltage working time under different sectors

Secto	0	1	2	3	4	5
T_1	$T_6=-T_z$	$T_1=T_x$	$T_2=T_y$	$T_3=T_z$	$T_4=-T_x$	$T_5=-T_y$
T_2	$T_1=T_y$	$T_2=T_z$	$T_3=-T_x$	$T_4=-T_y$	$T_5=-T_z$	$T_6=T_x$
T_0	$T_0(T_7)=T_s-T_1-T_6$					

where T_x, T_y and T_z can be expressed as

$$\begin{cases} T_x = \frac{4T_s U_\alpha}{3U_{dc}} \\ T_y = \frac{2T_s(U_\alpha + \sqrt{3}U_\beta)}{3U_{dc}} \\ T_z = \frac{2T_s(\sqrt{3}U_\beta - U_\alpha)}{3U_{dc}} \end{cases} \quad (13)$$

If $T_1+T_2 \geq T_s$, there will need time saturation processing, T_1 and T_2 need to be redefined as:

$$\begin{cases} T_1 = \frac{T_1}{T_1+T_2}T_s \\ T_2 = \frac{T_2}{T_1+T_2}T_s \end{cases} \quad (14)$$

3 Principle of DB-DTFC scheme

3.1 Structure of DB-DTFC scheme

 https://doi.org/10.1051/matecconf/202030604007

On the basis of control rate of DB-DTFC, $u_q^*(k)$ and $u_d^*(k)$ are selected. And then $u_\alpha(k)$ and $u_\beta(k)$ which are transformed from $u_q^*(k)$ and $u_d^*(k)$ can be input into the SVM module. SVM can synthetize the new voltage vectors by judging the basic voltage vectors selected by $u_\alpha(k)$, $u_\beta(k)$ and calculating the action time of the basic voltage vectors. In addition, the sector judgement module in SVM is similar to that of DTC. The control signal corresponding to the new voltage vectors synthesized by SVM drives the motor

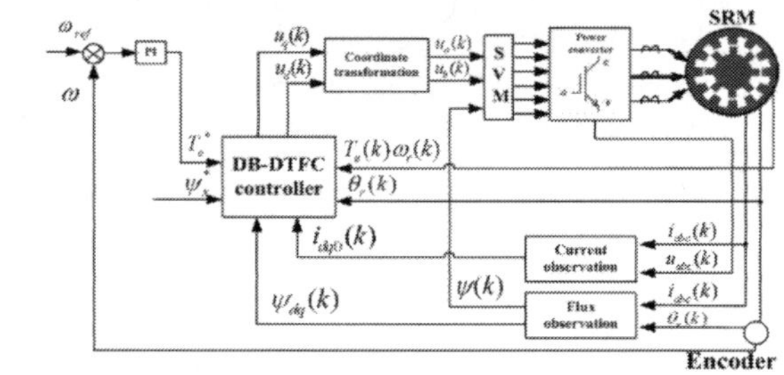

Fig. 3. Control block diagram of DB-DTFC

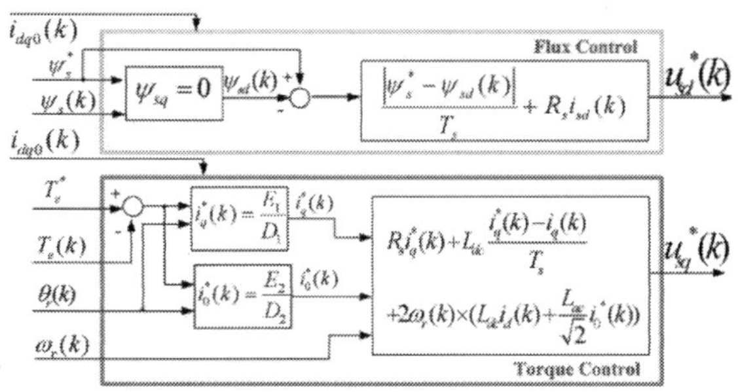

Fig. 4. The control rate of DB-DTFC

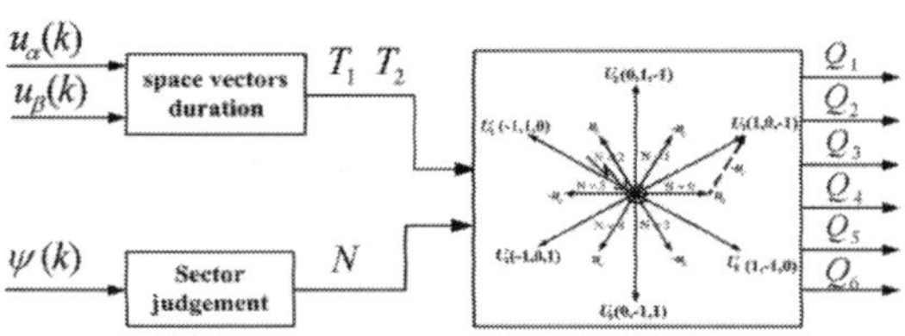

Fig. 5. Control block of SVM

3.1 Analysis of simulation results

The simulation results of proposed DB-DTFC and DTC-SVM will be analyzed and compared in this part. Relying on Matlab/simulink the simluation can be established based on the three-phase 12/8-pole SRM, and the rated voltage is 520V.

In order to assess the effect of different algorithms. The torque error T_{RC} can be defined as follow:

$$T_{RC} = T_{e-\max} - T_{e-\min} \tag{15}$$

$T_{e-\max}$ and $T_{e-\min}$ are the maximum and minimum values of torque respectively. Fig.6 is the waveform of 600rpm with 10N·m load torque, the optimization of T_{RC} is not obvious as that of Fig.7 and Fig.8 with high speed loading.

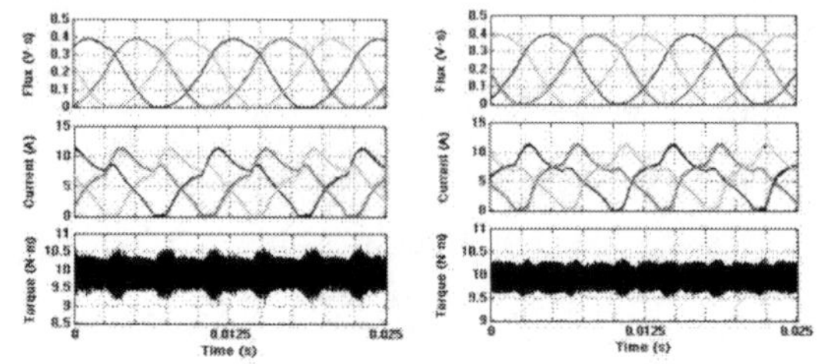

(*a*) (*b*)

Fig. 6. ω=650*rpm*, T_L=10*N·m* (*a*)DTC-SVM (*b*) proposed DB-DTFC

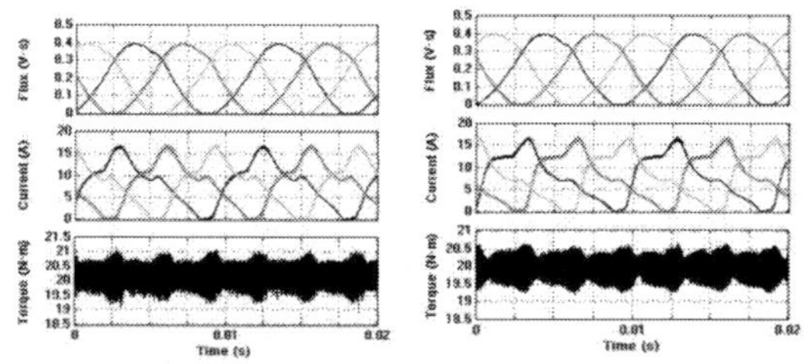

(*a*) (*b*)

Fig. 7. ω=800*rpm*, T_L=20*N·m* (*a*)DTC-SVM (*b*) proposed DB-DTFC

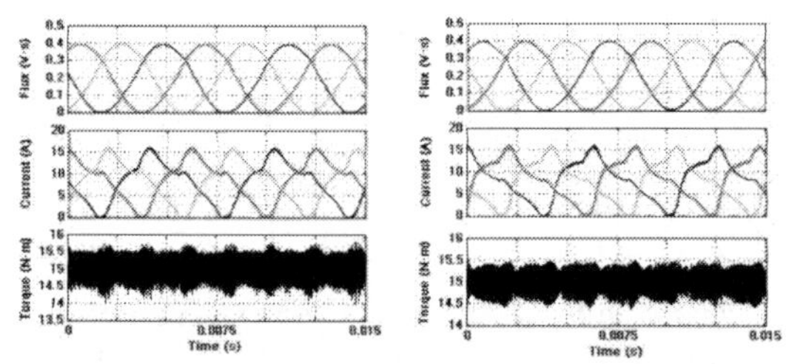

(*a*) (*b*)

Fig. 8. ω=1200*rpm*, T_L=15*N·m* (*a*)DTC-SVM (*b*) proposed DB-DTFC

Tab. 2 Torque ripple error comparison

Index	DTC-SVM T_{RC}	DB-DTFC T_{RC}
650rpm, 10*N·m*	1.65	0.81
800rpm, 20*N·m*	2.15	1.36
1200rpm, 15*N·m*	1.85	1.05

From the torque waveform, the two method both have good control to the torque. But the more excellent control performance of DB-DTFC can be seen than that of DTC. The torque ripple becomes smaller than that of DTC-SVM. Table 2 is a numerical comparison of torque ripple error about DTC and DB-DTFC. From table 2, the value of torque ripple error can be acquired in detail. Under different conditions, the torque ripple of

DB-DTFC decreases by 0.8$N{\cdot}m$ compared with DTC-SVM.

4 Conclusion

To solve the problem of switched SRM torque ripple, DB-DTFC is proposed based on $dq0$ coordinate system. DB-DTFC is based on DTC-SVM and introduces current vector into the control system. But the flux and torque control of DB-DTFC is different from DTC-SVM. Especially, decoupling flux and torque and adopting $dq0$ axis currents control respectively. Using i_q, i_0 control the torque and using i_d controls the flux. The flux control is similar to DB-DTFC based on IPMSM. But the zero phase current is taken into account in the torque control. DB-DTFC is compared with DTC-SVM by simulation on the stable state. The better performance of DB-DTFC is reflected in the torque ripple values. The simulation results can also verify the validity and feasibility of DB-DTFC.

This work is supported by: Special Innovation Projects for the Double-First Class Construction under grant no.CXXM2019SS005, Dalian Science Technology Innovation Fund Project under grant no.2018J12GX039 and National Natural Science Foundation of China under grant no.51407021.

References

1. A. Xu, C. Shang, J. Chen, J. Zhu and L. Han, *A New Control Method Based on DTC and MPC to Reduce Torque Ripple in SRM.* IEEE Access, **7**, 68584-68593(2019).
2. I. Husain, *Minimization of torque ripple in SRM drives.* IEEE Trans on Industrial Electronics, **49**, 28-39(2002).
3. R. B. Inderka, R. W. A. A. De Doncker, *DITC-direct instantaneous torque control of switched reluctance drives.* IEEE Trans on Industry Applications, **39**,1046-1051(2003).
4. C. Shang, A. Xu, L.Huang and J. Chen, *Flux linkage optimization for direct torque control of switched reluctance motor based on model predictive control.* IEEJ Trans on Electrical and Electronic Engineering, vol. **14**,105-1113(2019).
5. B. H. Kenny and R. D. Lorenz, "Stator- and rotor-flux-based deadbeat direct torque control of induction machines. IEEE Trans on Industry Applications, **39**, 1093-1101(2003).
6. J. S. Lee, C. Choi, J. Seok and R. D. Lorenz, *Deadbeat-Direct Torque and Flux Control of Interior Permanent Magnet Synchronous Machines With Discrete Time Stator Current and Stator Flux Linkage Observer.* IEEE Trans on Industry Applications, **47**, 1749-1758(2011)
7. M. Saur, D. E. Gaona Erazo, J. Zdravkovic, B. Lehner, D. Gerling and R. D. Lorenz, *Minimizing Torque Ripple of Highly Saturated Salient Pole Synchronous Machines by Applying DB-DTFC.* IEEE Trans on Industry Applications, **53**, 3643-3651(2017).
8. R. S. Dastjerdi, M. A. Abbasian, H. Saghafi and M. H. Vafaie, *Performance Improvement of Permanent-Magnet Synchronous Motor Using a New Deadbeat-Direct Current Controller.* IEEE Trans on Power Electronics, **34**, 3530-3543(2019).
9. W. Xu and R. D. Lorenz, *Dynamic Loss Minimization Using Improved Deadbeat-Direct Torque and Flux Control for Interior Permanent-Magnet Synchronous Machines.* IEEE Trans on Industry Applications, **50**,1053-1065(2014).
10. M. H. Vafaie, B. M. Dehkordi, P. Moallem and A. Kiyoumarsi, *Improving the Steady-State and Transient-State Performances of PMSM Through an Advanced Deadbeat Direct Torque and Flux Control System.* IEEE Trans on Power Electronics, **32**, 2964-2975(2017).
11. N. Nakao and K. Akatsu, *Vector control specialized for switched reluctance motor drives.* ICEM, 943-949(2014).
12. S. Kuai, H. Zhang, X. Xia and K. Li, *Unipolar sinusoidal excited switched reluctance motor control based on voltage space vector.* IET Electric Power Applications, **13**, 670-675(2019).
13. Y. A. Khan and V. Verma, *Novel speed estimation technique for vector-controlled switched reluctance motor drive.* IET Electric Power Applications, **13**, 1193-12038(2019)

Verification of Strain Transfer Characteristics of Distributed Surface-attached FBGs on Aircraft Wings

Zhen Ma and *Xiyuan* Chen*

Key Laboratory of Micro-Inertial Instrument and Advanced Navigation Technology, Ministry of Education, School of Instrument Science and Engineering, Southeast University, 210096 Nanjing, China

Abstract. In order to provide higher accuracy for transfer alignment on aircraft, a surface-attached FBGs is designed for distributed deformation measurement of aircraft wing. To verify the measurement accuracy of the surface-attached FBGs, a strain calibration method of the surface-attached FBGs applied with the wing strain measurement is proposed. Two Aluminium Alloy Test Plates(AATP) with different sizes were designed as the matrix material for tensile test. The test results show that AATP-A has better test effect. Through the two sets of tensile tests, the change of wavelength measured on the surface-attached FBGs is in good agreement with the change of strain value of strain gauge.The experimental results prove that the method based on surface-attached FBGs is feasible for the strain measurement of aircraft wing. The method is simple and reliable, which lays a foundation for the follow-up study of the full-field deformation measurement of aircraft wing.

1 Introduction

Structural health monitoring has always been an important research area in the field of aerospace. The deformation extent of aircraft wing directly affects the flight safety of aircraft. Airborne distributed measurement is an effective means for wing deformation monitoring. Fiber Bragg grating (FBG) has many advantages in distributed measurement of aircraft wing, among which the strain calibration technology of FBGs is a hot topic. For FBG strain calibration, researchers have done a lot of research work and achieved important results.[1-2] Tavares et al. [3] designed a method to calibrate tube packaged FBGS in a wide range by four-point bending test, and established the finite element model of plastic bending deformation of four-point bending test beam. Rothsch et al. [4] established a strain calibration device for FBG strain sensor according to the German VDI/VDE 2660 guide. The measuring accuracy of the device is high, but its structure is complex. Habel et al. [5] of Germany developed a calibration test device for FBGs. The unbiased digital image correlation (DIC) and electronic speckle interferometer (ESPI) are used for calibration measurement. The KALFOS (calibration of fiber optic sensors) facility designed by researchers consists of a load bearing facility and a temperature chamber. The experimental results show that the KALFOS method can analyze their strain transfer mechanism, and can easily observe the relative motion between sensors and samples which helps to validate the definition and quantitative description of component requirements in new or existing standards. This method has high accuracy. Lazarev et al. [6] designed a non-destructive calibration device for encapsulated FBG strain sensor using mechanical nano-motion sensor. Through the analysis of experimental data, the performance of FBG strain sensor is studied, the calibration curve of FBG strain sensor is given, which verifies the feasibility of this method. From these we can see that researchers have done a lot of research on calibration of tube-packaged FBGs, and the calibration technology is relatively mature on tube-packaged FBGs. However, there are few literatures about the calibration method of surface-attached FBGs, and there is no mature technical method.

Fig.1. Aircraft wing model

* Corresponding author: chxiyuan@seu.edu.cn

https://doi.org/10.1051/matecconf/202030605001

As shown in Fig.1, in order to study the strain transfer effect of surface-attached FBG in wing deformation measurement, we simulated the real wing structure design and designed the wing model. Several distributed fiber Bragg gratings are attached to the aircraft wing. The strain transfer characteristics of surface-attached FBGs are studied experimentally. Two specimens of different sizes were designed to verify the calibration method designed in this paper. The feasibility of surface-attached FBGs strain measurement is emphatically analyzed through comparative experiments.

2 Test equipment

As shown in Fig.2, in order to ensure the accuracy of fiber wavelength measurement, KL-530 optical fiber fusion splicer is used in the selection of test equipment to ensure the accuracy of fiber grating. The interrogation technique adopts SM130 interrogation technique made by U.S.A.. Electro-mechanical Universal Testing Machines (EUTM) is produced by Shenzhen Sansi Longitudinal and Horizontal Technology Co., Ltd. with a measuring range of 5000N. DH3818Y static strain gauge is used to read the strain value of strain gauge in the experiment at a frequency of 1Hz. Besides these, two aluminium alloy test plates (AATP) of different sizes are designed in this paper. As shown in Fig.3, AATP-A is 260 mm x 60 mm x 3 mm, and AATP-B is 200 mm x 35 mm x 8 mm. In order to transfer strain effectively, a fast-drying 3 ton epoxy adhesive was used to fix FBG and strain gauge. In the experiment, computer-1 was used to record the wavelength value, computer-2 was used to record the strain value, and computer-3 was used to record the tension change. Fiber Bragg Grating is sensitive to temperature change, so we conduct experiments under constant temperature environment. Under constant temperature, the change of wavelength value is proportional to the change of strain value.

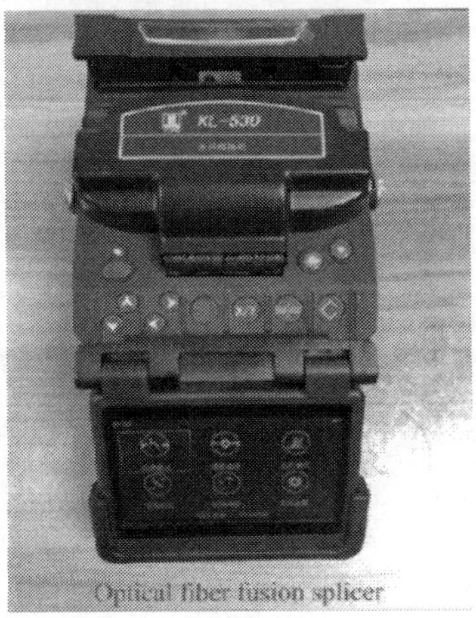

(a)

(b)

Fig.2. Test equipment

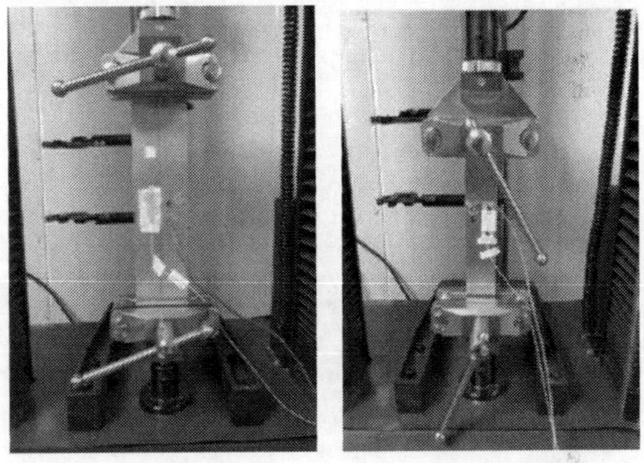

(a) AATP-A (b) AATP-B

Fig.3. Different sizes of AATP

3 Experimental methods and data analysis

3.1 Analysis of Strain Transfer Characteristics of Different AATP

In order to test whether the strain transfer effect of FBGs on different size specimens is consistent under the same force, we first carried out the tensile test of AATP-A. The test data show that the data obtained by FBGs on AATP-A specimen have good linearity. Then, we carried out tensile test on AATP-B specimen, the test data showed that the data obtained by FBGs on AATP-B specimen changed abnormally, and the test results were not as good as those on AATP-A specimen.

In order to study whether the abnormal changes of AATP-B data are caused by human factors, we conducted two experiments. As shown in Fig.4, in the first group of experiments, by observing the trend change of curve-1, we can find that when F> 1kN, the trend of curve-1 changes abnormally and the curve-1 will not continue to increase with the increase of force, it will jitter and slide. Meanwhile, the initial value of curve-1 is not 0, which means that AATP-B may have residual force at the initial time. On the premise of avoiding the possible interference factors in the first experiment, we conducted the second group of experiments on AATP-B. The trend of curve-2 shows that the reverse force appears in the first five seconds of tensile test. When F> 1.3 kN, the trend of curve-2 shows obvious non-linear change, and the value of force detected on computer-3 shows a similar decline of jitter.

Through the analysis of these two sets of test data, it can be concluded that the main factors affecting the abnormal change of the curve are the residual force caused by inadequate clamping force, slip and unreasonable clamping. Therefore, in order to improve the measurement accuracy and avoid the interference factors in the first two groups of tests, we need to pre-load specimen before the test to remove the residual stress. At the same time, the results of data analysis of two groups of AATP-B test show that AATP-A test measurement effect is better than AATP-B, so we choose AATP-A as the test material to verify the strain transfer effect of surface-attached FBGs.

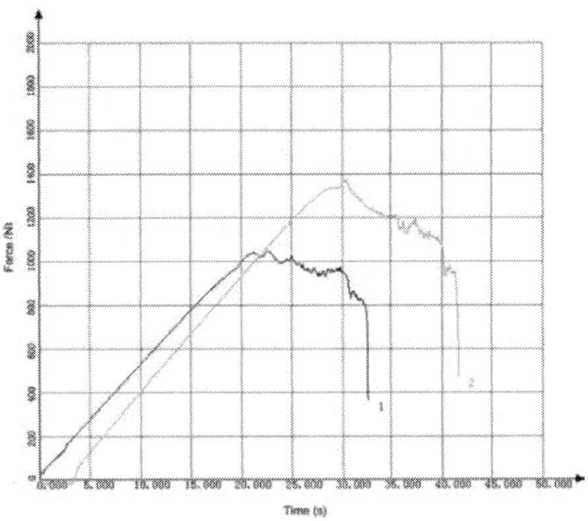

Fig.4. Curve of tensile test

3.2 Analysis of Strain Transfer Characteristics

Before strain transfer characteristic analysis test, in order to detect the maximum range on AATP-A and determine the optimal test method for AATP-A, this paper applied equal gradient increasing force on AATP-A until the force on AATP-A changed abnormally. Firstly, the force is applied to 300N at a speed of 50 N/s, then the pull force is maintained for 60s, and then the previous operation is repeated. As shown in Fig.5, when the force is greater than 1.5 kN, the value of the force measured on EUTM changes abnormally. For this reason, the maximum range of the force applied in this experiment is set to 1.5 kN.

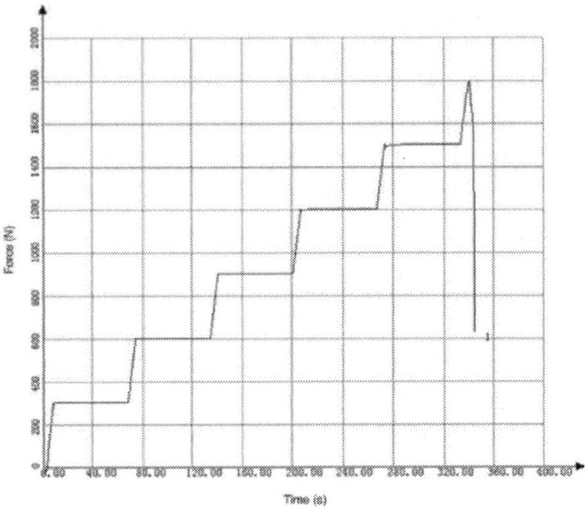

Fig.5. Maximum Range Test

In order to validate the strain transfer characteristics of surface-attached FBGs effectively, the strain transfer effect of surface-attached FBGs is validated by applying uniform increasing force. When applying tension to AATP-A by EUTM, it increases from 0 to 1.5kN at a speed of 50 N/s, then maintains for 60 seconds, and finally unloads AATP-A. The experiment was repeated twice, and then the data were observed and analyzed. The variation of wavelength values in two groups of tension tests is shown in Fig. 6.

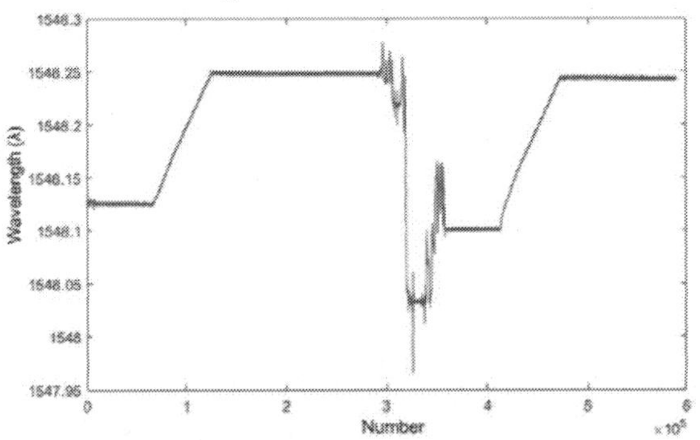

Fig.6. Wavelength value of tension test

From Fig.6, we can see that the working conditions of the two groups are the same, the initial values of the wavelength values in the two groups are different. The initial wavelength of the second group is less than the initial wavelength of the second group. When the tension is applied to the maximum value, the maximum wavelength of the two groups is basically the same.

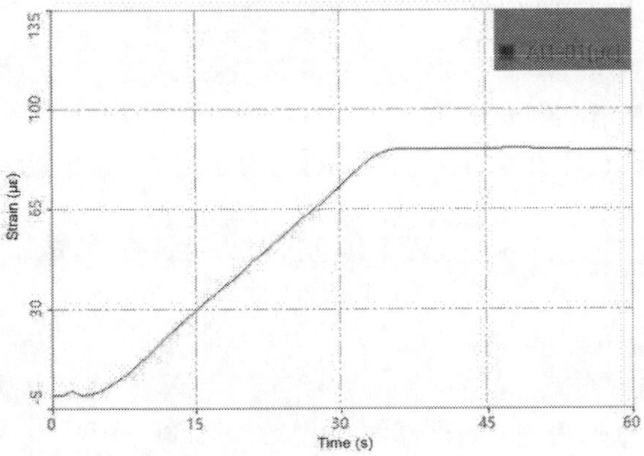

Fig.7. Strain value of the first tension test

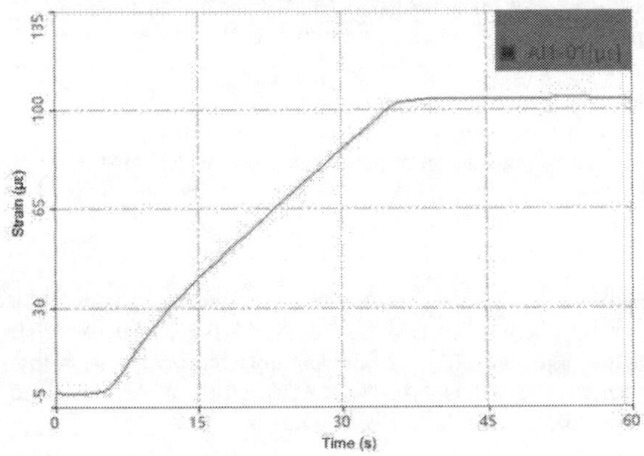

Fig.8. Strain value of the second tension test

In the meantime, we use strain gauges to obtain two sets of strain values. Fig.7 and Fig.8 show the changes of strain values measured by static strain gauge during two sets of tensile tests on AATP-A. Through two groups of experiments, we can find that the wavelength measured by the surface-attached FBGs and the strain value measured by the static strain gauge have good consistency. From Fig.7, we can find that the strain value fluctuates in a small range at the beginning. In the first group of experiments in Fig.6, the wavelength value fluctuates in a small range too. In this experiment, the static strain gauge itself has a high measurement accuracy. The experimental comparison proves that the change of wavelength has a good match with the strain value, that is to say, the surface-attached FBGs has a good strain transfer characteristics.

4 Conclusions

In order to verify the strain transfer characteristics of surface-attached FBGs, a strain transfer test verification method of surface-attached FBGs is designed. The test result of AATP-A is better than that of AATP-B. Then, AATP-A is used for tensile test to compare the changes of the measured wavelength and strain values. The test results show that surface-attached FBGs has good strain transfer effect, high measurement accuracy and good consistency with traditional strain gauge measurement methods. In addition, we also found that the surface-attached FBGs is greatly affected by noise. Therefore, in addition to strictly operating according to the experimental procedure, we also need to avoid the influence of temperature and noise on the test to improve the test accuracy. This experiment provides a reference for the application of surface-attached FBGs in the strain measurement of aircraft wings, and lays a foundation for the realization of the full-field deformation measurement of aircraft wings. We will further study the application of distributed measurement of surface-attached FBGs in transfer alignment.

Acknowledgements

This work is supported by the National Natural Science Foundation of China (No. 61873064 and No. 51375087), the National Defense Advanced Research Foundation of China.

References

1. Qiao, X. G., Zhang, L., Jia, Z. A., GAO, H., & MA, C. Research on FBG calibration demodulation system based on tunable fiber fabry-perot filter. Transducer and Microsystem Technologies, **27**.1, (2008)
2. Zhang, Z., Wang, Y., Sun, Y., Zhang, Q., You, Z., & Huang, X. Analysis and experimental study on the strain transfer mechanism of an embedded basalt fiber-encapsulated fiber Bragg grating sensor. Optical Engineering, **56**.1, (2017)
3. Richter-Trummer, V., Silva, S. O., Peixoto, D. F. C., Frazao, O., Moreira, P. M. G. P., Santos, J. L., & De Castro, P. M. S. T. Fibre Bragg grating sensors for monitoring the metal inert gas and friction stir welding processes. *Measurement Science and Technology* **21**.8 (2010)
4. Roths, J., Wilfert, A., Kratzer, P., Jülich, F., & Kuttler, R. Strain calibration of optical FBG-based strain sensors. Fourth European Workshop on Optical Fibre Sensors. *International Society for Optics and Photonics*, **7653**, (2010)
5. Habel, W. R., Schukar, V. G., & Kusche, N. Calibration facility for quality certification of surface-attached fiber optic and electrical strain sensors. *SENSORS*, 2011 IEEE. IEEE,(2011)
6. Lazarev, V. A., Leonov, S. O., Tarabrin, M. K., & Karasik, V. E. Fiber Bragg gratings strain measuring system and a sensor calibration setup based on mechanical nanomotion transducer. Optical Measurement Systems for Industrial Inspection X. *International Society for Optics and Photonics*, **10329** (2017).

Real-time estimation method of roll angle based on least squares recursion

Chen Jingsheng[1,*], *Li* Chuanjun[1], and *Hu* Peisen[1]

[1] School of Aerospace Engineering, Beijing Institute of Technology, Beijing, China

Abstract. The rotary missile stands a high overload during the launch and has to be powered up after launch, so it is necessary to achieve inflight alignment under high dynamic conditions. As a key technology of inflight alignment, the measurement method of roll angle has attracted more and more attention from researchers. The rotational speed of the rotary missile is very high, and most MIMUs cannot directly measure the roll angle. To solve this problem, this paper proposes a roll angle estimation method based on least squares method, analyzes its principle and derives the calculation procedure. Then on this basis, the roll angle estimation method based on least squares recursion is studied. The principle and calculation procedure of this method are deduced in detail. At last, the simulation experiment on MATLAB is carried out. The results show that this method is simple in calculation, high in accuracy and good in real-time performance, and has great application value.

1 Introduction

With the advancement of science and technology, modern warfare demands increasingly higher on the accuracy of weapon systems. For the guided weapons that use Strapdown Inertial Navigation (SINS) and Satellite Navigation (GNSS) integrated navigation, improving the initial alignment accuracy is an effective means to improve the system's strike accuracy. The GNSS/SINS integrated navigation system usually has to withstand high overload and high speed during the launching. At this time, both SINS and GNSS cannot work normally. Therefore, the SINS has to achieve inflight alignment under the stable or low-speed rotation conditions after the missile enters the controlled section[1].

Because of the high overload, SINS cannot be powered up at the moment the rotary missile is fired. During the launch, the missile is rolling in the cannon. The state and the roll angle of the missile is unknown after the launch, which makes the difficulty to the initial alignment of the SINS. According to the ballistic data of the short time after the missile exits the cannon, the position, velocity, yaw angle and pitch angle of the SINS can be initially set before the launch with the previously known launch conditions[2][3]. However, the roll angle cannot be directly measured, so the estimation of the initial roll angle has been widely concerned.

In order to solve this problem, this paper proposes a roll angle estimation method based on least squares method[4], using gyro data and GNSS data to estimate the roll angle. And on this basis, a real-time roll angle estimation method based on least squares recursion is proposed. This paper analyzes the principles and derive the calculation procedure. At last, a simulation experiment is carried out to verify the performance of the algorithm. The results of the experiment show that this method has high accuracy of roll angle and good performance in high rotation environment.

2 Roll Angle Estimation Based On Least Squares Method

This paper mainly studies on roll angle estimation method for SINS/GNSS integrated navigation system, and the functional composition of the system is shown as Fig.1.

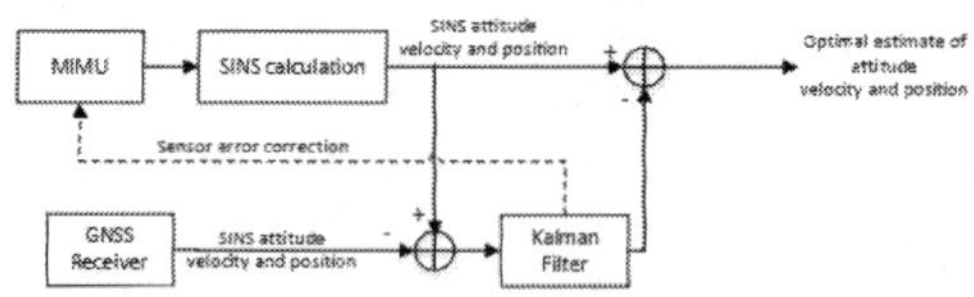

Fig. 1. Functional composition of the system

As we can see from the Fig.1, the SINS and GNSS calculate the attitude, velocity and position independently of each other, and then the results are integrated by a feedforward complementary Kalman filter to output the optimal estimate of the attitude, velocity and position. The results can help MIMU correct its own error and improve the accuracy. The placement of the sensors in MIMU is generally shown in Fig.2.

---

[*] Corresponding author: 1454341945@qq.com

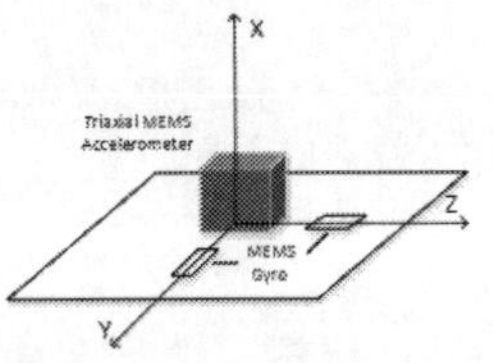

Fig. 2. The placement of the sensors in MIMU

During the flight of the missile, the Ox_1 axis of the missile body coordinate system and the Ox_4 axis of the missile body axis coordinate system keep coincident. The Oy_1 axis and the Oz_1 axis rotate with the missile body, and the Oy_4 axis keeps in the vertical plane. Oz_4 axis can be obtained from Ox_4 axis and Oy_4 axis by the right-hand system rule. The diagram of missile body coordinate system and missile body axis coordinate system is shown as Fig.3.

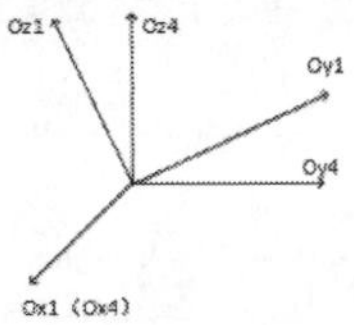

Fig. 3. The diagram of these two coordinate system

Therefore, we can get the conversion equation between the body axis coordinate system and the body coordinate system[5]:

$$\begin{cases} \omega_{x1} = \omega_{x4} \\ \omega_{y1} = \omega_{y4}\cos\gamma + \omega_{z4}\sin\gamma \\ \omega_{z1} = -\omega_{y4}\sin\gamma + \omega_{z4}\cos\gamma \end{cases} \tag{1}$$

In the formula, ω_{x1}、ω_{y1}、ω_{z1} are the three-axis components of the body rotational angular velocity in the body coordinate system. ω_{x4}、ω_{y4}、ω_{z4} are the three-axis components of the body rotational angular velocity in body axis coordinate system.

In the body axis coordinate system, the kinematics equation of the missile around the center of mass can be expressed as:

$$\begin{cases} \dot{\gamma} = \omega_{x4} - \omega_{y4}\tan\vartheta \\ \psi = \dfrac{1}{\cos\vartheta}\omega_{y4} \\ \dot{\vartheta} = \omega_{z4} \end{cases} \tag{2}$$

In the formula, γ is the roll angle, ψ is the yaw angle, and ϑ is the pitch angle. Substituting Equation (2) into Equation (1), the last two formulas of Equation (1) can be written as:

$$\begin{cases} \omega_{y1} = (\psi\cos\vartheta)\cos\gamma + \dot{\vartheta}\sin\gamma \\ \omega_{z1} = (-\psi\cos\vartheta)\sin\gamma + \dot{\vartheta}\cos\gamma \end{cases} \tag{3}$$

Organize it into a matrix, and get:

$$\begin{bmatrix} \omega_{y1} \\ \omega_{z1} \end{bmatrix} = \begin{bmatrix} \dot{\vartheta} & \psi\cos\vartheta \\ -\psi\cos\vartheta & \dot{\vartheta} \end{bmatrix} \cdot \begin{bmatrix} \sin\gamma \\ \cos\gamma \end{bmatrix} \tag{4}$$

In the formula, the values of ϑ、$\dot{\vartheta}$ and ψ can be approximately estimated by the GNSS velocity measurement data[6]. Substituting the corresponding values at each moment into Equation (4), a matrix equation of the Equation (5) is obtained.

$$\boldsymbol{Z}_{2n\times1} = \boldsymbol{H}_{2n\times2} \cdot \boldsymbol{X}_{2\times1} \tag{5}$$

Then the least squares estimation of X can be expressed as:

$$\hat{X} = \left(H^T H\right)^{-1} H^T Z \tag{6}$$

And X can meet the requirement to make:

$$J(\hat{X}) = \left(Z - H\hat{X}\right)^T \left(Z - H\hat{X}\right) = \min \tag{7}$$

Calculate the estimated value of $\sin\gamma$ and $\cos\gamma$ from Equation (6), then:

$$\gamma = \arctan\left(\frac{\sin\gamma}{\cos\gamma}\right) \tag{8}$$

The range of values is determined by Equation (9).

$$\begin{cases} \text{when} \sin\gamma \le 0 \text{and} \cos\gamma < 0, & 180^\circ \le \gamma < 270^\circ \\ \text{when} \sin\gamma \le 0 \text{and} \cos\gamma > 0, & 270^\circ \le \gamma < 360^\circ \\ \text{when} \sin\gamma \ge 0 \text{and} \cos\gamma < 0, & 90^\circ \le \gamma < 180^\circ \\ \text{when} \sin\gamma \ge 0 \text{and} \cos\gamma > 0, & 0^\circ \le \gamma < 90^\circ \\ \text{when} \sin\gamma > 0 \text{and} \cos\gamma = 0, & \gamma = 90^\circ \\ \text{when} \sin\gamma < 0 \text{and} \cos\gamma = 0, & \gamma = 270^\circ \end{cases} \tag{9}$$

Therefore, according to this method, based on the collected GNSS data, the corresponding pitch angle and yaw angle and their rate of change are estimated, and combined with the measurement data of the MIMU, the roll angle of the rotary missile at each moment can be estimated.

3 Real-Time Estimation Of Roll Angle Based On Least Squares Recursion

In the previous chapter, the roll angle estimation method based on least squares method was studied, but there exist some defects of this method. The least squares method has to be processed after all the data has been collected. When estimating the roll angle using the least squares method, it is necessary to let the missile fly for a period of time, and collect the MIMU and GNSS data during this time, and then use the least squares method to do the estimation, so that the measurement data cannot be processed in time. And each time using the least squares estimation, all the previously processed data need to be recalculated.

In order to solve this problem, this paper proposes a real-time estimation method of roll angle based on least squares recursion. This method can process the collected data in time after obtaining the gyro and GNSS data, without waiting for all the measurement data to be obtained after measurement. And when estimating the roll angle of the current time, the previous result can be utilized without calculating previously processed data again, and it saves the computation time very much.

It is assumed that the $k-1$ th time estimation result has been obtained when the least squares estimation is performed at the k th time.

$$\hat{\boldsymbol{X}}_{k-1} = (\boldsymbol{H}_{k-1}^T \boldsymbol{H}_{k-1})^{-1} \boldsymbol{H}_{k-1}^T \boldsymbol{Z}_{k-1} \tag{10}$$

In the formula, Z_{k-1}、H_{k-1} can be expressed as:

$$Z_{k-1}=\begin{pmatrix} Z(1) \\ Z(2) \\ \vdots \\ Z(k-1) \end{pmatrix},\ H_{k-1}=\begin{pmatrix} H(1) \\ H(2) \\ \vdots \\ H(k-1) \end{pmatrix} \quad (11)$$

Assume $M_{k-1} = (H_{k-1}^T H_{k-1})^{-1}$, then

$$\hat{X}_{k-1} = M_{k-1} H_{k-1}^T Z_{k-1} \quad (12)$$

When performing the least squares estimation at the k th time, $H(k)$ can be the estimated by GNSS velocity measurement data, and $Z(k)$ can be obtained from the gyro data. According to the least squares method, get Equation (13).

$$\hat{X}_k = (H_k^T H_k)^{-1} H_k^T Z_k \quad (13)$$

According to Equation (11), Z_k and H_k can be expressed as:

$$Z_k=\begin{pmatrix} Z(1) \\ Z(2) \\ \vdots \\ Z(k) \end{pmatrix}=\begin{pmatrix} Z_{k-1} \\ Z(k) \end{pmatrix},\ H_k=\begin{pmatrix} H(1) \\ H(2) \\ \vdots \\ H(k) \end{pmatrix}=\begin{pmatrix} H_{k-1} \\ H(k) \end{pmatrix} \quad (14)$$

Assume $M_k = (H_k^T H_k)^{-1}$, then

$$M_k = [\begin{pmatrix} H_{k-1}^T & H^T(k) \end{pmatrix}\begin{pmatrix} H_{k-1} \\ H(k) \end{pmatrix}]^{-1} \quad (15)$$

$$=[H_{k-1}^T H_{k-1} + H^T(k) H(k)]^{-1}$$

Substituting $M_{k-1} = (H_{k-1}^T H_{k-1})^{-1}$ into Equation (15), get Equation (16).

$$M_k=[M_{k-1}^{-1}+H^T(k)H(k)]^{-1} \quad (16)$$

After matrix inversion, get Equation (17):

$$M_k^{-1}=M_{k-1}^{-1}+H^T(k)H(k) \quad (17)$$

It can be seen from the Equation (17) that when M_{k-1} is definite, M_k can be obtained based on $H(k)$.

According to Equation (13) and Equation (14):

$$\hat{X}_k = M_k H_k^T Z_k = M_k \begin{pmatrix} H_{k-1}^T & H^T(k) \end{pmatrix}\begin{pmatrix} Z_{k-1} \\ Z(k) \end{pmatrix} \quad (18)$$

$$= M_k [H_{k-1}^T Z_{k-1} + H^T(k) Z(k)]$$

Pre-multiply the both sides of Equation (12) by M_{k-1}^{-1}, and get Equation (19) .

$$M_{k-1}^{-1}\hat{X}_{k-1} = H_{k-1}^T Z_{k-1} \quad (19)$$

Substituting Equation (19) into Equation (18),

$$\hat{X}_k = M_k [M_{k-1}^{-1}\hat{X}_{k-1} + H^T(k)Z(k)] \quad (20)$$

Substituting Equation (17) into Equation (20),

$$\hat{X}_k = M_k[(M_k^{-1}-H^T(k)H(k))\hat{X}_{k-1} + H^T(k)Z(k)]$$
$$=\hat{X}_{k-1}-M_k H^T(k)H(k)\hat{X}_{k-1}+M_k H^T(k)Z(k)$$
$$=\hat{X}_{k-1}+M_k H^T(k)[Z(k)-H(k)\hat{X}_{k-1}] \quad (21)$$

Therefore, Equation (17) and Equation (21) constitute the least squares recursion formula. In this formula, $\hat{X}_{k-1}$、M_{k-1}^{-1} are the results of the previous least squares estimation, $H(k)$ and $Z(k)$ can be obtained from the current measurement data. According to this method, measurement data can be processed in time, and the least squares estimation can be performed in real time without waiting for all the data to be obtained after measurement.

4 Simulation Experiment Of Roll Angle Estimation

Based on the last two chapters, Using MATLAB to simulate the roll angle estimation method based on least squares and least squares recursion[7]. The procedures of simulation are as follows:

(1) 5s after the missile was launched, its flight state is stable and the angle of attack and angle of side slip are small. At this time, the gyro data of the MIMU and the GNSS receiver velocity data are collected. The current time data of the GNSS is used to calculate the pitch angle and the yaw angle of the missile and their rate of change. Then roll angle estimation is started.

(2) Combined with the results of the previous roll angle estimation, using the collected gyro data of the MIMU and the GNSS data to perform the least squares recursion estimation.

Through simulation experiments, the real-time estimation values of $\sin\gamma$ and $\cos\gamma$ are obtained as shown in the figure below, and the obvious phase difference can be seen from the figures.

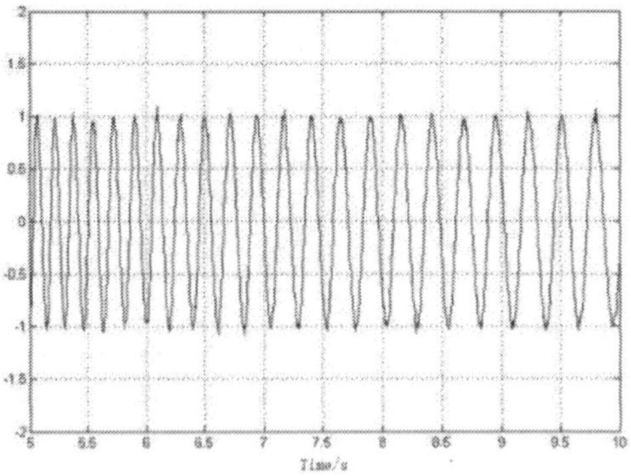

Fig. 4. The estimation values of $\sin\gamma$

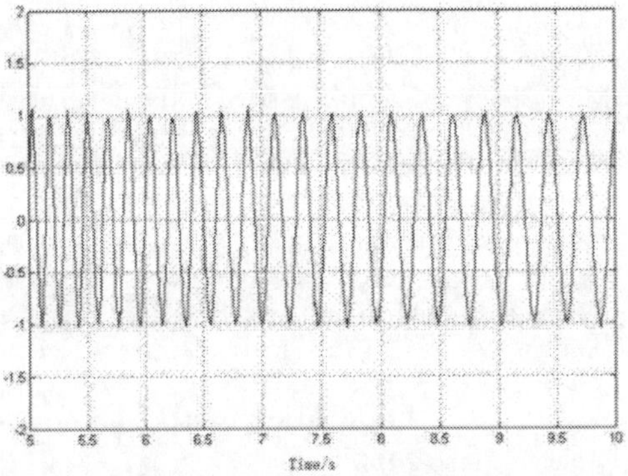

Fig. 5. The estimation values of cos γ

According to Equation (9), the roll angle and its error of least squares recursion estimation can be obtained and shown as follow figures.

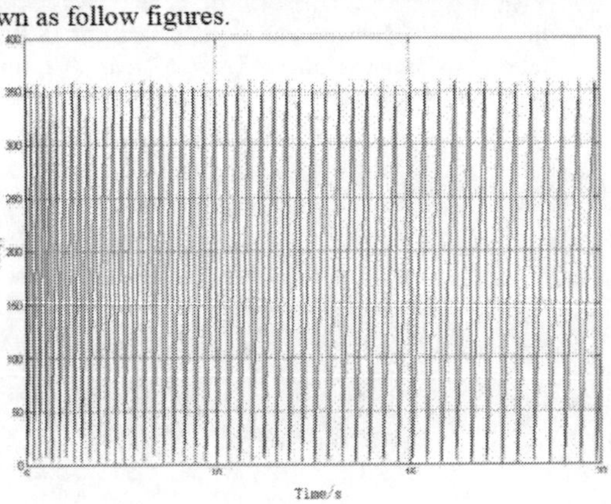

Fig. 6. Values of the roll angle real-time estimation

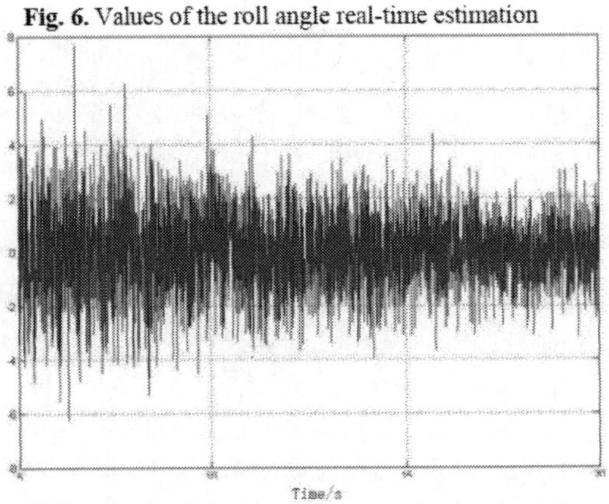

Fig. 7. Error of the roll angle real-time estimation

As can be seen from the Fig.6 and Fig.7, the roll angle error based on the least squares recursion estimation is stable throughout the flight time of the missile, and it fluctuates roughly within the interval of $\pm 4^{\circ}$ and the fluctuation amplitude becomes smaller and smaller, and finally remains at the $\pm 2^{\circ}$ interval. The result shows that the estimation accuracy of the roll angle meets the requirements of the coarse alignment of inertial navigation.

Using the gyro data of MIMU and GNSS velocity data collected in the simulation, the estimation based on least squares method is performed. Roll angle and its error are shown in the figures below.

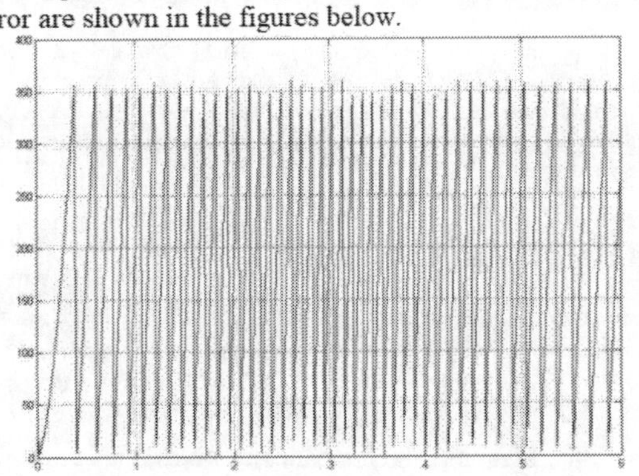

Fig. 8. Values of roll angle estimation

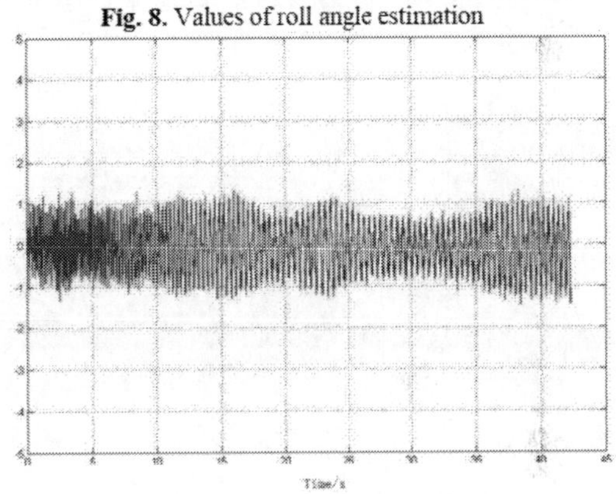

Fig. 9. Error of the roll angle estimation

It can be seen from the Fig.8 and Fig.9 that the estimated roll angle error based on the least squares method is relatively stable throughout the flight time of the missile, and is generally in the interval of $\pm 1^{\circ}$. The result shows that this method is of high precision and worth to be studied.

Through the simulation experiment, we can see that the roll angle error estimated by the least squares recursion is larger than the roll angle error estimated by the least squares method after the measurement. However, the error keeps fluctuating within the range of $\pm 4^{\circ}$ and the fluctuation amplitude is getting smaller and smaller. The estimation method based on least squares recursion has better real-time performance, and the data can be processed online for estimation. This is a great advantage of this method.

5 Conclusion

Based on the least squares method, this paper proposes a roll angle estimation method based on least squares recursion, and do the simulation experiment for this method. The simulation results show that this method is simple in calculation, good in real-time preforms and high in accuracy. It meets the precision requirements of the strap-down inertial navigation system for coarse alignment under the high speed conditions and has great application value.

This paper is supported by National Natural Science Foundation of China (61473039) and the key laboratory of dynamics and control of flight vehicle, Ministry of education, school of aerospace engineering, Beijing Institute of Technology.

References

1. Yang Denghong. Research on initial alignment method of GNSS/MIMU integrated navigation for rotating stable missile [D]. Beijing: Beijing Institute of Technology, 2015.
2. She Haoping, Yang Shuxing, Ni Hui. A New Algorithm for Estimating the Initial Rolling Angle of GPS/INS Combined Missiles in Air Alignment[J]. , 2009, 32(10): 1265-1270.
3. Ma Chunyan, Liu Li, Du Xiaojing, et al. Determination of Initial Rolling Angle of Terminal Guided Projectile[J]. Journal of Projectiles, Missiles and Guidance, 2005, 25(s4): 356-358.
4. Liu Yukuan, Li Chuanjun, Chen Jingsheng. Rolling Angle Estimation Method for GNSS/SINS Integrated Navigation System Based on Least Square Method. ICCSSE Proceedings Volumes, 2018, pp.330-333.
5. Qin Yongyuan. Inertial Navigation [M]. Beijing: Science Press, 2014.
6. An Liangliang, Wang Liangming. GPS/SINS combined attitude measurement method for high-rotation rockets[J]. Journal of Ballistics, 2016, 28(1): 39-44.
7. Jin Xuebo. Theory and Application of Kalman Filter——Based on MATLAB[M]. Beijing: Science Press, 2016.

https://doi.org/10.1051/matecconf/202030605003

PBL Program Producing Flying Robot in Mechanical and Aerospace Engineering Department

Susumu Hara[1,*], *Kazuhide* Kuroda[2], *Yu* Aoi[3], *Koki* Nakagami[1], *Koya* Hashizume[2], and *Seiichi* Hata[4]

[1]Nagoya University, Department of Aerospace Engineering, 464-8603 Nagoya, Japan
[2]Nagoya University, Department of Mechanical Systems Engineering, 464-8603 Nagoya, Japan
[3]Nagoya University, School of Informatics and Sciences, 464-8601 Nagoya, Japan
[4]Nagoya University, Department of Micro-Nano Mechanical Science and Engineering, 464-8603 Nagoya, Japan

Abstract. More and more mechanical and aerospace engineering departments of Universities and Colleges are adopting Project-Based Learning (PBL). One of the main purposes of PBL programs is the practical understanding of the relationship between classroom lectures on traditional theories, e.g., material mechanics, fluid dynamics, dynamics of machinery, and control engineering, and real phenomena of mechanical systems. Moreover, we need to take the use of recent digital manufacturing (e.g., 3D printers, CAD/CAM, and CAE) into account as recent educational content and methods. This paper introduces a novel trial of using PBL at Nagoya University. More specifically, the new program provides students with the experience of producing flying robots and introduces the above relationship and digital manufacturing.

1 Introduction

In recent years, many mechanical and aerospace engineering (MAE) departments of Universities and Colleges are adopting Project-Based Learning (PBL) [1], [2]. One of the main purposes of PBL programs is the practical understanding of the relationship between classroom lectures of traditional theories, e.g., material mechanics, fluid dynamics, and control engineering, and the real phenomena of mechanical systems. Moreover, we need to take the use of recent digital manufacturing (e.g., 3D printers, CAD/CAM, CAE) into account as recent educational content.

However, many MAE departments in Japan still use old-fashioned machinery for undergraduate lab classes due to budget constraints. In these lab classes, the integration of plural theories for manufacturing is not discussed. Generally, each lab class offers single experimental program for each traditional theory, e.g., an inverted pendulum control experiment program for understanding the control engineering theory. Subsequently, the technological gap between MAE departments and real industries may be enlarged because real industries emphasize the optimal integration "SURIAWASE (in Japanese)" of plural theories for high-quality high-efficiency manufacturing. Moreover, the recent increase in "passive engineers" in digital manufacturing is a concern. Digital manufacturing seems to be a black box for passive engineers. Taking these factors into account, the MAE department at Nagoya University (NUMAE) is now considering a new PBL program, "Design Practice." This program focuses on the following two elements: (i) The students consider the relationship between the integration of traditional plural theories and practical mechanical design, and (ii) the students experience the recent digital manufacturing effectively. The Design Practice program involves "Producing a Flying Robot" as a specific PBL theme. When designing a flying robot, the body strength and weight (material mechanics), aerodynamic shape (fluid dynamics), actuating system and control mechanisms/algorithms (dynamics of machinery and control engineering) all need to be considered. The integration of these considerations can lead to the development of a high performance flying robot. Simultaneously, parts of the robot, e.g., the wing rib, correspond to a good example of a digitally manufactured product.

The new program "Design Practice" begins in October 2019, and this paper reports on the plan for implementing the program with a focus on the development of a novel flying robot design for educational purposes.

2 Flying robots for the PBL program

The flying robot in the Design Practice program is a glider model without a power generation mechanism, as illustrated in Figs. 1 and 2. Figure 1 shows (a) the overview of the model and (b) the materials used to build the model. Although its length and wingspan are 1.20 m and 0.80 m, respectively, it weighs only 0.27 kg. The longitudinal direction core rod made of carbon pipe is lightweight, tough, and hard to break. These characteristics are well-suited for repeated experiments by beginners. It can fly by using an initial thrust force provided by a rubber tension based takeoff mechanism

* Corresponding author: haras@nuae.nagoya-u.ac.jp

https://doi.org/10.1051/matecconf/202030605003

(a) Overview.

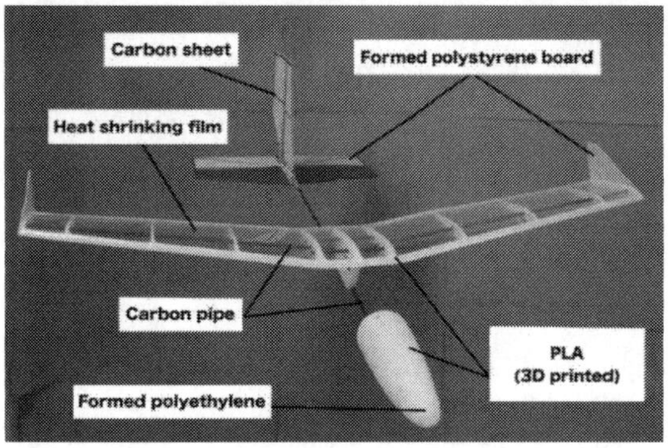

(b) Materials.

Fig. 1. Example of a flying robot for project-based learning.

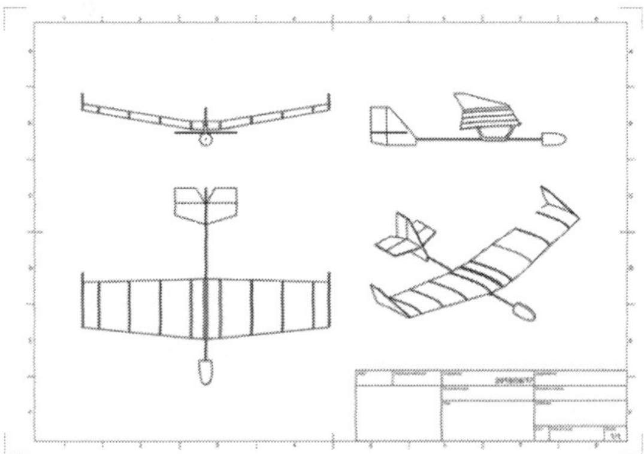

Fig. 2. Drafting of the flying robot.

(a) AHRS sensor.

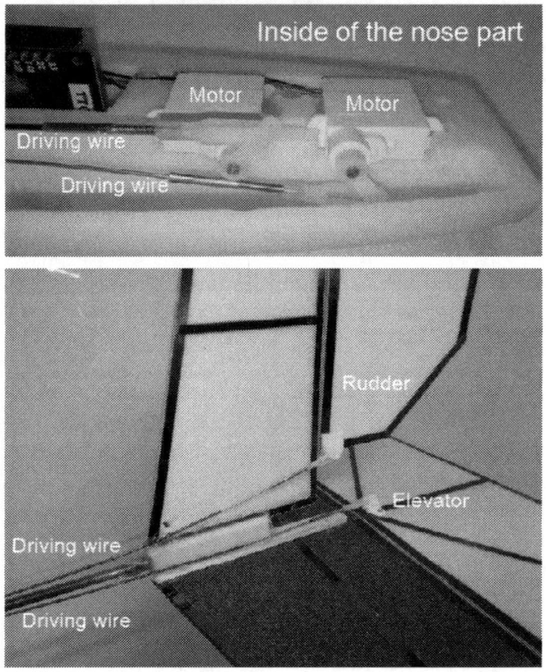

(b) Motors and driving wires (link mechanisms).

Fig. 3. Feedback control devices.

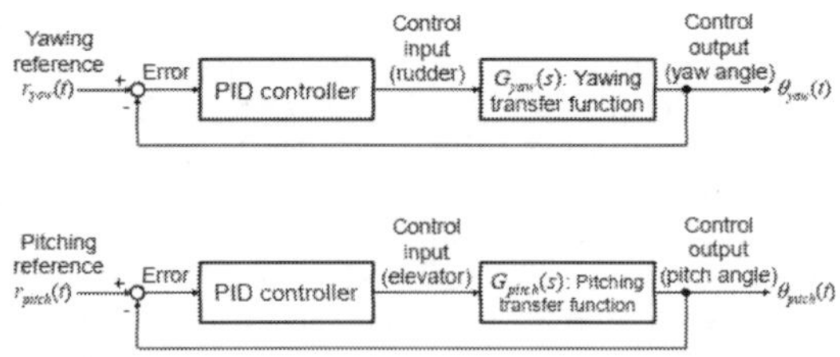

Fig. 4. Feedback control devices.

or a throwing motion from an operator. During flight, it can control its posture using two types of feedback control mechanisms for yawing and pitching controls. In this program, the general PID control algorithm is applied to the controllers, and their feedback gains are tuned by trial-and-error.

The control mechanism includes an attitude heading reference system (AHRS) sensor (Fig. 3a), a microcomputer (Espressif Systems Co., Ltd., ESP32-DOWDQ6), and an actuation system consisting of microservomotors and link mechanisms (Fig. 3b). The block diagram of the control mechanisms is shown in Fig. 4. The yawing and pitching references are set to 0 in the experiments. If the optimal feedback controls are realized, this can increase the flight distance. The states of the real experiments with and without the feedback control are shown in Fig. 5. From a comparison of both of the cases, it is easy to understand that the PID control realizes stable flight. The first author has released movies of this flying robot with and without feedback controls [3]. From this model, it is easy to understand the effectiveness of the control engineering theory. To experience digital manufacturing, a part of the wing is

 https://doi.org/10.1051/matecconf/202030605003

(0 seconds)

(0.33 seconds later)

(0.67 seconds later)

(1.0 second later)

(a) w/ feedback control (from the top to the bottom: 0 seconds, 0.33 seconds later, 0.67 seconds later, 1.0 second later).

(0 seconds)

(0.33 seconds later)

(0.67 seconds later)

(1.0 second later)

(b) w/o feedback control (from the top to the bottom: 0 seconds, 0.33 seconds later, 0.67 seconds later, 1.0 second later).

Fig. 5. The real experiments: (a): w/ feedback control, (b): w/o feedback control.

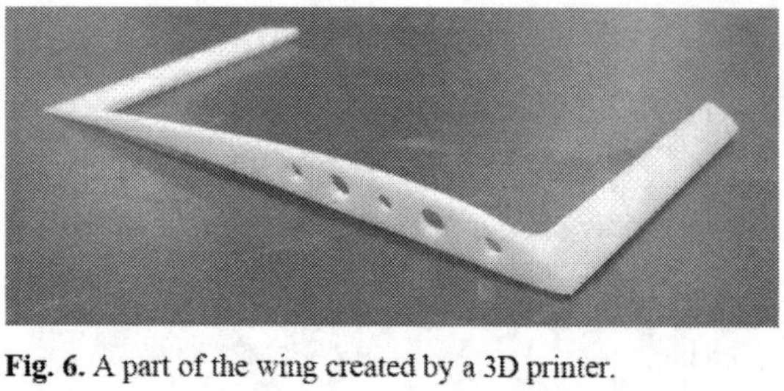

Fig. 6. A part of the wing created by a 3D printer.

Table 1. Major specifications of the flying robot.

Length	1.20 m
Wingspan	0.80 m
Wing area	1.41 X 10^{-1} m^2
Average chord length	0.175 m
Dihedral	10 deg
Angle of incident	5 deg
Sweepback angle	3 deg
Weight	0.27 kg

Table 2. The schedule of PBL program.

1st week	Guidance of "Design Practice"
2nd week	Guidance of "Flying Robot" and Safety Education
3rd week	Introduction of theoretical issues
4th–9th weeks	Production work (first half)
10th week	Midterm check
11th–13th weeks	Production work (later half)
14th week	Flying competition
15th week	Review meeting

created using a 3D printer, as shown in Fig. 6. The use of the 3D printer is also effective for maintaining the manufacturing pace of most of the students in comparison to when the manufacturing is completed by hand.

The major specifications of the flying robot are summarized in Table 1. One robot costs approximately USD 20, excluding the startup facilities such as a PC with software, the 3D printer, and a laser cutter.

3 PBL program

This program begins from the fall semester 2019 and will last 15 weeks (October–January, 135 min. per week). Its schedule is summarized in Table 2. One team consists of six students, and they will produce one flying robot with the members' cooperation. There are nine teams for one semester, and the advisers consist of one faculty member and two teaching assistants (graduate students). After two-weeks of guidance and safety education (in the first and second weeks), lectures on flight dynamics will be provided in the third week by using original documents based on typical textbooks [4], [5].

In this program, each team will design its flying robot by taking various trade-offs and constraints into account. These considerations are strongly related to traditional plural theories such as material mechanics, fluid dynamics, and control engineering. For example, the students have to balance strength improvements and saving weight. They also consider the relationships of the shapes of the control parts, control performance, and the actuating mechanism implementation. This characteristic is quite different from conventional single experimental programs mentioned in the introduction section. Therefore, the integration of plural theories is required in this program. Simultaneously, to avoid the difficulties of the integration of beginners, this program uses a sample robot produced by the teaching assistants a priori. However, the target of the program is to produce a flying robot that performs better than the sample.

From an educational point of view, the contents of the last two weeks (the 14th and 15th weeks) are very important. This involves a competition, from which each team can gauge the performance level of their robot. Moreover, the review meeting is also important as they can discuss the theoretical reasons why their robots performed well or poorly in comparison to the sample robot. Finally, all the students must summarize their activities, results, and considerations in a final report. They are evaluated not from their performance in the competition but through the final report.

Ideally, it is desirable that the students experience a comparison between a numerical simulation and the response of the flying robot. However, as a temporal restriction, it is difficult to include this option for undergraduate programs, and numerical simulation is expected to be used in graduate programs.

4 Conclusion

Taking recent experimental education circumstances in mechanical and aerospace engineering (MAE) departments into account, this paper introduced a novel project-based learning (PBL) program, Design Practice, which is beginning in October 2019 at Nagoya University. This program focuses on the optimal integration of traditional plural theories and the experience of recent digital manufacturing. We chose "Producing a Flying Robot" as a specific theme and discussed a novel flying robot design for educational issues. Its educational effect and subjects will be reported at another opportunity.

This work was supported in part by the Watanabe Memorial Foundation for the Advancement of New Technology and the DII Collaborative Graduate Program for Accelerating Innovation in Future Electronics, Nagoya University.

References

1. A. Kolmos, E. de Graaff, "Problem-based and project-based learning in engineering education," in *Cambridge Handbook of Engineering Education*

Research, (Cambridge University Press, Cambridge, 2014) pp.141-160

2. S.K. Esche and H.A. Hadim, "Introduction of project-based learning into mechanical engineering courses," in Proc. of the 2002 American Society for Engineering Education Annual Conference & Exposition, pp.7.755.1–7.755.13 (2002)
3. Twitter–"Control Systems Engineering Laboratory, Nagoya University." https://twitter.com/NU_CSEL
4. M. Makino, *Fundamentals of Aerodynamics*, (Sangyo Tosho, Tokyo, 1980) (in Japanese)
5. K. Rinoie, *Aircraft Conceptual Design–from Light Aircraft to Supersonic Transport*, (Corona Publishing, Tokyo, 2011) (in Japanese)

Research on Rapid Assembly Technology of Helicopter Transmissions

Wang Haiwei[1,*], *Li* Lulu[1], *Yuan* Sheng[2]

[1]Shaanxi Engineering Laboratory for Transmissions and Controls, Northwestern Polytechnical University, Xi'an, 710072, China

[2]Meteorological Bureau of Zhejiang Province, Hangzhou, 310000, China

Abstract. Research and application of rapid assembly technology can widely improve the design efficiency of helicopter transmissions. An assembly model of helicopter transmissions is presented by introducing Geometric Constraint Graph (GCG) method. A hierarchical model is obtained by multi-shrink decomposition. A corresponding data structure is generated according to the hierarchical model. CATIA based a parametric component database is built up. The Assembly modelling software is implemented by using VB. Finally, a helicopter transmissions example is taken as application and the rapid assembly for helicopter transmission system is achieved.

1 Introduction

Modern helicopter has developed into a high performance aircraft with heavy load, low weight, low noise, new technology, new materials, new equipment and advanced electronic equipment. The research of helicopter transmission system focuses on two aspects : transmission system component and transmission system configuration. Through component research to reduce the weight of the transmission system, improve the carrying capacity and enhance gear transmission performance [1]. The purpose of configuration research is to develop a new type of gear transmission system, transmission system layout and its rapid assembly design [2-3]. Configuration design is a design method based on the arrangement and combination of predefined components and parts. The configuration of helicopter transmission system directly affects the transmission form and transmission scheme and it is an important part of the scheme design stage. The design process is as follows: in a set of predefined components or parts, an assembly or combination that satisfies a series of requirements and follow a series of constraints is sought.

Assembly design is a crucial link in the life cycle of helicopter transmissions system design, the earlier the assembly problem is considered, the more obvious the cost, quality and time effect will be. In a broad sense, assembly process is the spatial embodiment of configuration design process. The rapid assembly of the helicopter transmission system is conducive to the rapid verification of the transmission system configuration, the analysis of its spatial structure and interference status, and the improvement of the design efficiency of the helicopter transmission system configuration.

Compared with traditional CAD assembly technology, virtual assembly technology has become the focus of research at home and abroad. Virtual assembly is to complete the assembly model of the designed parts on the CAD system and assemble different parts into a general assembly, its basic function is to define the relative position relationship between different parts through assembly constraints. After the assembly is completed, the spatial structure analysis and dynamic interference inspection between the parts are carried out to find the unreasonable parts design, so as to improve the design and complete the three-dimensional model design of the assembly. Boeing used virtual assembly technology to design and build a virtual prototype of a Boeing 777 with more than three million parts, allowing designers to walk through the virtual plane. Designers can call out anyone part which they want, and view and modify designs [4]. Abe [5] designed and developed a set of visual system for the assembly problem of mechanical parts, in the virtual environment, it supported designers to conduct assembly analysis and performance evaluation, new employees can systematic operation training when assembling machines, which greatly improves the efficiency. Wu [6] aimed at the problem of product model expression in the field of virtual assembly, put forward the assembly model compound expression method based on process. Gao [7] proposed a method based on the degree of freedom reduction to conduct constraint analysis. On the basis of identifying and understanding the designer's assembly intention and operation, the assembly constraint diagram of tree structure was established based on the assembly relationship, and the positioning and navigation of assembly in the virtual environment was realized through the analysis of spatial geometry.

* Corresponding author: whw@nwpu.edu.cn

In this paper, aiming at the problem of rapid assembly of helicopter transmissions system with complex spatial relations, a transmissions assembly model based on geometric constraint diagram was proposed, and a layered data structure including assembly constraint relation and assembly model was constructed, the assembly modeling of helicopter transmissions system was developed based on CATIA.

2 GCG based assembly model

The assembly model of transmissions system is a complex geometric constraint system, which organically combines various parts with certain assembly constraint relations. Because the helicopter transmission system has many parts and complex structure, the constraint relationship between the parts or components in the assembly model is very complex. An ideal assembly model system should be consistent in expression, maintenance and solution, so it is necessary to establish an effective assembly model expression method for helicopter transmission system.

2.1 Geometric Constraint Graph (GCG)

GCG method is the combination of graph theory and computer geometry [8-10]. The modeling of geometric constraint system based on GCG method can directly manage and express geometric entities and associated geometric constraints in two or three dimensional space, it has the characteristics of good expansion of geometric space and interpretability of geometry.

Taking a typical helicopter transmission system as an example, the process of recursion from its relational model to layered model is researched, and the assembly model of helicopter transmission system is further illustrated. The hierarchical model of transmission system is consistent with the layered structure of product assembly model in CAD system, this is beneficial to the modeling of assembly model in CAD system. At the same time with the help of the CAD system assembly design module function, complete the assembly of the transmission system. This provides the direction and foundation for the automatic assembly of helicopter transmission system in the scheme design phase.

Applying the structure of graph to express GCG is not only clear and straightforward, but also has good stability and high efficiency of graph correlation algorithm, so it is widely used [11-12]. The helicopter transmissions system is abstractly expressed as GCG, by analyzing the GCG, forming a separable and solvable sub-problem, and the solving strategy of sub-problem can be properly combined, which can effectively solve the assembly constraints of system.The established GCG is the relationship model of helicopter transmissions assembly. The mathematical expression of transmission system assembly model and GCG is as follows.

(1) Let S = (X, C, A), S represent a transmission assembly model; X is collection of parametric parts, such as the sun wheel, planet wheel and other parts; C is a collection of geometric constraints, such as angle constraints, distance constraints, vertical constraints, parallel constraints; A is the value of geometric constraint parameters, such as distance value, angle value etc.

(2) Let GCG = (V, E), GCG represents the geometric constraint diagram of the assembly model of the transmission system; V is the vertex in GCG and V = X, each vertex represents a geometry entity, a part of transmission system; E is the side of GCG and E = C, each side boundary represents a set of geometric constraints.

Fig. 1 is a configuration diagram of a certain type of helicopter transmission system, including the main reducer, intermediate reducer and tail reducer. The main reducer is divided into three levels of transmission, the first stage is the input level, which is reversing deceleration by the spiral bevel gear. The second stage is the parallel drive, which is decelerated by the spiral bevel gears. The third stage is planetary gear reducer transmission, the number of planetary wheel is three, and the total reduction ratio is 81. The intermediate reducer is installed on the base of the inclined beam, and is driven by a pair of spiral bevel gears to transfer the torque of the main transmission device to the tail reducer and change the direction of transmission. The angle of shaft intersection is about 58°, input speed is 4117r/min, and output speed is 3319r/min. The tail reducer is installed on top of the inclined beam and is driven by a pair of spiral bevel gears to reverse reduction. The angle of shaft intersection is about 105°, input speed is 3319r/min, and output speed is 1190r/min.

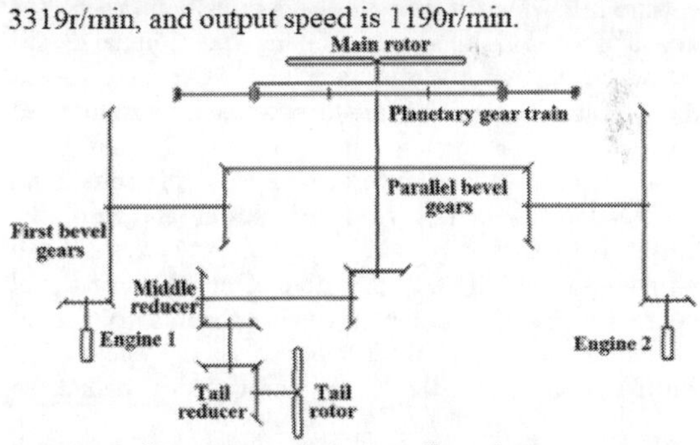

Fig. 1. A helicopter transmissions configuration

According to the composition of helicopter transmission system, GCG is introduced to express the assembly model of the transmission system, as shown in Fig. 2. There are two kinds of nodes in the GCG, the first type of rounded corner and square node represents parametric parts, such as sun wheel, planet wheel, transmission shaft, and each part contains modeling parameters. The second rectangular node represents the assembly constraint set between parts or components, for example, C1 represents the assembly constraint set between the sun wheel and the plant wheel.

The section encircle by the dotted line in Fig. 2 represents the nodes to be condensed and aggregated in GCG, corresponding to the basic components in the helicopter transmissions such as the planetary gear train. A component is an assembly unit that realizes a certain function or multi-function of a transmission system, a

https://doi.org/10.1051/matecconf/202030605004

component can be composed of multiple parts or a single part. The transmission system contains the following basic components: planetary gear train, parallel bevel gear transmission, spiral bevel gear pair, transmission shaft and other components.

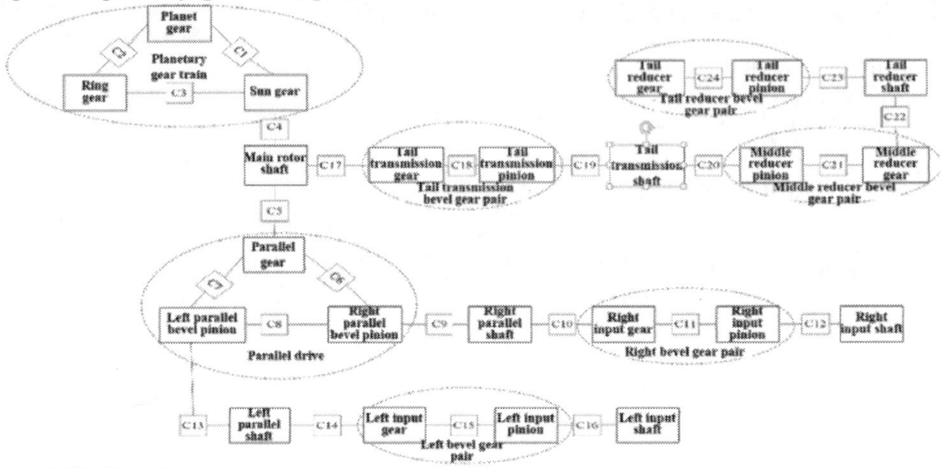

Fig.2. The GCG of the helicopter transmissions

2.2 Components condensed of GCG

Through the recursive assembly method, the components are condensed into a node, which changes the constraint distribution of the transmission and the assembly scale of the assembly system is also reduced. The constraints in the components such as planetary wheel become internal constraints and are separated from other external constraints, which is also the decomposition process of geometric constraint system. Fudos and Bouma[13-15] introduce the concept of cluster into GCG, they solved the problems of geometric constraints by graph-based construction method. In essence, the component is regarded as a group of parts cluster, if the remaining degrees of freedom in the part cluster are zero, the cluster is rigid, that is rigid body, if it is not zero, it is a pseudo-cluster, in the pseudo-cluster the rotational degrees of freedom around its axis of symmetry do not affect the assembly of the transmission system. The shrinking process of the basic components is called the first shrinking of the transmission system. Fig. 3(a) indicate the GCG by first shrink.

In the process of the first shrink, the constraints between planetary gear train, bevel gear parallel drive, bevel gear pair and the outside remain unchanged, and the constraints between the original parts become internal constraints. C1, C2 and C3 become internal constraints of planetary gear trains. C6, C7 and C8 become internal constraints of parallel drive, C15 and C11 become internal constraints of left bevel gear pair and right bevel gear pair, respectively, and C18, C21 and C24 become internal constraints of tail transmission bevel gear pair, mid-subtraction bevel gear pair and tail-subtraction bevel gear pair in turn. After the first shrink of the transmission system, a new cluster of parts forms a new node to replace, for example, the planetary gear train becomes the planetary gear train node.

Fig. 3(b) shows the secondary shrinking of the transmission system. It can depend on the transmission shaft for shrinking, such as the rotor shaft, planetary gear train, tail transmission bevel gear pair and parallel bevel gear drive shrink into the rotor shaft system, etc.

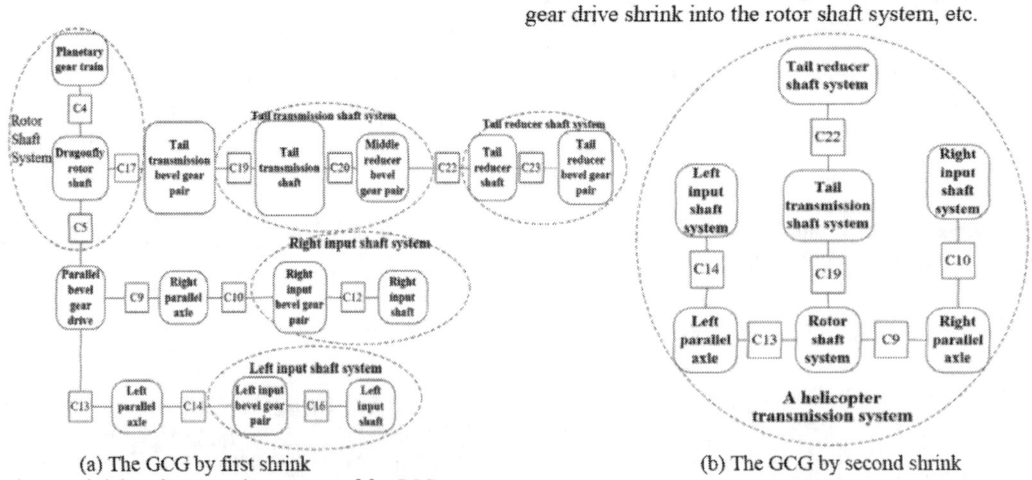

(a) The GCG by first shrink

(b) The GCG by second shrink

Fig. 3. The two shrink and aggregation process of the GCG

2.3 Layered model of transmission system

The layered model of the transmission system can be obtained from the process of the first shrinking to the second shrinking of the transmission system. Therefore, the assembly of a helicopter transmission system needs to be realized through its layered model, which is the result of GCG shrinking operation. Essentially, it is the inevitable result of decomposition of geometric constraint system. Fig. 4 is a layered model of helicopter transmission system. The layered model of the transmission system is divided into four layers, the first layer is the overall layer, which represents the general assembly of the helicopter transmission system; the second layer is the shafting layer, which is the result of the second shrinking, and the sub-assembly built with the transmission shaft as the core, such as the rotor shafting and the tail transmission shafting; the third layer is the component layer, which is the result of the first shrinking and expresses the basic components; the fourth layer is the part layer, It is the basis of basic components.

Fig. 4. The layered model of the helicopter transmissions

3 Data structure of assembly model

In the helicopter transmission system assembly model, nodes are relatively independent assembly units, such as shafting, components and parts. The node is taken as an object, the attribute element of the object is defined, and the corresponding data structure is obtained. Data structure should include assembly constraint relation, layered structure of assembly model and material attribute. The details are as follows.

(1) Ass (Assembly)

Name: the name of assembly

ID: unique identification of the assembly

Context: the ID number contains shaft system

Pointer: the constraint pointer point to shaft system

(2) SS (Shaft system)

Name: the name of shaft system
ID: unique identification of the shaft system
Context: the ID number contains component
Pointer: the constraint pointer point to component

(3) Com (Component)
Name: the name of component
ID: unique identification of the component
Context: the ID number contains part
Pointer: the constraint pointer point to part

(4) Par (Part)
Name: the name of part
ID: unique identification of the part
Parameter: part parameters
Material: material properties of parts

(5) Con (Constraint)
Name: the name of constraint
Type: the type of constraint
ID: unique identification of the part constraint
Value: constraint value

Assembly is the top node, it is expressed as the total assembly of the transmission system of the helicopter, contains Name, ID, Context and Pointer, the pointer describes assembly constraint information between the next layer. Shaft System refers to the sub-assembly after the second shrinking, such as rotor shaft system, etc. Shaft system that depend on shafts are used as sub-assembly. In fact, during the second shrinking, designers can specify partitioning shaft system interactively, but be careful that the components that connect each other are composed. Components are the basic units that make up transmission system, component can be a part or a combination of several parts to achieve a certain function, in interactive assembly, components in the component library can be directly called for combinational design. Part is the most basic unit; it is the basis of components. Constraint corresponding to CATIA assembly constraints, we can judge constraint type by specifying the constraint ID and discerning ID.

4 Realization of Rapid Assembly

To apply the aforementioned assembly technology and method, a software flow for rapid assembly of helicopter transmission system is constructed, as shown in Fig. 5. According to the layered model and data structure of helicopter transmission system, the corresponding assembly input files can be designed. Based on the secondary development of CATIA by VB, the corresponding interface was customized, the assembly input file was driven by program, and the corresponding assembly information was obtained, including properties of modeling parameters, assembly constraints among components and material. Finally, the component library and CATIA API interface function are called to realize the output of the three-dimensional assembly model of a helicopter transmission system. At this time, the model established is a pre-assembly model, which needs to be adjusted and modified by the designer. The formed CATIA file can be connected to the database to form a scheme, which is convenient for users to call and modify.

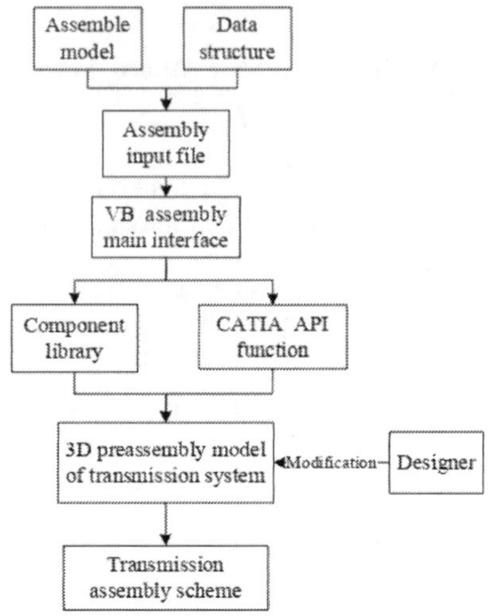

Fig. 5. The rapid assembly software process

According to the software flow framework of the transmission system rapid assembly, the bottom-up parametric assembly modeling design mode is adopted, and the secondary development of CATIA is carried out based on VB. The developed rapid assembly software can automatically generate the three-dimensional assembly of the transmission system by reading the basic parameters of the components and the position parameters of the parts in the CATIA environment according to the parametric data of the configuration results of the helicopter transmission system. The model is shown in Fig. 6.

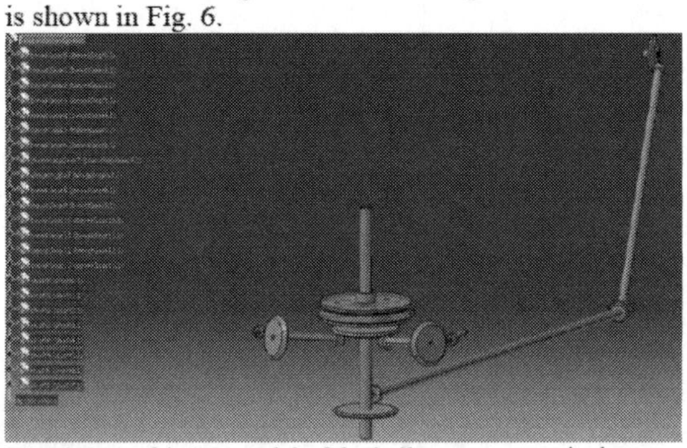

Fig. 6. Assembly 3D model of the helicopter transmissions

5 Conclusion

In this paper, geometric constraint graph is introduced to express the assembly model of helicopter transmission system. Through the development of a helicopter transmission system as an application example, it is proved that the assembly model of helicopter transmission system based on geometric constraint graph is an effective expression method, which can well achieve the consistency of expression, maintenance and solution. Combining with the assembly design function of CAD system, the layered model is corresponded to the feature tree of CAD system. Through parametric modeling and secondary development of CATIA, the rapid assembly of three-dimensional model of helicopter transmission system driven by program is realized.

Acknowledgements

This research is supported by the 111 project (no. B13044) and the concept research of new rotorcraft project (no.30103090101).

References

1. Lin He, Wang Sanmin, Dong Jincheng, *Journal of Aerospace Power*, **30**, 1 (2015) (in Chinese)
2. Bill R. C, NASA-TM-103276, 46th Annual American Helicopter Society Forum, Washington, D.C. (1990)
3. White G, Proceedings of Advanced Power Transmission Technology Symposium (1983)
4. Sui Aina, Wu Wei, Zhao Qinping, *Journal of System Simulation*, **12**, 4 (2000) (in Chinese)
5. Abe N, Wada A, Tanaka K, et al, Proceedings of ICAT' 97, Dec (1997)
6. Yang Rundang, Wu Dianliang, Fan Xiumin, et al, *Computer Integrated Manufacturing Systems*, **12**, 3 (2006) (in Chinese)
7. Gao Zhan, Zhang Shuyou, Gu Jiayin, et al, Chinese *Journal of Mechanical Engineering*, **13**, 6 (2002) (in Chinese)
8. Prouty R W, Krieger Publishing Company, Malabar, Florida (1998)
9. Johnson W, Yamauchi G, Watts M, NASA/TP, 213467 (2005)
10. Acree C W, American Helicopter Society 62nd Annual Forum, Phoenix, AZ (2006)
11. Shi Zhiliang, Chen Liping, *International Journal of Advanced Manufacturing Technology*, **32** (2007)
12. Kramer G A, *Computer Aided Design*, 25, 93 (1993)
13. Fudos I, Hoffmann C M, *ACM Transaction on Graph*, **16**, 2 (1997)
14. Bouma W, Fudos I, Hoffmann C M, et al, *Computer Aided Design*, **27**, 6 (1995)
15. Fudos I, Hoffmann C M, *International Journal of Computational Geometry & Applications*, **6**, 4(1993)

Aerodynamic effects of corrugation configurations in sweeping-pitching flight

Haibin Xuan[1], *Jun* Hu[1], *Yong* Yu[1,*] *Jiaolong* Zhang[1]

[1]School of Aerospace Engineering, Beijing Institute of Technology, Beijing 100081, P. R. China

Abstract. Some insects possess corrugated wings, which distinguish from the ordinary airfoils. It is important to research the corrugation effect on the aerodynamic performances. A series of corrugated wing models were designed based on former research in represent study to find out the underlying mechanisms. The effects of the corrugation pattern and inclination angle were studied using computational fluid dynamic (CFD) method during hovering flight at Reynolds numbers in the order of 10^4. The instantaneous aerodynamic forces and the vorticity field around the wing models were provided. The findings are as follows: (1) the results of this paper show that patterns of corrugation have different effect on aerodynamic performances. The corrugated wing like Corrug-1 changes the lift and drag very slightly compared with a flat-plate. The corrugation patterns like Corrug-2 and 3 of wing model reduce the lift and drag force. (2) the increase in the inclination angle has limited effect the aerodynamic forces. The inclination angles like corrug-3 and 4 produce almost the same forces.

1 Introduction

Biomimetic inspires engineers to find easier and more practicable approaches to solve technical problems [1]. In recent years, the insect-like macro air vehicles (MAVs) are gaining popularity due to their potential application in environmental monitoring and surveillance and security [2]. Many insects have unique flying capabilities; insects can move forward, flap up and down, plunge and sweep, especially the dragonflies and the locusts. In addition, the wing corrugations of some kinds of insects have been seen one of the remarkable features, which show the advantages of low mass and high stiffness compared to smooth wings [3]. Extensive studies carried out by many researchers to research whether the corrugations benefit aerodynamic performances. These studies could be summarized as steady gliding flight and unsteady flapping flight.

Experimental studies conducted by Rees, Rudolph and Buckholz [4-6] showed that insects have both chordwise and spanwise corrugations to enhance the stiffness, but have limited effect on the production of aerodynamic forces in the condition of steady flow. Rees [4] used flow visualization method to identify the field features around a corrugated and smoothed envelope through the corner points of the corrugated section. He noted that flow-separation vortices trapped within the corrugation valleys compared to the smooth wing. Conversely, Kesel [7] compared different wing cross-sectional corrugations of dragonfly with filled-smoothed airfoils, and pointed out that the corrugations of the dragonfly performs better than flat-plate wing. Hu and Tamai [8] conducted the experiment to investigate the flow behavior around a bioinspired corrugated airfoil compared with a traditional streamlined airfoil and a flat plate at the chord Reynolds number of Reynold number (Re) =34000 to explore the potential application of such bioinspired corrugated airfoils for micro air vehicle applications. They concluded that the corrugated airfoil has better performance over the streamlined airfoil and flat plate by preventing large-scale flow separation. Levy and Seifert [9] used (CFD) simulations to analyze the performances of corrugated and traditional smooth airfoil. They demonstrated that the corrugated airfoil outperformed the traditional streamlined airfoil when Re was in the range of 2000-8000. Chen and Skote [10] concluded that the profiled wing perform better than corrugated wing as the angle of attack (AoA) and the Re increasing for three dimensional gliding. Conversely, Bomphrey et al. [11] found that the corrugated wing generates marginally higher force coefficients than the smoothed wing when AoA up to 10°.

Meanwhile, researchers conducted simulations and experiment to investigate the effect of corrugation during unsteady flapping flight. Luo and Sun [12] studied ten representative insect wing models in sweeping motion with angle of attack 40°, the results of a corrugated wing model based on fruitfly and a smooth one showed that the corrugation had no effect on the aerodynamic forces. The reason might be that the separated flow or the size of leading-edge vortex exceeded the scale of the corrugation. Meng et al. [13] examined the aerodynamic effects of corrugation in insect wing models during hovering and forward flight using the method of CFD. They concluded that the time courses of aerodynamic

* Corresponding author: yuyong@bit.edu.cn

force of the corrugated wing are very close to those of the flat-plate wing.

Studies like above cited were mostly focus on the comparison of aerodynamic features between he corrugated and smooth wing or airfoil. The corrugation structures are little concerned. It can be seen that the effect of wing corrugation configurations for 3D simulations is still unclear, we performed 3D simulations using different corrugation configurations wing models during sweeping and pitching motion. The wing models used by former research [14, 15] to obtain deeper insights into the unsteady aerodynamics of flapping motion.

2 Computational technique

2.1 The flapping kinematics

It is common to simplify the laws of motion into mathematical function, two types of simplified laws, the trapezoidal and the sinusoidal are widely used. The idealized motion of 3D flapping wing consisted of pitching and sweeping [16], defined in Eqs. (1). The sketches of the pitching and sweeping motion profiles are shown in Fig. 1b.

$$\alpha(t)=90°-\beta\sin(2\pi ft), \quad (1a)$$

$$\varphi(t)=\Phi\cos(2\pi ft). \quad (1b)$$

b is set to 45°, the frequencies f of the plunging and pitching motions are the same (0.1Hz), ϕ has a value of 60° in the present study. The flapping flat plate was built based on New et al.'s experiment.[17]. This wing model is assumed to be a rigid plate, with a thickness of 2% chord length. Fig.1a depicts the pitching and sweeping axis's locations.

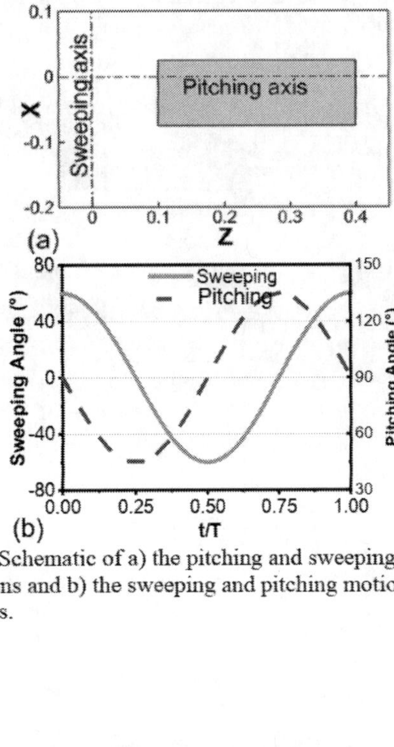

Fig. 1 Schematic of a) the pitching and sweeping axis locations and b) the sweeping and pitching motion profiles.

2.2 Solver settings and computational domain

Many literatures cited above stated that the airfoil corrugations cause the flow transition from laminar to turbulent. Levy and Seifert [9] stated that the flow pattern would be significantly different at higher Reynolds number and a transition from laminar to turbulent flow is expected. Hu and Tamai [8] stated that the corrugated wing that the protruding corners promote the transition of separated shear-layer flow from laminar to turbulent. The Reynold number in this paper is defined as: $Re=\rho U_{ref}\, c/\mu$, ρ is the liquid density, c is the chord length, μ is the dynamic viscosity, the reference velocity $U_{ref}=4f\Phi R_2$, f is the flapping frequency, and R_2 is the radius of second moment of area. Reynold number in this paper is in the order of 10^4. This paper used the improved delayed Detached Eddy Simulations (IDDES) to perform simulations. The unsteady RANS models are employed in the boundary layer, while the LES treatment is applied to the separated regions.

The modified k-ε of the IDDES is defined as Eqs. (2)

$$\frac{\partial \rho k}{\partial t}+\nabla\cdot(\bar{\rho}\tilde{U}k)=\nabla\cdot[(\mu+\frac{\mu_t}{\sigma_{k3}})\nabla k]+P_k-\rho\beta^{*}k\omega F_{IDDES} \quad (1a)$$

$$F_{IDDES}=l_{RANS}/l_{IDDES} \quad (1b)$$

The coupling of the pressure and velocity are handled using the coupled algorithm. The spatial discretization is carried out using a least square cell-based scheme. The pressure and momentum equations are solved with a second order scheme. All scaled residuals leveled off and reached 10^{-6}.

In the present study, we used sliding mesh to model the motion of the flapping wing [18], a hybrid grid method is used to generate the grid. The domain size of the 3D wing model for flapping flight is essential for CFD simulations. The computational domain is 20c. The cell size was increased by keeping a factor of 1.05 away from the wing and 1.1 for the outer boundary. The outer boundary is set as a pressure outlet, the wing edge was set with the condition of no-slip wall.

3 Results and discussion

In this section, two sets of results flapping flight of corrugated wing models are discussed: 1) The effect of corrugation profiles was discussed first, the aerodynamic forces and the vorticity around the wing were provided to get deep insight into the corrugations. 2) Three different inclination angles based on a special wing model were discussed in detail. The aims of this section are to find out the effects of wing corrugations on flapping wing performance, and the contributions of different corrugated patterns.

3.1 Effects of the corrugation patterns

Before later discussion, it is important to define the aerodynamic characteristics. The lift coefficient C_L and the thrust force C_D are normalized using following equations:

$$C_L=F_L/(0.5\rho U_{ref}^2S) \quad (3a)$$

$$C_D=F_D/(0.5\rho U_{ref}^2S) \quad (3b)$$

The grid and discretization time step influence the simulation results for unsteady simulations [19]. Three sets of grids were adopted to simulate for eliminating the mesh-error, this paper adopts the gird with 6×10^6 cells with a first layer of $0.0005c$ after comparison. Small time steps are more likely to accurately describe physical phenomena, but increase the demanded resources of the simulation. In this study, it was verified through calculations that discretization time-step $\Delta t = T/2000$ is the most suitable.

Fig. 2 shows the wing models used in this paper, the wing model is used in former research [20]. We modified the profile Corrug-1, replacing corrugated parts with smoothing plate. Two valleys and peaks were kept near the leading edge. Corrug-2 has the same valleys and peaks with Corrug-1 while has a camber rear part. Corrug-3 is covered with valleys and peaks. These four wing models have the same chord length and spanwise length.

Fig. 3 gives the time courses of lift and drag coefficients for three corrugated wings and one flat-plate wing in one flapping cycle. T in the figure indicates flapping cycle. The fifth flapping cycle was used to avoid the non-fully developed flow field. Three significant observations could be made. Firstly, the C_L differences of four wings are very small during downstroke while the differences are significant during upstroke. the second observation is that there is only very small difference between the results of different corrugated patterns. The last one is that C_L and C_D of flat-plate and corrug-1 wing models are mostly the same and symmetrical. The results presented in Fig. 3 show that the force coefficients were significantly asymmetric during the upstroke and downstroke for Corrug-2 and Corrug-3 (shown in Fig. 2). The drag curves have two positive peaks during downstroke and upstroke.

It is shown that the lift coefficients in Fig. 2a, overall, change little for different corrugation patterns during downstroke and upstroke. The first positive lift coefficient values of Corrug-1, Corrug-2 and Corrug-3 are 0.1%, 10.5 and 4.8% higher, respectively, than that of plate. For the drag coefficients, one noticeable discrepancy can be seen in Fig. 2b, the force coefficient curves have two positive peak values during the downstroke/upstroke. This finding is same as that reported by Lee and Lua [18] for flapping flight at low Reynolds numbers.

To further investigate the underlying mechanisms of the above aerodynamic effects of wing corrugations, the vorticity field around wings were provided in Fig. 4. At t=4.25T, it can be seen that the vorticity of the plate and corrug-1 have the most the same strength and shape. The valley near the leading-edge was filled with vortex, and the flow field region away the leading edge have the same flow characteristics. For both the corrug-1and corrug-2 cases, the vertex trapped in the valleys on the upper surfaces of the corrugated airfoils can be observed, as expected. There is no flow separation from the surfaces. At t=4.5T, there is no vortex attached to the plate, however, there is vortex remaining in the valley of the corrugated wing models. The valley was filled with vortex. One notable feature could be seen that there is a still vortices on the cambered surface. The cambered part can be seen as a large valley, which could trap the vortices. At t=4.75T, the flow field shows a symmetry with that at t =4.75T. However, the lift force coefficients show a decreasing trend for the corrugated wings. The reason may contribute the asymmetric of the wing model around the pitching axis.

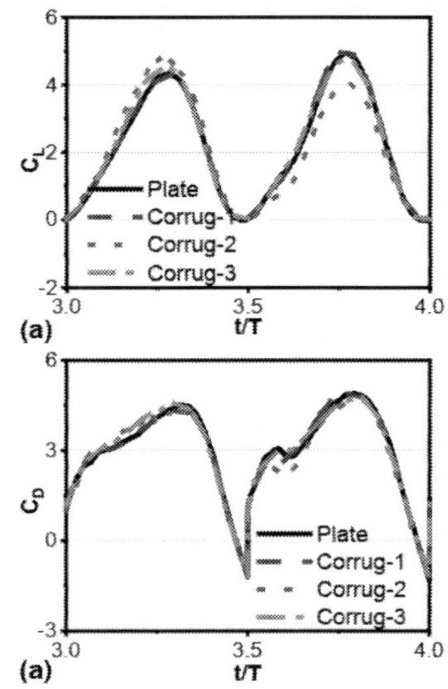

Fig. 2 The time histories of the aerodynamic forces coefficients, a) the lift coefficients and b) the drag coefficients

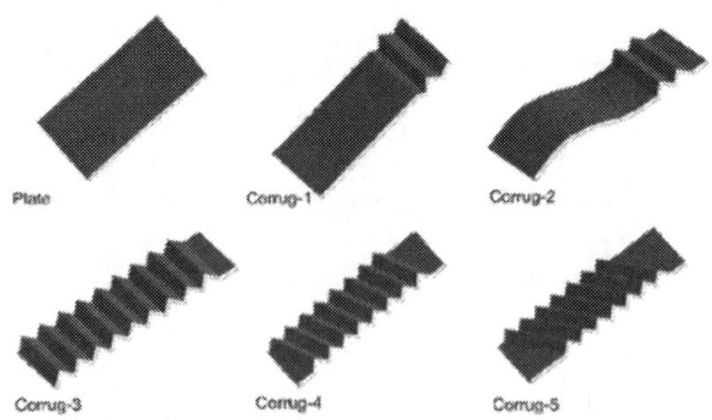

Fig. 3 The geometric sketches of the wing models

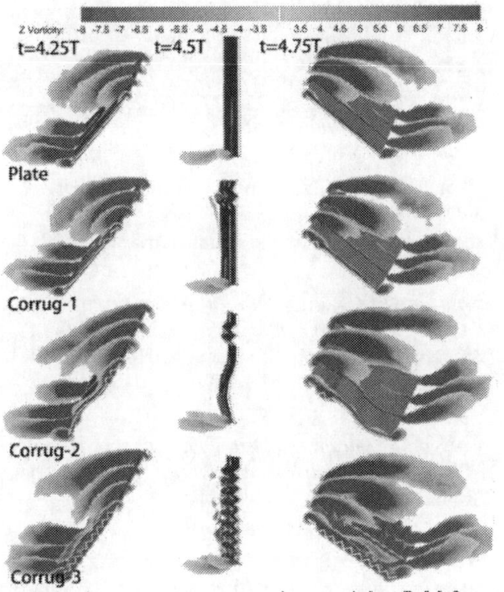

Fig. 4 The instantaneous spanwise vorticity field for different cases

3.2 Effects of the angles of the corrugation

The previous section discussed the effects of corrugation pattern changes. To further research the contributions of corrugation, this section focuses on a discussion about the effect of inclination angles. The different corrugated wing models are described is Fig. 3.

Overall, the lift and drag coefficient curves have similar shapes as those in the previous section (shown in Fig. 5). There are positive and negative lift coefficient peak values, while there are two positive drag coefficient peak values in the downstroke or upstroke. The peak values of the lift and drag coefficients are decreased with increasing tilt angle. The lift coefficient peak values change slightly during the upstroke. For the drag coefficients, one noticeable discrepancy can be seen, namely, the curve histories keep the same, as shown in Fig. 5.

Fig. 1 shows the pitching and plunging amplitudes for different corrugation patterns. Fig. 7 shows the time history of the aerodynamic force coefficients. The lift coefficient and drag coefficient vary with time. The results presented in Fig. 5 show that the force coefficients were significantly asymmetric during the upstroke and downstroke, which is the same with former section. The reason leads to the drag symmetrical shapes is still unclear.

Fig. 6 shows the flow field around the flapping wing in terms of the spanwise vorticity. At t = 4.25T, which reaches to the minimum AoA in one cycle, the wing rotates clockwise around the pitching axis and continues moving downward at the same time. At this moment, there are vortices attached to the upper surface and shed from the trailing edge into the wake. The lift and drag reach the maximum.

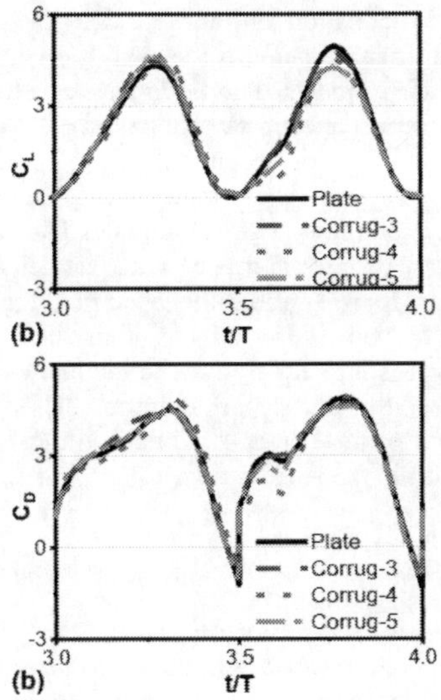

Fig. 5 The aerodynamic forces, (a) lift coefficients and (b) drag/thrust coefficients.

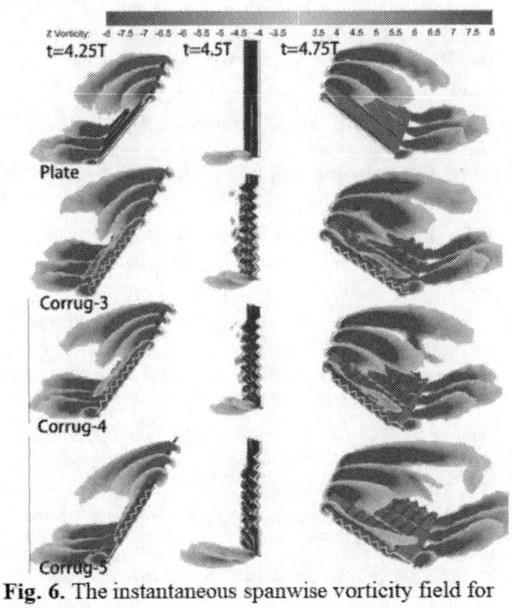

Fig. 6. The instantaneous spanwise vorticity field for different cases

At t=4.5T, the former LEVs and trailing edge vortex have successfully separated from the wing. Some relatively small vortices have already formed in the valley of the corrugated wing while that of flat-plate not. The following wing movement is similar to the previous movement in a timely symmetric manner; as a result, the symmetry phenomenon affects the integrated aerodynamic coefficients.

4 Conclusion

The effects of the wing corrugations were studied using CFD method during hovering flight at Reynolds numbers in the order of 10^4. Different corrugation patterns and inclination angles were discussed using the instantaneous aerodynamic forces and the vorticity field around the wing models. The findings are as follows: (1) the results of this paper show that patterns of corrugation have different effect on aerodynamic performances. The corrugation like corrug-1 changes the lift and the drag very slightly compared with a flat-plate. The corrugation patterns like Corrug-2 and 3 of wing model reduce the lift and drag force (during the upstroke, the lift peak values, 10.5 and 4.8% lower than that of plate, respectively). (2) the increase in the inclination angle has limited effect the aerodynamic forces. The inclination angles like Corrug-3 and 4 produce almost the same forces (the difference is less than 1%).

References

1. Zhang, S., Y. Sunami, and H. Hashimoto, *Deformation behavior of dragonfly-inspired nodus* structured *wing in gliding flight through experimental visualization approach.* Scientific Reports, 2018. **8**(1): p. 5751.
2. Shyy, W., et al., *Aerodynamics, sensing and control of insect-scale flapping-wing flight.* Proceedings of the Royal Society A: Mathematical, Physical and Engineering Science, 2016. **472**(2186): p. 20150712.
3. Hord, K. and Y. Liang, *Numerical Investigation of the Aerodynamic and Structural* Characteristics *of a Corrugated Airfoil.* Journal of Aircraft, 2012. **49**(3): p. 5вЂ“101.
4. Rees, C.J.C., *Aerodynamic properties of an insect wing section and a smooth aerofoil* compared. Nature, 1975. **258**(5531): p. 141-142.
5. Newman, B.G., S.B. Savage, and D. Schonella, *Model tests on a wing section of an Aeschna* dragonfly, in *Scale effect in animal locomotion*, T.J. Pedley, Editor. 1977. p. 445-477.
6. Rudolph, R., *Aerodynamic properties of Libellula quadrimaculata L.(Anisoptera: Libellulidae), and the flow around smooth and corrugated wing section models during gliding flight.* Odonatologica, 1978. **7**(1): p. 49-58.
7. Kesel, A.B., *Aerodynamic characteristics of* dragonfly *wing sections compared with technical aerofoils.* Journal of Experimental Biology, 2000. **203**(Pt 20): p. 3125-35.
8. Hu, H. and M. Tamai, *Bioinspired Corrugated* Airfoil *at Low Reynolds Numbers.* Journal of Aircraft, 2008. **45**(6): p. 2068-2077.
9. Levy, D.-E. and A. Seifert, *Simplified dragonfly* airfoil *aerodynamics at Reynolds numbers below 8000.* Physics of Fluids, 2009. **21**(7): p. 071901.
10. Chen, Y.H. and M. Skote, *Gliding performance of 3-D corrugated dragonfly wing with spanwise variation.* Journal of Fluids and Structures, 2016. **62**: p. 1-13.
11. Bomphrey, R.J., et al., *Flight of the dragonflies* and *damselflies.* Philosophical Transactions of the Royal Society B: Biological Sciences, 2016. **371**(1704): p. 20150389.
12. Luo, G. and M. Sun, *The effects of corrugation and wing planform on the aerodynamic force production of sweeping model insect wings.* Acta Mechanica Sinica, 2005. **21**(6): p. 531-541.
13. Meng, X.G., L. Xu, and M. Sun, *Aerodynamic* effects *of corrugation in flapping insect wings in hovering flight.* The Journal of Experimental Biology, 2011. **214**(3): p. 432-444.
14. Flint, T., et al., *Validated Unsteady* Computational *Fluid Dynamic Analysis of an Oscillating Bio-Inspired Airfoil.* Applied Mechanics and Materials, 2015. **799-800**: p. 698-706.
15. Liu, Z., et al., *Artificial insect wings with* biomimetic *wing morphology and mechanical properties.* Bioinspiration & Biomimetics, 2017. **12**.
16. Lee, Y.J., et al., *A quasi-steady aerodynamic model for flapping flight with improved adaptability.* Bioinspiration & Biomimetics, 2016. **11**(3): p. 036005.
17. Lua, K.B., T.T. Lim, and K.S. Yeo, *Scaling of* Aerodynamic *Forces of Three-Dimensional Flapping Wings.* AIAA Journal, 2014. **52**(5): p. 1095-1101.
18. Lee, Y.J. and K.B. Lua, *Wing–wake interaction: comparison of 2D and 3D flapping wings in hover flight.* Bioinspiration & Biomimetics, 2018. **13**(6): p. 066003.
19. Sun, M. and J. Tang, *Unsteady aerodynamic force* generation *by a model fruit fly wing in flapping motion.* Journal of Experimental Biology, 2002. **205**(Pt 1): p. 55-70.
20. New, T.H., et al., *Effects of Corrugated Aerofoil Surface Features on Flow-Separation Control.* AIAA Journal, 2013. **52**(1): p. 206-211.

AUTHOR INDEX

EDP Sciences – Web of Conferences
17, Avenue du Hoggar
Parc d'Activité de Courtabouf
BP 112
F-91944 Les Ulis Cedex A
France

ISBN 978-1-7138-0292-1

CURRAN ASSOCIATES INC.
proceedings.com